*Texas Weather*

We are blessed with a habitat best described as a "sea of air," the most compelling evidence of which is a preponderance of clouds of various shapes and consistency.

George W. Bomar

# TexasWeather

University of Texas Press, Austin

In memory of E. L. Bomar, my grandfather and my inspiration to be a student of Texas weather, and dedicated to all Texas volunteer weather observers.

Copyright © 1983
by the University of Texas Press
All rights reserved
Printed in the United States of America

First Edition, 1983

Requests for permission to reproduce material from this work should be sent to Permissions, University of Texas Press, Box 7819, Austin, Texas 78712.

Library of Congress Cataloging in Publication Data

Bomar, George W.
  Texas weather.

  Bibliography: p.
  Includes index.
  1. Texas—Climate.  I. Title.
QC984.T4B67  1983  551.69764  82-23898
ISBN 0-292-78052-4
ISBN 0-292-78053-2 (pbk.)

# Contents

# Preface

From the numbing blizzards that pound the Panhandle every winter to the enduring heat that sears all sections of the state a few months later, Texas in the meteorological sense truly is the "Land of Contrast." Perennially the state perseveres through bombardments of hail, high winds, and flash floods as well as the threat of being struck on its coastal flank by a hurricane or some other intense marine storm. Moreover, in any given year, sections of Texas suffer from the effects of a tornado, a snowstorm, a hail-bearing thunderstorm, or a sand or dust storm. In many years some portions of the state experience destruction or severe damage from an untimely freeze, a drought, or a lengthy spell of rain. Assuredly, no two years are alike, for the community that reeled one year from a dry spell likely received ample rain the following year, while a not-too-distant neighboring locale that endured a disastrous hailstorm one spring experienced relative tranquility during the following year's peak storm season.

Such diversity in weather conditions fosters a profound and abiding interest in Texas weather by residents, tourists, and newcomers alike. Because a particular severe storm, such as a hailstorm or a tornado, is a rather rare event for a specific locality, public reaction to it is as much that of fascination as of fear. Longtime inhabitants often pridefully recall witnessing at least a few bizarre weather events that have enhanced the state's reputation for having a

weather scene with a propensity for altering itself in a matter of minutes. Those who visit in the Lone Star State may carry home stories of how they successfully outlasted some of nature's more dangerous actions. Because growing numbers of Americans are discovering the economic, recreational, and climatic advantages of living in a "Sun Belt" state like Texas, timely, relevant data and information on the state's weather are needed.

This book attempts to describe—graphically, pictorially, and numerically, where practical—those weather events that distinguish Texas as a land of climatic disparity. It is also the intent of this work to explain why such diversity exists and to present—as precisely as current technology will permit—projections on the immediate future. This book will not suggest unique solutions or responses to those elements of the Texas weather that often irritate or torment. Instead, it furnishes fundamental knowledge with which readers can analyze and estimate nature's threat and thereby ascertain the course of action to be taken to preserve life and possessions.

Substantial advancements in our understanding of the ways by which the weather affects the populace and the environment have been realized over the past thirty years. However, very little effort has been invested in applying this wealth of knowledge to the peculiarities of Texas' brand of weather. It is hoped that this volume will serve many of the needs of users in technical fields, such as engineering, building and highway construction, and disaster preparedness, as well as be a handy reference for users whose occupations and avocations fit into such categories as farming and ranching, gardening, outdoor sports, and recreation.

Since this book is designed to address a reading audience of varied concerns, I recognize that none of the aspects of Texas weather should be treated here in great depth with accompanying mathematical and physical reasoning. Numerous other books are available to augment the reader's understanding of such topics as climate-human relationships, weather modification, environmental pollution, and the geometry of our solar system. Few, if any, readers will find all chapters to be equally informative and useful. Nevertheless, I hope that enough information is given to stimulate the serious readers to determine for themselves the probable range of weather types they are likely to experience at a given locality.

Most of the chapters are introduced by intentionally vivid accounts of some of Texas' most catastrophic weather tragedies. This is not intended to unnecessarily frighten the reader. Rather, it is to stress that coexistence with the whims of Mother Nature depends in part upon a comprehensive awareness of the potential hazard posed by many different weather elements. It is also meant to underscore the fact that present-day technology cannot be relied upon fully to give early warnings of imminent adverse weather. My goal here is to sharpen the readers' awareness of the injury that various forms of weather can induce and to prod them to exercise precautionary measures.

The introductory chapter is aimed to acquaint readers with the general nature of our atmosphere and to furnish tips on how to meaningfully monitor weather conditions. Part of Chapter 1 is devoted to volunteer weather observers, whose loyal service to the government over the years has proven to be invaluable in attempts to draw definitive conclusions about our erratic weather. Chapter 2 presents an elemental description of the role of fronts, those phenomena popularly and justifiably regarded

as the prime instigators of most of the changes in the weather we experience from day to day. Chapter 3 deals with the occurrence of rainfall, by far the most common form of precipitation in Texas and looked upon as the spice of our weather.

Beginning with Chapter 4, emphasis is given to specific ingredients that compose the medley of weather events that distinguish Texas' climate as the most diverse of any state in our nation. Chapter 4 consists of a lengthy treatise on the hurricane—its distinctive traits and its previous and projected impacts on the citizenry of Texas. Because the thunderstorm is a year-round phenomenon that begets both beneficial (rain) and harmful offspring (hail, high winds, and tornadoes), it is the subject of the most exhaustive section of this book (Chapter 5). Chapter 6 is devoted to the tornado, the descendant of the thunderstorm. Readers are taken from one extreme to the other in Chapters 7 and 8—from a summer scene afflicted by intense heat and drought to a winter cursed by bitter cold and treacherous snow and ice. Chapter 9 focuses on the role of the wind, with special emphasis given to the bedeviling mixture of strong winds and dust. Finally, Chapter 10 is concerned with enabling the readers to do more than lament the kind of weather with which they have been cursed; it supplies lists of safety rules and procedures that are designed to help mitigate undesirable aspects of our weather.

Practically all the data used originated with the U.S. Weather Bureau and, since October 1, 1973, its successor, the National Weather Service. Unless otherwise stated, maps, charts, and other references to "normal" weather conditions incorporate data for 1951–1980, a period regarded by climatologists as being sufficient to suggest what is normalcy. Use of maps in this book should be tempered by the realization that weather-observing stations with sufficiently lengthy weather histories are not evenly distributed throughout Texas; therefore, some regions may have a greater concentration of "point-source" data than others. For specific points (especially the major metropolitan areas of Texas), the reader should consult the appropriate appendixes.

# Acknowledgments

Numerous individuals and organizations were helpful in the preparation of the manuscript and in providing much of the visual material contained in this book. I recognize the encouragement and assistance of Charles Arthur, Barbara Spielman, and others on the staff of the University of Texas Press. I also acknowledge the assistance of the National Severe Storms Laboratory (NSSL) for its critical review of and suggestions on the content of the chapters dealing with thunderstorms and tornadoes. I am particularly grateful to Vincent Wood of NSSL for his encouragement and assistance and to Dr. Edwin Kessler, Bob Davies-Jones, Steve Nelson, Don Burgess, and Donald MacGorman of the NSSL—all of whom reviewed parts of the manuscript and supplied many of the vividly descriptive photographs of tornadoes, lightning, and hail. I appreciate the photo of the hailstorm thin section supplied by Nancy Knight of the National Center for Atmospheric Research.

I am also indebted to the National Weather Service (NWS) and the Texas Department of Water Resources (TDWR) for access to the voluminous reservoir of climatological data so essential to this work. My appreciation is given to David Owens, of the Austin office of the NWS, for his helpful suggestions on the chapter concerning the operations of the NWS, to the TDWR's Weather & Climate Section, and to analyst Jack Johnson for the use of the satellite photographs and the wind roses displayed throughout the book.

I also wish to acknowledge the helpful review of Dr. James R. Scoggins, head of the Department of Meteorology at Texas A&M University, and Dr. Donald Haragan, head of the Atmospheric Science Group at Texas Tech University. Gratitude is also given for the interest, encouragement, and assistance of the Office of State Climatologist, headed by Professor John F. Griffiths, and the Department of Meteorology at Texas A&M University. A special thank-you goes to Jack Lawler of the Texas Department of Public Safety for the use of numerous photographs of damage scenes wrought by severe local-storm phenomena.

Appreciation is extended to several friends and associates whose help in proofreading the manuscript was invaluable: E. R. and Faye Simmons, Pat O'Dell, Mary Furey, and Kay Crawford. Pat also supplied some of the photographs of snow and ice scenes. Most of all, my sincere appreciation is given to Charlcie O'Dell and to my wife, Judy, for their assistance in typing the manuscript.

*Texas Weather*

# 1. *The Vital Signs of Our Temperamental Atmosphere*

Though the calendar hinted that the crisp, cool days of Indian summer were just around the proverbial corner, the simmering heat that gripped the plains and prairies of Texas one day in September more forcefully argued that Texas' long and torrid summer was far from over. Temperatures in the middle of the afternoon eased near 100°F (38°C) from the southern High Plains eastward into the prairies and plateaus of central Texas. To the cotton farmer adjusting the controls of his irrigation system or the oil-field worker inspecting one of a multitude of seesawing pumps, the day's weather was virtually identical to that of so many other days that had preceded it throughout a hot and nearly rainless summer. Within a few hours, however, an unmistakable omen appeared on the far northern horizon, signaling an abrupt and drastic alteration in the weather. A bank of ominous dark clouds spread rapidly southward, yielding flashes of lightning, rumbles of thunder, and dashes of rain. Abruptly, the unfamiliar breath of Old Man Winter could be felt whipping through the High Plains and across the Cap Rock. The wind lurched into the north almost instantaneously, and quickly the temperature nosedived. Readings that had lofted far into the 90s in midafternoon stood in the 50s as the sun vanished behind the western horizon. Winds picked up in speed as the evening grew longer, and a persistent overcast supplied intermittent bursts of light rain throughout the night. For the next three days, a chilly northeast wind retained temperatures in the 50s even in daytime. To longtime inhabitants of the flatlands of western Texas, the abrupt and marked change in the weather was due to none other than a bona fide "blue norther."

The fact that summerlike heat had given way suddenly to a wintry chill in mid-September surprised no one. What proved to be startling—and devastating as well—was a second abrupt shift in the weather that reintroduced torrid temperatures. Once the wind veered from the northeast back into the southwest, temperatures skyrocketed. The three-day spell of chilly, damp, and windy weather, capped by a rapid warm-up, placed too much stress on tens of thousands of acres of burgeoning cotton. This time around, the culprit was not the usual untimely pounding by pea-to-marble-size hail. Rather, a war between frigid and simmering air masses that originated in such diverse areas of the globe as the North Pole and the Mexican Chihuahuan Desert was responsible for the $40–50 million loss wrought on the cotton crop in a 25-county area of the southern and central High Plains.

No item of human interest is as much the subject of more insipid conversation on street corners and in coffee shops than the vagaries of the weather. Because of its diversity, severity, and—at times—its unpredictability, Texas' brand of weather has produced a disproportionate share of disappointments, both minor and major. It has also been the object of many jokes and exaggerated claims. With very few exceptions, it is the focal point of considerable and immediate daily interest.

The distinctive shades of heat and cold, of drought and dampness, do far more, however, than serve as topics of popular conversation. They mold and shape the citizenry to fit the environment; the never-ending skirmishes between competing masses of polar and tropical air hold the populace hostage. Of course, the degree of bondage is not nearly as acute now as in the past when nearly all our ancestors lived directly off the land, fishing, hunting, caring for their herds, and literally raising cane. Today we live in predominantly centrally heated or cooled environments. Yet, our dependency on the behavior of the atmosphere is never more clearly understood than when too little rain shrinks the water supply to threatening levels or when too much heat or cold limits the production of food and fiber. Quite possibly, one of our most consoling thoughts is the realization that the fortitude and vitality that allowed our ancestors to persevere are the same hidden resources required in a modern era still plagued by storms, flood, and drought.

## Our Atmosphere—A Gaseous Envelope

In a genuine sense, we live—and thrive—in one gigantic greenhouse whose transparent ceiling encircles the globe at an altitude of 8–10 miles. Within this gaseous envelope is a mixture of gases that protects all things that live from the deep cold and the lethal radiation of outer space. This habitat is far from self-sustaining, however. Rather, life on Earth is at the mercy of a colossal, incandescent cauldron of gas known as the sun. Without heat energy from the sun, life on this planet could be maintained for only a few fleeting moments. Still, without our atmosphere's capacity for transforming and distributing the energy from the sun around the globe, life in the forms we know today could not flourish.

## A Collection of Permanent Vapors

The atmosphere is made up of a uniform mixture of permanent gases known as dry air, which contains varying amounts of other materials, such as water vapor and organic and inorganic impurities. Four gases—nitrogen (78% by volume), oxygen (21%), argon (1%), and carbon dioxide (0.03%)—account for more than 99% of pure, invisible, and odorless dry air. Certainly oxygen is the most crucial gas for the sustenance of animal life, whereas carbon dioxide is vital for the plant world. However, carbon dioxide is also of much climatic significance mainly because it effectively and selectively absorbs appreciable amounts of radiation emitted by Earth that would otherwise be lost to space. If it did not have this capability, nighttime temperatures would be markedly lower. Of greatest importance to our weather and climate, however, is the presence in the atmosphere of water vapor, also an invisible and odorless gas that is highly variable in amount but usually accounts for about 3% or 4% of the total volume of air. Its significance far outweighs its percentage contribution to the total volume of air, for it not only provides the ingredients for clouds and precipitation but also absorbs certain types of solar and terrestrial radiation. Water vapor also possesses the unique characteristic of being able to change its state from solid to liquid to gas while still part of the atmosphere. It is while water vapor is undergoing a transition—from vapor to liquid to form clouds, for example—that it serves as a major source of atmospheric energy.

In the lowest 50–60 miles of Earth's atmosphere, the composition remains somewhat constant. The concentration of ozone ($O_3$) increases with altitude to a maximum 15 miles above the surface. Ozone is an important regulator of the types and amounts of solar energy that reach the land and water surfaces of Earth, in that it acts to shield terrestrial life from the lethal effects of ultraviolet radiation emanating from the sun. Hundreds of dust particles per cubic centimeter fill the atmosphere and play an important role in the formation of clouds and precipitation by acting as nuclei upon which atmospheric moisture collects to

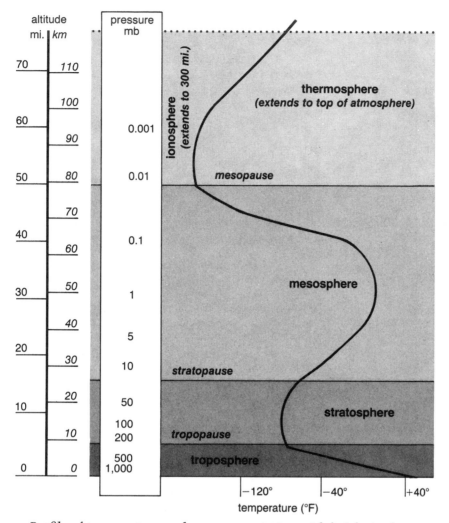

*1. Profile of temperature and pressure variation with height in the atmosphere.*

form droplets. It is the presence of these myriads of submicroscopic dust particles, along with certain molecules of gas, that give us the blue color of the sky and the brilliant red hues of sunsets by selectively scattering the sun's rays. Though some of this dust is washed to Earth's surface by the rainfall it helps to generate, the atmosphere's supply of dust particles is constantly being replenished. Of increasing concern these days is the massive introduction of many impurities, especially those that result from the burning of fossil fuels, which are decidedly harmful to humanity. Because of the state's blossoming population—and with it an explosive growth

of industrial activity and use of automobiles—atmospheric pollution has become a major headache for many Texans living in the largest metropolitan areas.

### A Succession of Layers

The atmosphere consists of four fairly distinct layers that are differentiated mainly on the basis of how the temperature varies with elevation. The layer adjacent to Earth's surface, and the sphere in which virtually all of humanity operates, is the *troposphere*. Extending to about 8–9 miles above the ground, it is the domain within which variations in the weather are most pronounced. This is so because the tropo-

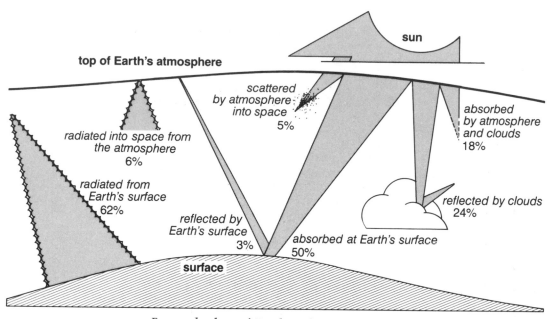

*2. Energy budget of Earth and its atmosphere.*

sphere contains about three-fourths of the atmosphere's total mass and practically all of its water vapor (and clouds). Throughout the troposphere, the upper limit of which is called the *tropopause*, the temperature decreases with increasing height.

Unlike the troposphere beneath it, the *stratosphere* exhibits very little, if any, change in temperature with increasing height. Relatively warm temperatures may be found near the top of this layer, due to the concentration of ozone, which serves as a highly efficient absorber of solar energy. Above the top of the stratosphere, at an altitude of about 16 miles, is the *mesosphere*, where temperature increases, then decreases with greater elevation. Near the upper limit of the mesosphere, at an altitude of about 50 miles, is where most meteors burn and disintegrate. Part of the uppermost portion of the atmosphere, known as the *thermosphere*, is a layer from 50 to 300 miles high known as the *ionosphere*. Particles that make up this slice of the atmosphere reflect certain radio waves. While changes in the density and composition of the upper layers of Earth's atmosphere conceivably affect the weather near Earth's surface, our greatest concern is with the behavior of the lowest layer of the at-

mosphere, for the troposphere is the sphere of our weather.

Fueled by the Sun

An immense and continuous stream of energy from the sun enters the envelope of air surrounding Earth. Energy gained from the visible segment of solar radiation (also known as *insolation*) supplies the fuel necessary for the multitude of processes that make up Earth's weather and climate. Of course, since Earth is not warming up or cooling off substantially, it then must return about as much energy to space as it receives from the sun. However, some parts of Earth—the tropics and subtropics—collect more solar energy than they give back to space, while other parts—most notably the polar regions—give off much more than they receive directly from the sun. This continual transfer of heat energy, both horizontally and vertically, keeps the whole atmospheric machine in balance. It is this exchange that sets up a variety of temperature conditions.

Only about one-fifth of the insolation is absorbed by Earth's atmosphere, while one-half is absorbed, either directly or indirectly, by the surface of Earth. The remainder is reflected back into space by clouds and Earth's surface or is scattered elsewhere in the atmosphere. Heat energy is transferred between Earth's surface and the air, or from one portion of the atmosphere to another, by one of three processes: (a) convection, (b) conduction, or (c) radiation. The processes of condensation and evaporation of water also bring about exchanges of heat energy between various surfaces and the atmosphere. These various means of transferring heat energy maintain a fairly stable global-heat budget over relatively long periods of time. The amount of insolation received at any point on Earth depends upon such factors as the output of energy from the sun, the distance between Earth and the sun, the angle at which the sun's rays strike Earth's surface, the duration of the daylight period, and the constituency of the atmosphere itself. The varying angle of the sun's rays hitting Earth's surface is responsible for the unequal distribution of solar energy over this planet, which in turn determines the seasons of the year and the degree of variability of our weather (see fig. 4).

Temperature: The Atmosphere's Most Prominent Trait

Temperature is the main indicator of the uneven distribution of incoming radiation from the sun. The word *temperature* is a relative term indicating the degree of molecular activity, or simply the coldness or hotness, of a substance. The more rapid the movement of molecules, the higher the temperature. To measure the degree of coldness or hotness, an arbitrary scale is used. In the United States, temperature is commonly expressed in degrees Farenheit (°F), where the boiling point of water at sea level is 212°F and the freezing point is 32°F. Another temperature scale, one that is used in most of the world, is Celsius (formerly known as Centigrade). In this system of units the boiling point is 100°C and the freezing point is 0°C. One scale may be converted to the other by the following formulas:

$$°F = 32° + 9C/5$$
$$°C = (°F - 32°) \times 5/9$$

Values of temperature throughout this text are expressed in degrees Fahrenheit, with degrees Celsius included parenthetically (except in the appendixes).

Various kinds of temperature statistics are used to define the climatology of an area. The most frequently used value is the daily mean, or average, temperature. It is computed by taking the lowest and highest readings for a 24-hour period, summing them, then dividing that sum by two to get the mean. Actually, hourly readings of temperature measured throughout the 24-hour period would serve as a better basis for deriving a daily mean temperature, but there are fewer than fifty points in Texas where hourly temperature readings are measured and recorded. The difference between the minimum and maximum temperatures for the 24-hour period is the diurnal (or daily) range. Most weather forecasts include the diurnal range by providing predictions of overnight low and afternoon high temperatures for a given locale. The mean monthly temperature is derived by adding the daily means and dividing by the number of days in the month. Average monthly low and high temperatures denote annual variations in temperature from one season to another. Finally, average annual temperatures, both minimum and maximum, are of interest. These values consist of the average values of mean monthly temperatures.

*The Statewide Temperature Pattern.* The temperature patterns that highlight Texas weather are influenced to a large extent by the amount of insolation reaching the surface, a quantity of no small value owing to the fact that Texas' range in latitude (from 26°N in the extreme south to 36°N in the northern fringe) places it on the equatorial side of the mid-latitude regions. Yet, its subtropical latitude is not the only controlling factor related to the receipt of solar radiation. The Gulf of Mexico has a profound bearing upon weather throughout Texas—and especially in the coastal plain—because prevailing winds throughout much of the year blow from the sea onto the land. Cold spells usually last no more than a few days at a time near the Texas coastline because of the warming effect of Gulf waters once the winds shift from the north back into the southeast. Still another—albeit less pronounced—in-

*3. The four great physiographic regions of Texas, the ten climatic divisions, and other major geographic sections.*

fluence on the variation of temperature across Texas is the presence of mountain barriers. The sheltering effect of mountain ranges is sometimes felt in the Trans Pecos in winter. Cold polar or Arctic air surging southward out of the central Great Plains sometimes piles up on the lee side of such ranges as the Guadalupe, Chisos, Davis, Delaware, and Chinati mountains, so that the Rio Grande Valley from the Big Bend upstream to El Paso is spared the stiff north winds and plummeting temperatures that accompany the cold-air intrusion elsewhere in the state.

*Diurnal Variation in Temperature.* Day-to-night variations in temperature across Texas are almost always appreciable. With the setting of the sun, the amount of incoming solar radiation quickly drops, and the outgoing terrestrial radiation increases markedly. Evaporation virtually stops and, if Earth's surface cools sufficiently, condensation in the form of dew or frost occurs. Depending upon the amount of moisture in the air, the difference between daytime high and nighttime low temperatures may be as much as 30°–40°F (17°–22°C). On some days in Texas, particularly in winter, the air may be so laden with moisture that the diurnal range in temperature is only a few degrees. Most of the time, the temperature will bottom out at daybreak, the point at which absorbed incoming solar radiation begins to counteract the removal of heat energy by radiative processes that have been going on all night at Earth's surface. Though incoming solar radiation ordinarily reaches a maximum at noon, daytime high temperatures usually occur at mid or late afternoon, a few or several hours before sundown. This lag is due mainly to the capacity of the atmosphere to store heat.

The daily range in temperature is nearly always larger on clear days than on cloudy ones because an overcast sky reduces substantially the escape of terrestrial radiation at night. Conversely, with clear skies, a maximum of solar radiation penetrates the atmosphere to Earth's surface during the day and at night the outgoing radiation is not inhibited by clouds. The amount of moisture in the air is also a major determinant in day-to-night temperature fluctuations. Because of the abundance of moisture in the air in areas along and near the Texas coastline, diurnal ra $\ldots$es in temperature are much smaller than in areas that are more continental (drier). For instance, on a typical summer day, the diurnal temperature range at Galveston may be from 80°F (27°C) to 86°F (30°C), while up on the High Plains the temperature at Amarillo may vary from a low of 65°F (18°C) to a daytime high of 94°F (34°C).

Soil type also has a profound bearing on the daily range of air temperature. The air above a sandy, loosely packed soil likely will have a greater diurnal range in temperature than that over a tightly-packed clay soil. Hence, frost is more likely in a sandy area than in one dominated by clay. Also, daily temperature ranges are larger over dry soils than over wet ones. Furthermore, a snow cover induces the air temperature to fall lower at night.

One other localized temperature variation that has become increasingly evident in some of Texas' largest metropolitan areas is the urban heat-island effect. On many nights, variations in the heat energy budgets of the large cities and the surrounding countryside may be as much as 3°F (2°C) to 6°F (3½°C). Indeed, on some nights that are especially clear and calm, the temperature at a business intersection in an urban area may be 10°F (6°C) warmer than at the same elevation on the outskirts of the urban area. Certainly, the large amounts of concrete, asphalt, and brick are responsible for storing a lot of heat energy, which prevents nighttime cooling from reducing the temperature as much as in open, grassy areas of the nearby countryside. The heat generated by industry, auto traffic, and residences, along with the introduction of many kinds of pollutants into the atmosphere, contributes to greater retention of heat by the lower atmosphere at night in cities.

The Four Seasons

Just as day-to-night variations in the amount of solar radiation reaching Earth's surface affect the short-term surface energy, so also do variations in insolation from season to season affect the annual march of temperature. The perennial shift in temperature reflects the gradual increase in solar energy from midwinter to midsummer, and the gradual diminution from midsummer to midwinter. The consistent angle at which Earth tilts toward and away from the sun provides the rhythm of the changing seasons so familiar to all. Because the ground is slower to conduct heat than the atmosphere above it, the air temperature usually lags from one to two months behind the periods of minimum and maximum solar radiation. Thus, even though the sun's rays are striking the Texas atmosphere at the greatest angle in mid-December, the state usually does not sustain its coldest weather until sometime in January or February. This lag is also responsible for the peak heating period in summer coming in July or August, not with the onset of summer in mid-June. On the following pages is a description of the impact that the Earth-sun configuration has upon the seasonal variation of temperature across Texas. Emphasis is given during winter to the occurrence of abnormally cold weather—the type that results in hard frosts and freezes. The opposite condition—a searing heat wave with its consequent drought—is treated extensively in Chapter 7.

*4. Because Earth tilts on its axis, there are large seasonal changes in the distribution of solar energy over Earth's surface; (a) at the time of the two equinoxes, the sun's rays are vertical at the Equator and days and nights are equal over the whole Earth; (b) at the time of the solstices, the noon-time rays of the sun have their greatest poleward displacement, 23½° north or south.*

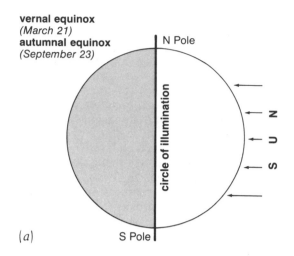

**vernal equinox**
(March 21)
**autumnal equinox**
(September 23)

N Pole

circle of illumination

N

U

S

(a)   S Pole

*Spring.* By the time Earth has tilted on its axis so that the sun's rays are striking the Equator directly from above, an event known as the "vernal equinox" (because night and day are of equal length), Texas' weather has already begun to thaw from the winter cold. By the first day of spring, which takes place on or about March 21, the southern half of Texas will have seen its last freeze of the cold season. Along with the turbulence that makes spring the most uproarious season of the year, humidities increase, as do daytime temperatures. By the middle of spring (April), daytime high temperatures vary from the low 70s in the Texas Panhandle to the low 80s in the southern extremity. Of course, cold fronts push through Texas with much vigor at this time of the year, and differences between low and high temperatures for several days after frontal passage usually are substantial. But once a norther "wears out," a process that often takes only a couple of days, strong southerly flow enriches the atmosphere with abundant moisture and much warmer temperatures.

At some point in early or mid-spring, if it did not occur in late winter, the season's last freeze signals the inauguration of the "frost-free" period. Actually, for agricultural purposes, the interval between the last killing frost of spring and the initial killing frost of autumn is the most useful indicator of the growing season. Often, though, those final or initial frosts coincide with the last or first freeze occurrences. The length of the freeze-free season decreases with an increase in latitude, the only exception occurring in the mountains of the Trans Pecos, where the high elevations often experience an uncommonly late freeze. A last spring freeze on or about April 22 in the northern Panhandle marks the start of a freeze-free period that will commonly last about 185 days. Near the center of the state—in the vicinity of Brownwood—the freeze-free period lasts about 225 days, beginning with the average last spring freeze on March 30. In some years in the southern tip of Texas, the freeze-free period has no beginning or ending.

Freezing temperatures are particularly perilous in spring when they occur later than usual in the season. In mid or late spring, an untimely freeze may catch field crops in the seedling stage, or trees and shrubs that are budding or blossoming. The hard freeze that gripped the southern half of Texas on the morning of March 2, 1980, was especially nocuous because it struck peach and plum trees that had budded just a week earlier in the midst of 80°F (27°C) heat. About half of the peach crop, valued at more than $3 million, was lost in several counties in the Texas Hill Country. Morning temperatures in the low 20s also destroyed sprouting cotton, grain sorghum, and vegetables in the coastal bend section of the state.

*Summer.* Summer is invariably hot in Texas. On the majority of days in June,

**winter solstice in Northern Hemisphere**
*(December 22)*

**summer solstice in Northern Hemisphere**
*(June 22)*

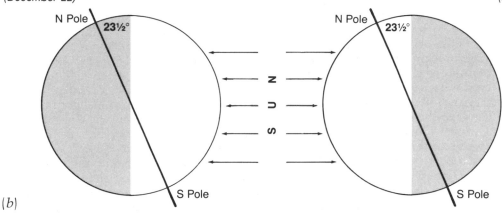

(b)

July, and August, temperatures in every sector of the state—except for some of the offshore islands—peak somewhere in the 90s or low 100s. Readings usually drop enough during the night to provide some relief from the heat, particularly in the higher elevations of the High Plains and Trans Pecos. In these two westernmost regions the lower atmosphere is almost always dry enough to allow sufficient nighttime cooling so that temperatures around dawn dip into the 60s. In fact, in the mountainous portions of the Trans Pecos, temperatures frequently drop into the 50s around daybreak before climbing back into the high 80s and low 90s by midafternoon. Farther east, however, enough moisture from the Gulf is usually present to prevent temperatures from dropping much below 75°F (24°C) before sunrise. Such moisture can also cause daytime conditions to be nothing short of stifling.

The contrast in temperature across Texas from one day to the next is much less pronounced in summer than at any other time of the year. Summer—the onset of which occurs on or about June 22—is the most constant of the four seasons in that day-to-day variations in morning low and afternoon high temperatures are seldom ever appreciable. The constancy of summer weather is due largely to the absence of weather-changing frontal systems from the higher latitudes. Cool fronts often ease into the northern fringe of Texas in summer, but they infrequently penetrate very far.

Even when they do, the polar air that is infused into the state behind the front hardly ever causes drops in temperatures of more than 5°–10°F (3°–6°C). In addition, high relative humidities—ranging from noontime readings of 45% in North Central Texas to 65% in the Upper Coast (*see Appendix F-4*)—allow fluctuations of only 20°–30°F (11°–17°C) between morning low and afternoon maximum temperatures. Morning lows in midsummer (July) range from the upper 60s in the northern High Plains to the upper 70s in more than half of the remainder of Texas. Daytime highs, on the average, are in the 90s statewide.

On a seasonal basis, summer temperatures normally attain a peak from mid-July to mid-August in all but the two westernmost regions, where hottest temperatures often occur in late June or early July. With few exceptions, hottest daytime temperatures day in, day out are found in Southern Texas near the Rio Grande between Del Rio and Laredo, along the Rio Grande between El Paso and Big Bend National Park, or in the northern Low Rolling Plains within the triangle bounded by Childress, San Angelo, and Gainesville.

*Autumn.* As on the first day of spring, the onset of autumn occurs when the sun's noon rays are directly overhead, or vertical, at the Equator. The time of the "autumnal equinox" varies a few hours from year to year on or about September 22. Almost always, the state's weather will have begun to cool down from the torrid summer by the time the first "official" day of autumn arrives. One or more polar cool fronts will have penetrated most, if not all, of the state, and the drier continental air that ensues will have brought about a wider range between morning low and afternoon high temperatures. Relative humidities statewide often are significantly lower in midautumn (October) than those that typified summer. During October, morning low temperatures range from the low 40s in the northern reaches of the Panhandle to the mid 60s in the extreme southern sector of the state, while afternoon highs vary from the low 70s in the extreme north to the mid 80s in the Lower Valley.

With the season's inaugural frost arriving in the Panhandle in the last week of October and elsewhere in the northern half of Texas at some time in November, the growing season (or "frost-free" period) comes to an abrupt end. Frost may be of one of two varieties: (a) advection, or air-mass, frost, which occurs when surface temperature drops to the freeze mark because of the influence of a passing cold front; and (b) radiation frost, which results on clear, nearly still nights when ice crystals form on cold objects. Air-mass frosts—known commonly as freezes—can be especially hazardous when they occur prematurely in autumn before plants, including many kinds of crops, have had time to make the necessary seasonal physiological adjustments. Damage to plant life caused by a radiation frost is different from that exacted by a widespread freeze. Plants killed by a general freeze may be only partially damaged by a radiation frost, which tends to be more spotty, though the economic impact of one may be as great as that caused by the other. For instance, a berry or fruit crop may be eradicated by one hard frost, but the plants themselves may not be harmed to the extent that they cannot produce again at a later time.

*Winter.* On or about December 22, with the South Pole inclined 23½° toward the sun, the winter solstice, or first day of Texas' coldest season, arrives. Because of Earth's orientation at this time of the year, the intensity of light from the sun is at a minimum because of the longer path through Earth's atmosphere (*see fig. 4*). On the average, both day and nighttime temperatures steadily drop to a minimum sometime in late December or in January. In midwinter (January) temperatures range in the morning from the low and mid 20s in the northern High Plains to the low 50s in the lower Valley (*see Appendix B-5*). Daytime highs vary from the upper 40s in Texas' northernmost extremity to the upper 60s in the far south (*see Appendix B-7*). Day-to-day deviation of low and high temperatures from normal is often drastic, however. In nearly every winter, the temperature will dip to near 0°F (−18°C), or even lower, on one or a few occasions in the northern fringe of the Panhandle. In fact, subzero temperatures have abounded in the Texas Panhandle, and in other areas of northern Texas, on more than a few occasions over the past several decades. A reading of −10°F (−23°C) or lower can be expected somewhere in the High Plains once every two years.

It is in winter, for the most part, that temperatures dip to levels low enough to pose the greatest risk to the agricultural sector of the Texas economy. Freezing temperatures are most numerous in the colder half of the year in the Texas Panhandle, where an average of two out of every three mornings features low readings at or below 32°F (0°C) (*see Appendix B-2*). Freezes are least common in the southern extremity of the state, where an average of one freeze is experienced every winter. Even farther north, in a strip of coastal plain closest to the Gulf that includes Houston, Victoria, and Corpus Christi, the number of freezes in the typical year is less than twelve.

Nonetheless, whenever growing plants are subjected to freezing temperatures for any appreciable length of time, damage or even death is virtually inevitable. Consequently, every few years a mass of cold polar or Arctic air will plunge temperatures enough to harm the winter crop in the southern quarter of Texas. For instance, the hard freeze that gripped all of Texas on January 9–12, 1962, totally destroyed the unharvested citrus crop in southernmost Texas except for a small quantity salvaged for juice processing. Growers sustained a loss amounting to $9.5 million, while vegetable producers suffered a loss of $8 million. While temperatures hovered around the −15°F (−26°C) level in the Panhandle, the deep freeze held the Lower Valley in its grasp for 65 hours. Readings on the tenth dipped to 10°F (−12°C) at Rio Grande City, while Brownsville recorded a minimum of 19°F (−7°C), the lowest temperature there so far in the twentieth century. The freeze was so severe and persistent that young citrus trees suffered extensive damage and older trees had their bark split. The state's flax crop was virtually totally destroyed, and almost all of the oat crop was lost.

Another extraordinarily frigid blast of Arctic air caused very heavy crop damage in Southern Texas on January 2–3, 1979. That cold wave forced temperatures throughout the Lower Valley below the freeze level for nearly twelve hours and resulted in a citrus fruit loss of approximately $25 million. Many other kinds of crops also sustained extensive damage because temperatures remained too low for too long. (Critical temperatures for various crops are 27°F (−3°C), avocado leaves; 26°F (−3°C), oranges; 25°F (−4°C), grapefruits and young cabbage; 24°F (−4°C), citrus leaves, mature cabbage, lettuce, and most other winter vegetables; 22°F (−6°C), citrus twigs.)

The average "return" period for a major freeze like that which hit the Lower Valley in January 1979 is once every six years. Severe freezes, like that which ravaged Texas' southern extremity in January 1962 when temperatures drop into the teens in the Lower Valley, occur about once every ten to twelve years. Rural agricultural areas are nearly always several degrees colder on nights of major freezes than are nearby cities. As a rule, temperatures throughout the Lower Valley average about 3°F (2°C) colder than the "official" National Weather Service temperature measurement made at Brownsville Airport. Hence, if the reported low temperature at Brownsville is 26°F (−3°C), then many of the crop-growing areas in the Lower Valley sustained readings in the neighborhood of 23°F (−5°C), which is the threshold value for a major freeze. A low of 22°F (−6°C) at Brownsville suggests readings in the teens elsewhere in the Lower Valley, or a "severe"

freeze in which intense and extensive damage is done to most citrus and winter vegetables.

Weather records reveal that the decade leading up to the turn of the twentieth century contained several extremely severe freezes. At Brownsville in February 1895, freezing temperatures prevailed for six straight days, with readings bottoming out in the upper teens and low 20s. During that time, more than 6 inches of snow fell in the Lower Valley. The Lower Valley also endured a disastrous freeze in February 1951, when temperatures again dipped into the teens and low 20s. It was on that occasion that a heavy glaze struck the state's southern tip, with ice accumulations amounting to 1–2 inches in many places.

## Monitoring the Atmosphere's Behavior

With the same degree of regularity that characterizes the passing of the seasons, a ritual is repeated hundreds of times every day of the year from the lofty peaks of the Trans Pecos and the interminable plain of the semiarid Panhandle to the verdant woodlands of steamy East Texas and the marshy lowlands of the coastal bend: A person ambles a short distance from a house or business establishment to what appears to be a haven for honey bees—a white, wooden, enclosed structure standing at eye level in an open field. Once the hinged door that faces north is opened, the observer peers inside, then picks up a pad and pen and scribbles down a few notes. Clasping a black measuring stick, he or she leaves the little white house momentarily and walks a few steps away to a silver canister, into which the observer thrusts the stick, then withdraws it. He or she removes the funnel that caps the cylindrical container and empties its contents. Back at the shelter, one additional notation is made. Finally, satisfied that all systems are set for the next round of measurements, the observer strolls from the scene, leaving behind yet another few bits of information on the vital signs of our ever-changing environment.

The proud tradition of cooperative weather observers faithfully measuring the elements and vagaries of nature continues.

It is these bits and pieces of data that enhance our understanding of the disposition of the sea of air that incessantly swirls around us. For more than a century, hundreds of volunteer public servants have dutifully logged their observations and then filed completed reports to appropriate government authorities. The practice of noting the weather for the government has been

*5. A standard NWS instrument shelter—ideal for housing thermometers used by the cooperative observer; the structure protects the instruments from direct sunlight and yet its vented sides allow free circulation of air to give an accurate reading of temperature.*

maintained within some families for several generations. More importantly, some locales in Texas have complete weather histories spanning fifty years because of the tireless efforts of an individual who unfalteringly managed to collect temperature and precipitation data without missing a single day (*see Appendix F-9*).

Due to the praiseworthy efforts of these volunteer observers, a vast storehouse of weather data has been accumulated on the climate of Texas. The availability of these data has had profound effects on the lifestyles of many in our mobile society. Prior to their arrival, outsiders contemplating a visit or move to the state can readily tap this storehouse for information on what to expect from the weather—and hence what to wear or plan for. With climate often a crucial consideration in many industrial endeavors, a decision maker can be advised of typical or prevailing weather conditions in some sector of Texas before a commitment is made to invest. Obviously, an analysis and understanding of weather are of critical importance to successful agricultural practices. Adequate planning for developing water resources and for highway design and construction also must be based upon reliable climatological data. Unquestionably, many facets of the Texas economy have prospered greatly from a keen awareness of the tendencies and limitations of various facets of Texas weather. The volunteer weather observer has contributed in no small way to that prosperity.

The History of Weather Observing

The first known weather annals for Texas were kept by Spanish settlers, who established and maintained numerous missions in the western, southern, and eastern sectors of the state from 1682 until the early decades of the 1800s. Anglo-American settlers, who later displaced the Spaniards and colonized the same areas, also kept records of significant weather developments in the decades leading up to the American Civil War. The earliest Texas weather record on file in the national climatic archives in North Carolina consists of data recorded at Camp Nacogdoches in September 1836. Most of the documentation, found mainly in logs, diaries, and journals, is piecemeal and deals only with severe or other unusual weather events. Still, these weather records have proven to be of considerable value to climatologists seeking to understand general weather trends in the pre–Civil War era.

An organized approach to collecting and recording weather information was first begun with the establishment of numerous military outposts on the Texas frontier between 1840 and 1860. The earliest such military weather station was maintained at Fort Houston in Anderson County, near the present site of Palestine. As at many other military outposts, weather observations lasted until the outbreak of the Civil War. In some instances, the weather annals were maintained after the war by civilian observers who lived near the outpost and assumed the task of making the observations and logging the data on a daily basis. Indeed, continuity of record keeping was preserved by these nonmilitary personnel to the extent that some locales (e.g., Fort Stockton and Fort Davis) have a nearly complete weather history spanning more than a century (see Appendix F-10).

Civilian weather observers were also at work prior to the Civil War in areas without military bases. Weather annals containing rainfall data as early as 1846 were kept by an inhabitant of San José Island (the original name restored by the legislature in 1973 to the land long known as St. Joseph Island), a settlement offshore from Rockport. A settler at Fredericksburg began a systematic record of both temperature and precipitation in August 1849. Texas A&M University, known at the time as the Texas Agricultural and Mechanical College, began temperature and rainfall measurements in May 1882 and, except for a few brief breaks, College Station has a virtually complete weather history covering a whole century. A nearly complete weather history is also available for the state's capital city; both temperature and precipitation measurements were begun in Austin in January 1856 and, except for five missing months in 1882–1883, the record is complete for that 126-year interval. No other locale in Texas is known to have as thorough a weather record as that for Austin, although more than a dozen other sites have an almost complete record covering the period from the Civil War era to the present (see Appendix F-11).

The Cooperative Weather Observer

When Congress enacted legislation establishing the U.S. Weather Bureau in 1890, weather observations were being made on a daily basis at seventy-eight locations within Texas, with volunteer civilian observers making more than three-fourths of all the recordings. Today rainfall is measured on every day of its occurrence at slightly more than six hundred locations within the state, and private citizens (cooperative observers) account for about 90% of those observations. Readings at about two of every three cooperative-observing stations are made around daybreak (at 7 A.M. or 8 A.M.) on every day of the year. Nearly all the remaining stations collect weather data prior to sunset, or around 5:00 in the afternoon. Every one of Texas' 254 counties has at least one weather station where daily precipitation is recorded and most have two or more. With very few exceptions, minimum and maximum temperatures are measured as well as daily precipitation in each county seat.

Each of the more than six hundred cooperative-weather stations in Texas is equipped with a standard nonrecording rain gauge. About half of the stations also are furnished with an instrument shelter that houses a set of special thermometers used for measuring daily extremes in temperature (see fig. 5). A second network, consisting of about two hundred "recording" rain-gauge stations, uses more sophisticated rain-measuring devices to record the rate as well as the amount of rain that falls. In all, some type of rainfall data is gathered on a daily basis at more than eight hundred different locations in Texas.

Daily data on precipitation and temperature obtained by the cooperative weather observers are recorded on forms supplied by the National Weather Service (NWS). These forms are submitted, usually at the end of every month, to the NWS and the National Climatic Center (NCC) in Asheville, North

Carolina, for processing and editing. The data are published on a monthly basis by the National Oceanic and Atmospheric Administration (NOAA), an agency of the U.S. Department of Commerce, in one of several forms: *Climatological Data: Texas*, a periodical containing daily measurements of rainfall, snowfall, minimum and maximum temperatures, evaporation, and soil temperature; and *Hourly Precipitation Data*, a publication that furnishes precipitation totals on an hour-by-hour basis for those cooperative stations in Texas equipped with recording, weighing-bucket rain gauges. The reader may obtain these data by subscribing to the publications mentioned above; the annual subscription fee is but a few dollars, and inquiries and subscription payments should be sent to National Climatic Center, U.S. Department of Commerce, Federal Building, Asheville, North Carolina 28801.

Countless numbers of other serious weather observers not on the government payroll regularly have logged data on various aspects of the weather's behavior in their communities. With rather primitive instruments like coffee cans and yardsticks, and for periods extending over several decades, these veteran weather historians have noted faithfully in their own personal records the occurrence of nearly every rain shower, hailstorm, and blizzard that beleaguered their locales. Few, if any, of these "unofficial" observations ever find their way into government documents. The volumes of notes kept over the years sometimes are handed down to younger members of the family. Undoubtedly, many records made by these citizen observers have been displaced or destroyed.

One such self-motivated weather historian who keenly watched the skies in all seasons and consulted a limited array of simple weather instruments was Elmer Bomar. A farmer of rich blackland soil near McKinney, Texas, for nearly all of his eighty-eight years, he rarely let a day pass without noting the range in day and nighttime temperatures, as well as the occurrence of every form of precipitation. Invariably, he could pinpoint with striking accuracy the time of passage of a nocturnal winter cold front by listening for the grating sound made by a rusty wind vane, which always lurched abruptly with the shifting of the wind. By correlating this sound with the striking of an old clock in the living room of his two-story home, this astute weather watcher could then assess the precise time of frontal passage no matter what time of the night it came to pass. His personal weather annals for the decades of the 1950s and 1960s give fresh insight to the impact that drought and excessive rains had on the operations of a small-scale farmer in north Texas. Undoubtedly, an untold number of other longtime Texas residents also observed the weather with similar meticulousness and steadfastness.

The National Weather Service

Profound changes in the way we observe and predict the weather have been made since money was first appropriated by Congress nearly a century ago to run the U.S. Weather Bureau, now the National Weather Service, at that time an element of the Signal Service. Greatest strides in improving forecasting skills—as well as more extensive and detailed weather observations—have been made in recent decades, particularly since the end of World War II. Today a massive computer at the National Meteorological Center in Suitland, Maryland, is used to digest nearly fifty thousand weather observations made each day by land stations, ships, and aircraft. Methods for disseminating these processed weather data have been radically transformed in the past quarter of a century: raw material for forecasters today consists of satellite photos, analyzed surface and upper-air weather charts, radar images, coded teletype messages, and computerized charts containing projections of storm movement and growth. Weather forecasts generally are more reliable today than ever before, not just because the quantity of useable weather data has swelled phenomenally, but also because techniques for handling the data and formulating projections have undergone considerable revision and modernization. No longer are daily forecasts dependent largely upon surface weather observations; emphasis now is given to detecting more subtle

*Jan 1976*

Jan was a cold dry month Had 1 in of Rain on the 24th Coldest day was the 8th 8 dg. Warmest day was 23thd 75dg.

*Feb 1976*

Feb was a cold Dry month Just 75/100 in of Rain in all. Coldest day the 7th 18dg. Warmest day was the 16th 82 dg.

*MARCH 76*

March was a warm dry month Just 2 days of Freezing weather 32 dg on the 6-13th Warmest day was the 2 day - 85 dg. Drauth Broken on the 7th Had 3 94/100 in of Rain for the month

6. *An excerpt from the personal weather annals of Elmer Bomar.*

changes that constantly occur higher in the atmosphere and that have a major impact on conditions at the surface.

*The NWS Surface Observatory.* Each hour on the hour day or night a trained observer at many of the eighteen NWS stations in Texas measures and records the following weather information: sky condition (cloud types as well as amount of cloud cover and the height above ground level of each cloud layer), visibility, air temperature, dew-point temperature (from which the relative humidity may be computed), barometric pressure, wind speed and direction, and such weather occurrences as rain, thunder, snow, fog, or dust (if they exist at or near the station). The weather charts the reader sees on television or in the newspaper are based upon these routine observations (*see fig. 9* for symbols used in describing and plotting the weather). The data are transmitted electronically, most often by teletype and facsimile receiver, to other NWS stations and numerous other facilities (such as radio and television stations) where the data are required on a continual basis. Subsequent to their collection, these data are published by the U.S. Department of Commerce as a monthly periodical called *Local Climatological Data,* which the reader may obtain for any or all of the NWS stations in Texas (or any other state for that matter) by subscription.

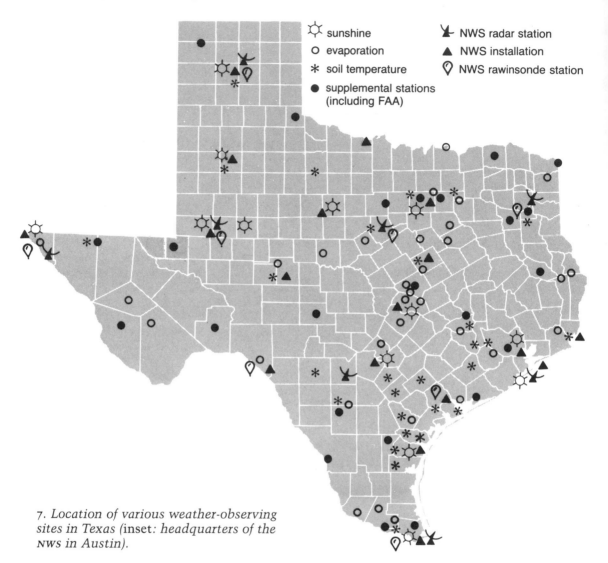

☼ sunshine        🦅 NWS radar station

○ evaporation      ▲ NWS installation

✳ soil temperature   ⑨ NWS rawinsonde station

● supplemental stations
   (including FAA)

7. *Location of various weather-observing sites in Texas (*inset: *headquarters of the* NWS *in Austin).*

*The NWS Upper-Air Station.* In addition to the hourly weather observations made at the surface, conditions aloft in the atmosphere are monitored twice daily at several of the NWS installations through use of a rawinsonde (weather balloon). An instrument package borne by a large helium-filled balloon is released each morning around dawn and each evening about dusk from eight different locations within Texas (*see fig.* 7). The instrument package, consisting of temperature, pressure, and humidity sensors, collects data as it is carried through the atmosphere to altitudes of 50,000 feet or higher and transmits the data back to a receiving unit on the ground. By tracking with precision electronic equipment the movement of the balloon and the instrument package as they ascend, the direction and speed of winds at all levels of the atmosphere may be gauged. Without the upper-air data from the weather balloon, the weather analyst's capability in forecasting imminent weather would be seriously impeded.

*Supplemental Surface Weather Stations.* Sets of surface weather observations virtually identical to those collected by the NWS are obtained on most hours of each day by trained Flight Service Station personnel who work at twelve Federal Aviation Administration (FAA) offices located at airports around Texas. Data from them complement those furnished by Texas' eighteen NWS stations and enable the weather analyst to acquire a better view (known as a "synoptic" view) of prevailing weather conditions and patterns throughout the state.

To secure additional weather data each hour from locales that are remote from manned operations, the NWS operates several Automatic Meteorological Observing Stations (AMOS) in western and southern sectors of the state. These automated stations measure and transmit to a nearby NWS office such data as temperature, wind, pressure, and amount of precipitation. Periodically, a human observer will make additional readings (such as visibility, sky

*8. A meteorologist consults a collection of dials on a console to determine the temperature, pressure, humidity, and wind movement at one of the eighteen NWS installations in Texas.*

## 9. Symbols used in describing and plotting the weather.

**weather systems**

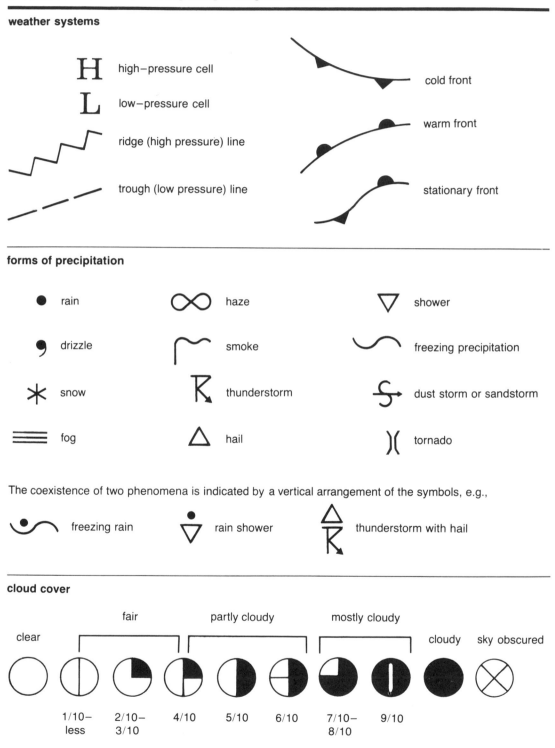

H   high–pressure cell

L   low–pressure cell

ridge (high pressure) line

trough (low pressure) line

cold front

warm front

stationary front

**forms of precipitation**

● rain

∞ haze

▽ shower

drizzle

smoke

freezing precipitation

✳ snow

thunderstorm

dust storm or sandstorm

≡ fog

△ hail

)( tornado

The coexistence of two phenomena is indicated by a vertical arrangement of the symbols, e.g.,

freezing rain

rain shower

thunderstorm with hail

**cloud cover**

clear    fair    partly cloudy    mostly cloudy    cloudy   sky obscured

1/10–
less    2/10–
3/10    4/10    5/10    6/10    7/10–
8/10    9/10

The proportional filling of the circles denotes the  number of tenths of the sky covered by clouds.

## cloud forms

The primary symbols corresponding to cloud classifications are as follows:

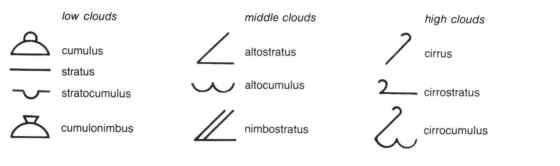

| low clouds | | middle clouds | | high clouds | |
|---|---|---|---|---|---|
| ⌂ | cumulus | ⟋ | altostratus | ⟋ | cirrus |
| ▬ | stratus | | | | |
| ⏑ | stratocumulus | ⏝ | altocumulus | ⤳ | cirrostratus |
| ⌂ | cumulonimbus | ⫽ | nimbostratus | ⤳ | cirrocumulus |

## wind

The direction and speed of the wind are indicated by an arrow such that the shaft shows the direction from which the wind is blowing and the barbs give the speed in knots. The short barb indicates 3–7 knots (3–8 mph); the long barb, 8–12 knots (9–14 mph); and the filled triangle, 48–52 knots (55–60 mph), e.g.,

westerly wind at 23–27 knots
(26–31 mph)

northeasterly wind at 3–7 knots
(3–8 mph)

south-southeasterly wind at
53–57 knots (61–66 mph)

Occasionally, the wind may be indicated on weather maps by simple arrows that denote wind direction by pointing in the direction to which the wind blows; the length of the arrows is proportional to wind speed, e.g.,

a light southeasterly breeze

a moderate southwesterly wind

a vigorous northerly wind

## a plotting model

The general arrangement of symbols describing the weather at a specific weather-observing site is shown in the diagram below. In the case shown, a thunderstorm occurred during the time between the present and the foregoing observation; the amount of rain was 1.65 inches. At the time of this observation, a shower was in progress, one-half of the sky was covered with clouds, a cumulonimbus cloud was still present, the wind was blowing from the north at 20 knots (23 mph), the barometric pressure was 1,013.5 mb (29.92 inches), the air temperature was 73°F (23°C), the dew-point temperature was 66°F (19°C), and the visibility was 7 miles.

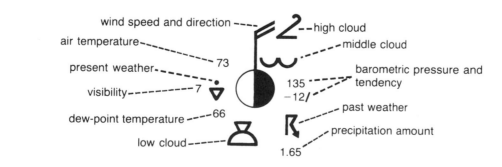

condition, and certain weather occurrences) from some of these AMOS sites and send them along with those data obtained automatically. Currently, the NWS obtains data each hour from AMOS installations at Galveston, Del Rio, Marfa, Guadalupe Pass, Sanderson, and Junction. Yet another valuable source of hourly weather information consists of reports filed by trained observers working for airlines in such cities as Paris, Temple, Longview, and Laredo. With reliable weather data coming in each hour from more than forty points in Texas, weather analysts are supplied with sufficient information to identify in timely fashion those weather systems (such as cold and warm fronts) often responsible for abrupt changes in Texas weather. Without these supplemental sources of data, an important atmospheric "trigger," such as a slow-moving front, could not be so readily located and anticipated.

Often, it is not enough just to know how much rain has fallen and where. There is an additional need, particularly by those who till the soil, to know how quickly water is evaporating from the soil or how much is draining off underground. Although it is sparse, a network of thirty-one soil-temperature stations in Texas gauges daily rates of soil temperature (*see* fig. 7). From these data, values of evapotranspiration, which is the loss of moisture from the soil supporting and nurturing a crop, may then be calculated. NWS collects and analyzes soil-temperature data and issues specialized agricultural weather advisories that are tailored to assist farmers in improving efficiency. Soil-temperature data are published each month in *Climatological Data: Texas* by the NCC in Asheville, North Carolina.

The rate at which water evaporates from the surfaces of many of the state's lakes and reservoirs is gauged at thirty-five locations by the NWS (*see* fig. 7). A shallow circular pan about four feet in diameter is used each day to measure the amount of water drawn off by the combined action of sun and wind. Pan evaporation measurements are highly useful in assessing the impact of sunshine and wind movement on the depletion of water from a surface water reservoir. As with soil-moisture data, evap-

oration data are published monthly in *Climatological Data: Texas*.

In this energy-conscious era, marked by a veritable explosion of ventures involving the use of solar energy, the need for reliable sunshine data has been heightened considerably. Unfortunately, though solar radiation can be measured, the number of data-collection points in Texas is small, mostly because fairly expensive and complex instruments are required. The duration of sunshine most often is expressed as a percentage of the total number of minutes the sun could have shown if clouds had not obscured it. The reader can secure daily sunshine data through subscription to *Climatological Data: Texas*.

*The Weather Radar Network.* No tool is of greater value to the professional weather watcher in recognizing the presence of severe storm phenomena than the radar. Although little more than a radio transmitter and receiver, the weather radar has assumed monumental importance in observing and warning of potential and impending weather disasters. Its greatest advantage is in its ability to detect precipitation between weather observing stations, which in Texas may be separated by one hundred miles or more. The structure, evolution, and motion of storm cells can be monitored continuously, thus allowing trained meteorologists to formulate warnings and other advisories on hailstorms, tornadoes, other severe thunderstorms, and ice storms.

Virtually all of Texas is under constant radar surveillance. Network radar units (models WSR-57 and WSR-74s), stationed at Amarillo, Midland, Stephenville, Hondo, Longview, Galveston, and Brownsville (*see* fig. 7), provide an electronic view of precipitation patterns in practically every corner of the state every hour of the day. In addition, less sophisticated local-warning radar units are operated at NWS offices in Lubbock, Wichita Falls, Abilene, San Angelo, Waco, Austin, Victoria, and Corpus Christi; these secondary systems are used on an as-needed basis to support warning and forecast programs. Furthermore, many television and some radio stations in the state own and operate weather radars. Just recently, radar units with the capability of

10. *Recorded messages containing a variety of weather information are broadcast continuously by the NWS via the NOAA Weather Radio.*

casts of current and anticipated weather conditions on a 24-hour basis from each of about thirty locations around Texas, including the sites of the state's eighteen NWS offices. These broadcasts, consisting of taped weather messages repeated every four to six minutes and routinely revised at least three or four times every day, are tailored to the needs of people living within the receiving area.

In emergency weather situations, each NWS office interrupts routine weather broadcasts and substitutes special weather advisories, such as watches and warnings. When conditions grow ominous, especially in heavily populated areas, the NWS meteorologist often will dispense with taped messages and instead "go live" with up-to-the-minute statements on the developing storms. When a special weather advisory, such as tornado or severe thunderstorm warning, is issued, a signal sent from the NWS office will activate those weather radio receivers specially equipped with an alarm. Sounding of the alarm (or siren) notifies the listener that a weather emergency exists; by depressing a lever or pushing a button, the listener then may hear the warning message and take appropriate measures.

Maintaining a Home Weather Station

Without such sophisticated gear as radars and weather balloons, NWS meteorologists would not have the reliable weather data they need to devise short-range forecasts. But these electronic monitoring stations, few in number, sample merely a small portion of the atmosphere. The vast majority of Texans reside in areas distant enough from an NWS office to have a climate notably different from that measured constantly by NWS sensors. Hourly NWS readings seldom match those measurements logged by amateur observers in other sections of a metropolitan area in which an NWS observatory is located. Most every urbanite can testify of those occasions when the NWS in his or her city reported a morning low temperature of 38°F (3°C), when the thick layer of frost covering his or her lawn at sunrise on the same day correctly suggested that the neighborhood sustained a much greater chill. This discrepancy is often due to the fact that official NWS temperature readings

projecting storm echoes in an assortment of colors have been added to some stations' electronic repertoires. These illustrate the intensity and size of very severe storm cells with a clarity heretofore unavailable to the television viewer.

*The NOAA Weather Radio.* Most urbanites and a growing number of rural residents can obtain first-hand much of the data collected by the NWS through a government-operated radio network known as the NOAA Weather Radio. When the National Oceanic and Atmospheric Administration (NOAA), the parent agency of the NWS, first instituted the network of radio stations that continuously broadcast weather reports, the Weather Radio system was designed to speed warnings of natural disasters and other national emergencies to the general public and emergency action units. In the past few years, this network has been expanded to provide detailed broad-

*11. To measure the amount of snowfall choose an area where drifting of snow has been eliminated as much as possible. (Texas Department of Highways and Public Transportation)*

are made by instruments mounted about five feet above ground level, where the temperature may be substantially higher than at ground level. At other times, the news media, using the "official" measurement made at the NWS installation, reports that the city received more than an inch of rain from a summer afternoon thundershower, while most residents elsewhere in town saw not a single drop.

A longing to have more "personalized" weather measurements prompts some folks to purchase an inexpensive, farm-type rain gauge and mount it on a fence post to measure the amount of rain that fell *in their yard*; the value derived is far more meaningful to that person than that reported by the NWS at its airport installation on the other side of town. Too often, however, the readings obtained from homemade weather stations are far from accurate because the observer has not taken care to properly position the instruments. The following is intended to help the amateur weather observer set up an inexpensive, yet reliable and properly exposed, home weather unit.

*Collecting Precipitation.* The most important piece of weather information for nearly any amateur weather observer is the amount of rainfall. The most common type of rain gauge in use by the public today is the fence-post variety consisting of a glass tube having gradations of 0.1 inch. If properly exposed, it can give the observer a reasonably approximate measure of how much rain fell since the gauge was last consulted. Obviously, if precision is critical to the observer, a larger, more sophisticated rain-collecting device is a must. One simple and inexpensive, but reasonably accurate, rain gauge can be made from a juice can. Mount the can on a large stake several feet above the ground and use a yardstick to measure the amount of rain. Some plastic rain gauges available in hardware and other stores afford measurements in hundredths of an inch, but this type of gauge is vulnerable to cracking in freezing weather. A more expensive type of rain gauge—that used by all climatological stations in Texas—is an excellent choice, for it not only provides precise measurements (in hundredths of an inch) but also is easy to install and maintain, and it is very durable.

In addition, this "8-inch standard National Weather Service" rain gauge will collect as much as 20 inches of rain before overflowing. If observers desire to have a truly elaborate home weather station, they may obtain a remote-reading rain gauge (one that allows them to read the amount of rainfall using a resettable counter in their homes or offices) or a tipping-bucket rain gauge (a recording device complete with strip chart that indicates rainfall intensity as well as amount for a period lasting as long as one week).

It is extremely critical that the observer's rain gauge be properly exposed. The mouth or top of the gauge should stand at least three feet above ground level. Be sure not to use a container with sloping sides; ideally the container should be a perfect cylinder. Position the gauge as far away as possible from trees, fences, buildings, or any other obstructions that might have a bearing on how much water falls into the gauge. Remember that good exposures are not always permanent; the growth of vegetation, as well as other alterations to the surroundings, can transform an excellent exposure into an undesirable one in a relatively brief spell of time.

Of considerable interest during the cold season is the amount of snowfall observed, especially in those regions of Texas where snow accumulations are common. The amount of snow that collects in a rain gauge is not indicative of the actual snow accumulation because some of the snow catch may melt in the gauge before the observer is able to make a reading. Rather, the weather observer should select a flat, smooth surface (such as a concrete slab) away from an obstruction, where drifts often occur. To get a representative snowfall reading, it is advisable to make at least three measurements—by thrusting a ruler or yardstick through the snow until the measuring device makes contact with the underlying surface—in different locations and then average the values. As a very general rule, the depth of snow can be converted to equivalent inches of rain by dividing that depth by ten. Yet, in some instances when the air is especially dry or moist, more or less than 10 inches of snow is required to equal one inch of rain. It is

best for the observer to allow the snow catch to melt in the gauge before measuring the liquid content. You may want to remove the gauge from its support and melt the snow catch in the house to obtain a highly accurate measurement of liquid precipitation.

*Measuring the Temperature.* The second most important element of the weather for an observer to monitor is temperature. Since it is dependent upon the amount of wind and sunshine, as well as precipitation when it occurs, the temperature of the air is the single best indicator of overall weather conditions in a particular locale. By using temperature data collected each day in conjunction with observations of precipitation, the amateur observer can be well on the way toward quantifying the effects of weather on such domestic enterprises as farming, gardening, and yard work.

Be certain not to locate a thermometer where it will be struck at any time of the day by direct sunlight. This precludes positioning the instrument in most places on the east, south, or west sides of the home. Ideally, thermometers should be located away from a large structure, such as a house, for by radiating heat or sheltering the instrument from the wind, it contributes to bias in the temperature reading. As a result, a thermometer that is on the north side of a house, in the shade and protected from the prevailing southerly breeze in the summer, will give readings likely to be too low. Too, a home—especially one made of brick or stone—absorbs a great amount of heat from the summer sun, such that in the evenings the house emits some of that heat, which in turn can contribute to an unrepresentative temperature reading.

If possible, place your thermometer(s) in a shielded but well-ventilated structure, such as the instrument shelters used by cooperative weather observers for the NWS (*see fig. 5*). These may be purchased from a specialized weather instrument manufacturer, although the cost may be prohibitive even for an earnest weather watcher. In that event, you may elect to build your own shelter. You should remember that such a homemade structure should be made of wood and painted white (to allow a minimum amount of solar heat to be absorbed by the device); it must protect the instruments contained within it from sunlight; and it should provide for sufficient ventilation: the sides should be louvered to permit air to circulate freely. Make certain also that the ground over which the shelter is located is representative of the surrounding area; avoid installing a shelter on a steep slope or in a sheltered hollow (unless, of course, you want temperature data for such an environment). Keep the shelter as far away from any obstruction as is practicable and displace it as much as possible from extensive concrete or paved surfaces. The thermometers should be positioned in the shelter at a height of about five feet—or approximately eye level.

*Registering the Pressure and Wind.* Of lesser value to the amateur weather watcher than precipitation or temperature data are readings of atmospheric pressure. Yet, the pressure of one's immediate environment can be a good source for supplemental information, for by carefully observing changes in the barometric pressure, a weatherperson can become more adept at anticipating imminent changes in the weather (e.g., an approaching cold front). Aneroid (or dial) barometers that hang on the wall inside the home are usually satisfactory for gauging important fluctuations in atmospheric pressure. It is not necessary to place your barometer outdoors—particularly in an instrument shelter—to get a representative pressure reading of your environment.

To have a record of wind movement you will need both a wind vane (to measure wind direction) and an anemometer (to measure wind speed). Precision wind-recording instruments may be purchased from most manufacturers of meteorological equipment. Some of the more expensive varieties provide a digital readout of continuous measurements of wind speed and direction. The industrious observer may elect to forego the expenditure of a sizable amount of money for a wind-measuring device and instead construct one. You can fashion a homemade wind vane, using a few pieces of wood and a metal plate (such as a license plate) for a fin, and obtain a highly satisfactory indication of the direction from which the wind is blowing. A homemade wind sock resembling those used at airports also suffices as an accurate indicator of wind direction. As with most other types of weather equipment, wind instruments should be as far removed as possible from any obstructions.

One very important weather observation that can be made without an instrument is the determination of the condition of the sky. The ability to identify various cloud types helps the observer to be better informed about on-going weather in his or her locale. Once an observer develops the skill of recognizing differences among high, middle, and low clouds, he or she can anticipate with greater confidence what type of weather is impending and how soon it will occur. The reader will find in Chapter 2 a detailed description of the major classifications, or genera, of cloud types in terms of their height and composition.

Signals from Mother Nature

For several millennia people have used their practical knowledge of the weather to grow or otherwise secure food, defend family and country, sail the seas, and capture energy. Seasoned mariners use their knack of scanning the distant horizon to ascertain if it is timely to venture out to sea. Fruit growers note the shift in the wind and ready their smudge pots for use on a night sure to be one accompanied by a killing frost. This intuitive feel for weather on the part of many veteran weather watchers has given rise to numerous credible and practical observations that are both helpful and enlightening:

A shift in the wind invariably means a significant change in temperature; most often, a southerly wind brings warmer temperatures, while a northerly wind causes the temperature to fall.

Dew rarely forms on a night when skies are overcast.

Frost can occur even though "reported" temperatures go no lower than the upper 30s.

The humidity declines notably when winds veer from a southerly direction to a westerly or northerly direction;

conversely, in most areas of Texas, a wind swinging into the east or southeast means an increase in moisture.

A shift toward inclement weather (rain or even snow) sometimes is forewarned by a ring, or halo, around the moon.

Clearing skies are in the offing when dark clouds grow lighter and lift to high altitudes, while the barometer reveals the air pressure to be rising rapidly.

Whereas the atmosphere usually manifests certain clues which, when properly read by the astute observer, provide tips on impending weather developments, so too the behavior of many elements of the animal world suggests a predictable change in the weather. Hunters have long been aware of the fact that prey react consistently to major fluctuations in the weather pattern. Adept fishers can testify of the effect of weather on the volume of the catch. While some of the observations formulated over the years by veteran weather watchers amount to little more than thinly disguised "old wives' tales," numerous practical observations made from the animal world contain considerable veracity:

Insects are more active prior to a storm because the warm, moist air prevalent beforehand is more comfortable to them.

Joints in the human body are more likely to ache before a change to colder, wetter weather (due to low pressure in advance of the storm allowing the gas in one's joints to expand and create pain).

Frogs croak more often before a storm, for the air is more humid at that time, thus allowing them to stay out of the water longer without their skin drying out.

Birds fly lower before a storm arrives, for lower pressure signaling a forthcoming storm creates pain in their pressure-sensitive ears; the lower they fly, the less the pain.

Much remains to be learned about the forecasting potential inherent in the activities of nature. With our complex and sophisticated array of modern instruments, we have made tremendous strides in obtaining a wealth of information about the vagaries of Earth's atmosphere. At the same time, we have been made aware of the fact that the atmosphere is fundamentally unpredictable to a significant degree and that this innate inconsistency is independent of the amount of knowledge we gain about the atmosphere's behavior or the extent to which we devote our resources to the forecast problem. We should not despair, however, for we are not yet near the limit of our ability to predict the weather. Still more improvements in forecasting methodology are in the offing.

At the present time, we must be resigned to the fact that existing weather prognostic capability permits reasonably accurate forecasts for periods not much longer than 24–48 hours. The fundamental predictability of Earth's atmosphere, in terms of accuracy and detail, varies inversely with lead time. To say it another way, forecasts for longer and more distant periods of time become less specific and less reliable.

Some help in unraveling the complex puzzle of how our atmosphere reacts to stimuli from within and beyond itself may come from a closer scrutiny of how other constituents of the animal world respond to nature's seemingly whimsical behavior. More likely, however, our ability to anticipate nature's next move on the weather front will be enhanced once we attain greater skill in using efficiently and expeditiously the myriads of weather data now being amassed. With improved computer technology and a growing awareness of the inner workings of our intricate atmosphere, giant strides in formulating and communicating better and longer-range forecasts may become as nearly inevitable as the rising of tomorrow's sun.

# 2. Fronts: Conflict in the Skies

The ground hog had just resumed his slumber in the wake of his perennial pronouncement that winter still had six weeks to run. Skeptics had begun to scoff again at those who placed credence in such a primitive method of weather forecasting when, just one day after the little animal appeared from its subterranean habitat to see its shadow, summerlike heat gripped the southern section of Texas. Indeed, the temperature along the Rio Grande almost hit the 100°F (38°C) mark on the afternoon of February 3 in that year of 1899. However, as so often happens in late winter and early spring, the weather scene in Texas underwent a sudden turnabout. With nothing but strands of barbed wire separating the plains and prairies of Texas from the frozen tundra of the north polar region, a ponderous mass of glacial air poured through the state, forcing temperatures to unparalleled depths. By the time the Arctic air settled in over Texas on the morning of Lincoln's Birthday, the scales on some thermometers were nearly inadequate to gauge the intensity of the extreme severe cold. Readings bottomed out below 0°F (−18°C) in virtually all of the northern two-thirds of Texas, while, in the usually tepid southern extremity of the state, temperatures skidded just short of 10°F (−12°C). No corner of the state—not even the partially insulated offshore islands—escaped the bone-numbing cold of this cold snap.

The climatologist who at that time wrote about the epic "blue norther" of February 1899 offered the most classic of understatements: a cold-weather spell of "marked intensity" that made the month as a whole "very unfavorable for farming operations." The bluest of all Texas northers destroyed all but those winter vegetables that had been heavily protected. Damage to fruit and vegetable crops was extensive and severe. The total dollar loss amounted to "many thousands of dollars," no small sum in pre-twentieth-century terms. The extraordinarily large mound of Arctic air produced a barometric pressure reading of 31.06 inches (of mercury) at Abilene. Its chill lowered the temperature to −23°F (−31°C) at Tulia in the southern Texas Panhandle, while morning minimums from −10°F (−23°C) to −15°F (−26°C) were common in the Red River Valley. Due largely to this historic cold front, the average statewide temperature for the shortest month of 1899 was more than 11°F (6°C) below normal.

## The Confrontation of Air Masses

Listen to any longtime resident of Texas talk about the weather, and you will be struck by the fact that the word *front* is a key element of his or her vocabulary. This is because most notable and abrupt changes in the weather—like that of late winter 1899—come about as a consequence of the passage of fronts. There usually is no need

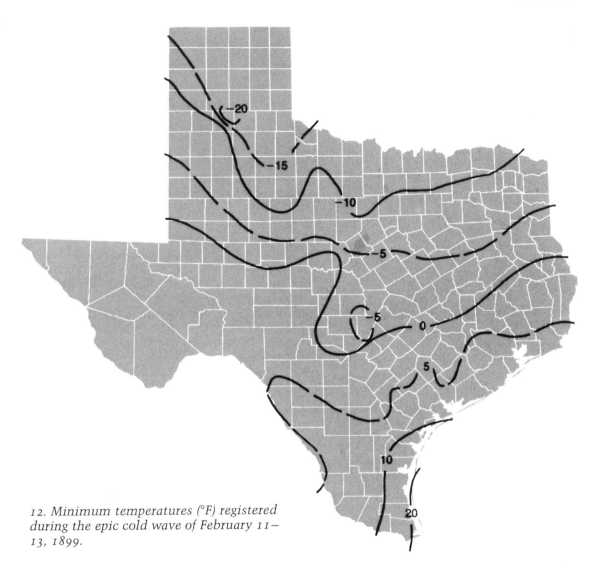

*12. Minimum temperatures (°F) registered during the epic cold wave of February 11–13, 1899.*

for a weather expert to inform us that a front has moved through a locale, for its arrival is often signaled by something dramatic, such as a clap of thunder, rain, or a sudden, chilling wind too harsh to be ignored. A front is not some thin, discrete line like that depicted on most television weather charts. Instead, it is a pronounced transition zone that lies between two masses of air having different temperatures and, hence, differing densities. Confront one air mass with another that is cooler and heavier, for instance, and the clash that ensues takes place in a narrow zone separating the . two antagonists. How pronounced the

change in temperature is within the frontal zone depends upon the season of the year and, more specifically, the source region of the air mass that is advancing.

Sometimes, particularly in winter, these fronts—or transition zones—will be as narrow as 15–25 miles, though more commonly they are much thicker than that. In summer, when fronts tend to be weak and diffuse, the boundary zone may be more than 100 miles across. Though weather charts depict fronts as two-dimensional features, they are in fact three-dimensional, as shown in fig. 13, which is a sketch of a vertical cross section through a cold frontal zone. This illustration reveals an unmistakable signature of the type of cold front popularly referred to in Texas as a "norther." Ahead of the advancing cold front, temperatures vary little over great distances; this is denoted by isotherms—lines connecting points of equal ("iso") temperature ("therm")—that are essentially parallel to the ground. Within the frontal zone, however, temperatures change substantially over short distances, as denoted by the isotherms that run vertically and close together in the figure. Behind the frontal zone and deeper into the advancing cold air mass, temperatures continue to fall but not quite as dramatically as within the frontal boundary itself. Another aspect of a cold frontal zone is the fairly uniform temperature throughout the frontal boundary from top to bottom.

## Types of Air Masses

Few events are as welcome to Texans baked by an unrelenting summer heat wave as a fresh surge of cool Canadian air. There is something invigorating about a sudden shift in the wind that brings in the flow of air that is noticeably cooler and drier, the most prominent characteristics of a vintage north Canadian air mass. While the relief may have seemed slow in coming to those residents who long for the crisp days of autumn, the infiltration of Canadian polar air takes place rather quickly. The time required to transport the cool air more than a thousand miles across the U.S.-Canadian border and as far south as the northern limit of the Gulf of Mexico pales in comparison to the time taken for the air mass to form. An air mass is an extensive portion of the lower atmosphere typified by a fairly uniform distribution of temperature and moisture, and it comes about only after air has resided in a particular area within or near the northern polar region for many days—and sometimes for several weeks.

The types of air masses that invade Texas are designated by the names of their source

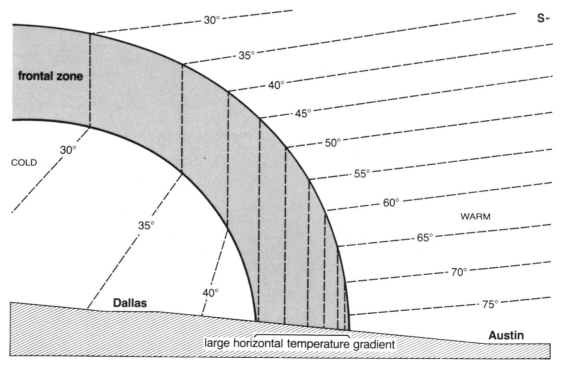

*13. Vertical cross section showing a large horizontal temperature gradient (depicted by isotherms—dashed lines) within a frontal zone moving southward through North Central Texas on a winter day.*

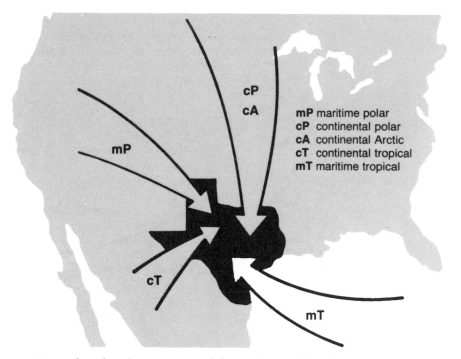

*14. Typical paths of movement of the various cold and warm air masses that invade Texas.*

regions. Essentially, the face of Earth consists of low and high latitudes that are either continental or oceanic (or maritime) in nature. Accordingly, fronts entering Texas usher in one of the following four varieties of air masses: (*a*) continental polar (or Arctic), designated as cP (or cA); (*b*) maritime polar (mP); (*c*) continental tropical (cT); or (*d*) maritime tropical (mT) (*see fig. 14*). Many of the cold fronts that invade Texas in winter usher in continental polar or Arctic air, though the former is much more common than the latter. True continental Arctic air reaches Texas only occasionally and is responsible for the especially frigid spells that force the temperature near or below 0°F (−18°C) in the Panhandle and that cause freezes in the Lower Valley. By contrast, maritime polar air, because its source region is the relatively warmer waters of the northern Pacific Ocean, is almost always milder than its counterpart that develops over the more frigid interior continental regions of Canada. Penetration of maritime polar air is more common in Texas in spring and autumn than that of

continental polar air. Intrusions of tropical air are common throughout the year in Texas, though the eastern half of the state is enveloped more often by maritime tropical air than is western Texas.

Without the migration of these various types of air masses across Texas, fronts would not be so prominent a topic of discussion. More importantly, Texas would be without most of the sudden, drastic fluctuations in wind, cloudiness, temperature, humidity, and precipitation that make the climate of the state so varied, colorful, and unpredictable. Because basically the same types of air masses enter Texas intermittently throughout most of the year, there are distinct patterns of weather associated with them that an experienced observer of the sky can, on most occasions, readily identify.

Cold Fronts

The movement of a cold front is marked by the displacement of warm and, usually, moist air by colder and, usually, drier air. Cold fronts almost always enter Texas from

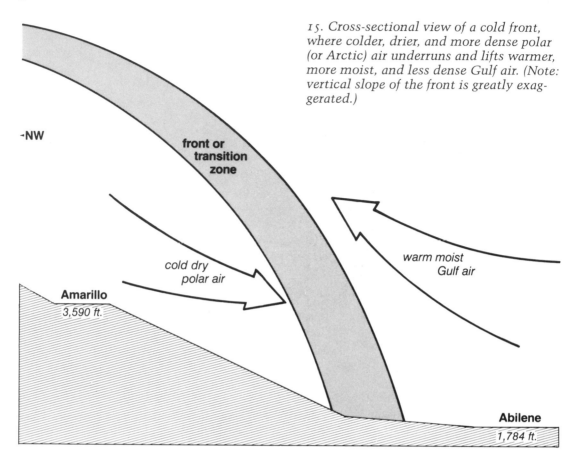

15. *Cross-sectional view of a cold front, where colder, drier, and more dense polar (or Arctic) air underruns and lifts warmer, more moist, and less dense Gulf air. (Note: vertical slope of the front is greatly exaggerated.)*

the northwest or north (much less often from the west or northeast), and their passage is invariably typified by a shift in the wind from the south or east. Since the incoming cool air is more dense than the air it is supplanting, it wedges underneath the mild or warm air and lifts it (*see fig. 15*). This uplifting of the warmer, moisture-laden air leads most of the time to the formation of a zone or band of clouds. This distinctive zone of cloudiness associated with a cold front is very apparent on photographs taken above Earth's atmosphere by a weather satellite (*see fig. 16*).

The kind of weather produced by a cold front depends on the stability and the moisture content of the two interacting air masses—especially that of the air mass being replaced—as well as the degree to which the incumbent air mass is forced to rise. Cold fronts are designated on weather charts as solid lines (colored blue) with barbs pointing in the direction toward

which the front is moving (*see fig. 9*). Most of the time when a cold front whips into Texas, the mild or warm air that is forced to ascend ahead of it flows in a fairly narrow zone almost parallel to the incoming front. In the colder half of the year, many cold fronts entering Texas have fairly steep slopes and they move rapidly; any precipitation associated with their movement is most often short-lived and confined to a zone 25–50 miles either side of the front. If a cold front moves rapidly, thereby forcing the resident air to rise vigorously, then one or more bands of large convective clouds, or thunderstorms, form in the frontal region. A fast-moving cold front may so disturb the air in advance of it that one or several lines of intense thunderstorms, or squall lines, may either precede or accompany it. By contrast, in late spring and early autumn, and occasionally in summer, a cool front has difficulty penetrating the state. Steering winds aloft are weak, and,

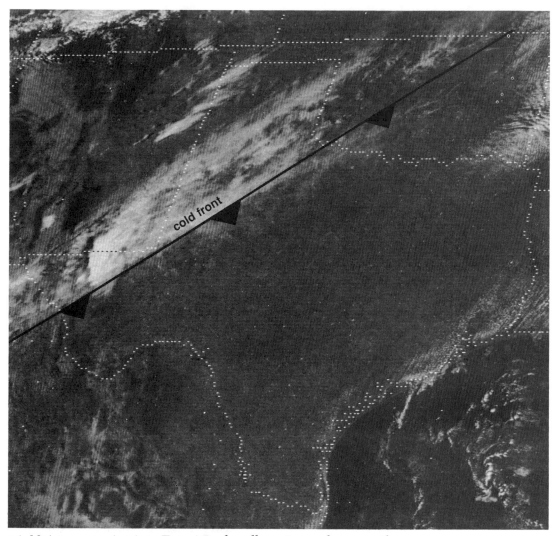

cold front

*16. Moisture pouring into Texas' Panhandle region and up over the leading edge of an advancing cold air mass appears as strands of clouds that mark the frontal boundary. (National Oceanic and Atmospheric Administration)*

because the dome of cool air is shallow, warm moist air surges up over the cool air and extends for many miles behind the cold front. In this instance, cloudiness caused by the front is much broader, and precipitation produced by the clouds usually lasts longer and extends far behind the cold front.

Cold fronts traversing most or all of Texas are most common in late winter and early spring. Sometimes during these seasons, the fronts push through the state

without triggering significant precipitation, a happenstance that may be attributed to the absence of enough moisture in the warm air mass being replaced by the invading cool air. The average number of cold fronts for most sections of Texas is seven or eight in April, while six cold fronts normally infiltrate the state in each of the months of January, February, and March (*see Appendix F-5*). Average frequency of cold fronts progressively drops off from north to south across Texas in the summer, a reality known too well by heat-beleaguered Texans who bemoan the fact that cool fronts entering the northern perimeter of the state often lose their "punch" before they get very far into the state. Whereas at least a half-dozen cold fronts customarily enter the Panhandle in July and August, no more than two of them ever reach the southern quarter of Texas in those months. In some years, no cold fronts penetrate Southern Texas or the Lower Valley at any time in summer.

Warm Fronts

The progression of warm air through Texas from the south usually produces weather events that are more subtle than those that accompany cold fronts. Still, the signs of an imminent warm front are unmistakable to the knowledgeable observer. During the cooler half of the year in Texas, one of the first clues of an impending warm-frontal passage is the migration at high levels in the atmosphere of thin, wispy, ice clouds known as cirrus. As the leading edge of warm, and nearly always moist, air approaches, the thin strands of cirrus clouds thicken gradually into an overcast, which within hours lowers to the extent that light precipitation begins falling. The precipitation, which is nearly always rain, may intensify as the warm air at the surface draws nearer to the human observer. Because the rain falls through the shallow layer of cool air about to be replaced by the warm front, the cool air often becomes saturated and fog develops. Once the warm front passes, the fog dissipates, the sky lightens, and the precipitation ceases. Well behind the warm front, in the midst of the warm, moist tropical air, only scattered middle clouds are seen.

Unlike the cold front, the passage of a warm front at the surface is the culmination of the series of progressively worsening weather events described earlier. Warm moist air streams northward up over the shallow dome of cool (or cold) air long before the leading edge of the warm air mass is felt at the surface. This is because the invading warm air is less dense than the air it is trying to supplant and it overrides the shallow cool air that hovers near the ground

(*fig. 17*). During its ascent, the warm air cools and condensation in the form of cloudiness takes place. Slowly the incoming warm air erodes away the edge of the cool air, or the cool air mass retreats, allowing the warm air to advance at the surface as well as aloft. Distinguishing marks of the passage of a warm front in Texas are a wind shift, where winds veer from the north or northeast into the east or southeast, a sharp increase in humidity (usually expressed on a weather chart as sizable jumps in the dew-point temperature), and sometimes notable rises in air temperature. Of course, changes in the amount and type of cloud cover, and any resulting precipitation, are sure tip-offs to the arrival of a warm front.

Warm fronts, like cold fronts, may generate thunderstorms. The type and amount of precipitation that precedes the arrival of a warm front at the ground depend upon the moisture content and stability, or lack of it, within the oncoming warm air mass. Also a factor is the slope of the frontal zone between the residual cool air next to the surface and the overriding warm, moist air. If the warm air surging northward out of the Gulf of Mexico is highly unstable (and it often is in autumn, late winter, and early spring), then towering thunderheads may form. Most of the time an Earth-bound observer cannot discern these thunderclouds because a low overcast obscures them, but their presence is unmistakable. Downpours that give substantial amounts of rainfall usually are the fallout from these large thunderstorms that are not uncommon in the coastal plain during winter. Occasionally, warm, moist air may slide up over a shallow dome of cold air as far inland as the Texas Panhandle and, with the aid of some upper-atmospheric disturbance, form thunderstorms that release a combination of rain, sleet, and snow. Thunder-snow showers, the result of precipitation shed by a thunderstorm falling through a subfreezing atmosphere, are not so rare in far northern Texas in winter.

Pronounced warm fronts whose passage can be discerned are confined mostly to winter and spring in Texas. When they occur, they are more distinguishable in the

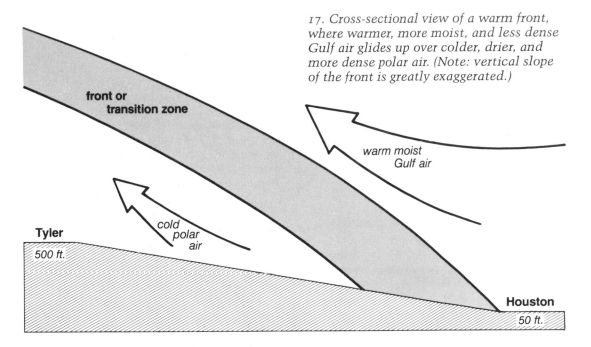

*17. Cross-sectional view of a warm front, where warmer, more moist, and less dense Gulf air glides up over colder, drier, and more dense polar air. (Note: vertical slope of the front is greatly exaggerated.)*

eastern portion of the state than in any other sector. On the average, about two warm fronts push through North Central and East Texas during April, the month that is the most active of any in the year in terms of warm-frontal movement. Warm fronts ushering in moisture-laden Gulf air rarely permeate the Trans Pecos, and only one, on the average, manages to reach the Panhandle in each of the three winter months. Clearly defined warm fronts are almost extinct in summer in Texas, an absence explained by the fact that warm, moist air almost constantly covers the state at this time of the year. Warm fronts appear on weather maps as solid, red lines to which are appended half circles "pointing" in the direction to which the warm air is moving (*see fig. 9*).

Stationary Fronts

One sure sign of approaching summer in Texas is the diminishing number of cool fronts that sweep down out of the central Great Plains states through the heart of Texas. Many of the cool fronts that penetrate the Panhandle and Red River Valley fail to reach interior portions of the state in May, June, July, and August. Rather, these

fronts "bog down" somewhere south of the Panhandle or Red River and either retreat later as a warm front or dissipate altogether. Nonetheless, these feeble fronts, though they become stationary, often trigger bountiful rains. Actually, they never completely stall; instead, they oscillate back and forth over short distances, and these gyrations frequently send off small waves in the flow of the lower atmosphere. These minor waves sometimes can be recognized by the clusters or bands of rain and shower areas that they produce as they surge along on a path often parallel with the front. In truth, these seemingly dormant weather systems are more appropriately called "quasi-stationary." If the air to the south of the nearly stationary front is tropical in origin and rich in moisture and if the front lingers in the same area for several days, the precipitation is likely to be persistent and steady, with flooding a common consequence. A quasi-stationary front that lay along the Red River for several days in October 1981, as depicted in fig. 18, allowed bands of heavy rain to form that caused extensive flooding in much of the region bounded on the north by the Red River and on the south by Abilene and Fort

Worth. These massive areas of persistent rain moved very little over a period of four days and thus dumped rains of 5–15 inches that caused more than $105 million in flood damage to property.

The Panhandle is affected by more quasi-stationary fronts every year than any other section of the state. During each of the months of May, June, and July, an average of ten fronts will penetrate part of this northernmost sector of Texas, then stall out and ultimately vanish or reverse themselves and slip out of the area a day or so later as warm fronts. A sizable amount of the year's precipitation traditionally collected in this northern fringe of the state comes from thunderstorms instigated by these almost stationary, oscillating fronts. On the other hand, quasi-stationary fronts are seldom seen in the state's southern extremity in June and July, principally because very few, if any, fronts ever survive a southward trek across Texas in the year's hottest period. Fronts are least likely to go virtually stationary during October through March. These months make up part of the transitional seasons of spring and autumn, when the circulation pattern over Texas is strong. Most cool fronts entering from the north or northwest push completely through the state without slowing down appreciably.

## Weather Systems of the Upper Atmosphere

Few people other than pilots and meteorologists have much interest in the goings-on in the upper layers of Earth's atmosphere. After all, the element of weather that interacts with humanity is the series of fronts and low- and high-pressure cells that continually progress just above Earth's surface. Yet, the perturbations in the flow of air high in the atmosphere are inextricably linked with the weather systems that manifest themselves in the lowest few feet of the atmosphere. To be able to anticipate when, where, and at what speed fronts will enter and exit Texas, for instance, demands that one be aware of changes that are constantly taking place at ten, twenty, even thirty thousand feet up in the atmosphere. It is not within the purview of this book to explain in an elaborate way the sophisti-

cated nature of the horizontal and vertical flow of air in our atmosphere. The reader may consult one of a number of fine meteorological textbooks to gain insight into the forces (e.g., gravity, pressure gradient, friction, the Coriolis factor) that control atmospheric motion. Rather, it is the intent here to describe in an elementary way how low- and high-pressure cells and their related elements in the upper atmosphere interact to provide those living in Texas with highly varied brands of weather.

Atmospheric pressure is simply the weight of all the air above the level at which the pressure is being measured. Obviously, then, pressure is greatest at Earth's surface, even though at that level the atmospheric pressure varies from spot to spot. Air pressure decreases rapidly with height, especially in the lowest 10,000 feet of the atmosphere (see fig. 19). In this nearest-to-Earth layer, the pressure typically drops about 3 millibars, which is the equivalent of about 0.10 inch of mercury, for each 1,000-foot increase in elevation.

The rate of pressure variation horizontally—or across the surface of Texas from one end of the state to the other—is much less than that in the vertical direction. Nonetheless, these relatively minor differences are vitally important in that they explain why wind speed and direction differ from one area of the state to another. Surface atmospheric pressure is depicted on standard weather maps by isobars, which are lines connecting points having the same pressure. Air pressure usually is given to the public in "inches of mercury": when the weather report states that the barometric pressure is 29.90 inches of mercury (Hg), it is meant that the atmospheric pressure is sufficient to support a column of mercury 29.90 inches high. On the other hand, meteorologists choose to quantify air pressure in units known as millibars (mb), whose conversion to inches is given by the expression 1 inch Hg = 33.86 mb.

A low-pressure area is one in which the pressure is lower than that of surrounding areas at the same elevation. It is also referred to simply as a "low" or a "cyclone," and it is designated as an "L" on a weather map (see fig. 9). If there is no recognizable center of low pressure but rather an elon-

*18. Bands of heavy rainstorms, supplied with moisture from the remnants of Hurricane Norma out of the Pacific Ocean, persisted for four days along a stationary front in northern Texas in October 1981 and caused extensive flooding. (National Oceanic and Atmospheric Administration)*

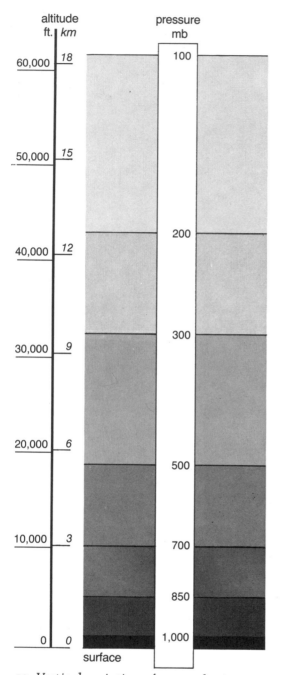

altitude
ft. | km

pressure
mb

60,000 | 18    100

50,000 | 15

40,000 | 12    200

30,000 | 9     300

20,000 | 6     500

10,000 | 3     700

             850

0 | 0     1,000

surface

*19. Vertical variation of atmospheric pressure (measured in millibars) over central Texas on a typical day in spring or autumn.*

gated region of low pressure, the feature is called a "trough." Conversely, an area with a higher pressure than its surroundings is called a high-pressure cell. It is also identified by the term "high" and "anticyclone." If there is no identifiable center of high pressure but only an elongated pattern of higher pressure, the feature is known as a "ridge." Remember that the usage of these terms applies to pressure features at the same elevation. Quite often, the distribution of pressure aloft differs considerably from that at sea level.

As a general rule, high-pressure systems that enter Texas in autumn, winter, and spring are much stronger than those that prevail in summer. They consist of colder air with brisk winds on the periphery but calm conditions at the core. Occasionally in winter, high-pressure cells with a central pressure of 30.50 inches or higher will migrate into Texas. When these huge domes of polar air follow in the wake of a cold front, they bring strong winds (which rotate in a clockwise—or anticyclonic—direction) and rapidly plunging temperatures. As the center of the high approaches a particular locale, winds slow down and eventually cease to blow when the core of the high rests overhead. The clockwise flow of air around the western edge of a summer high (known commonly as the "Bermuda high") is weaker and warmer, more stagnant, and likely to produce haze. Most of the time, high pressure connotes fair weather, although if a trough of low pressure in the upper atmosphere is affecting Texas the weather can be cloudy and wet even though high pressure prevails at the surface.

Low pressure, on the other hand, usually spawns foul weather. As with highs, low-pressure cells are more intense in autumn, winter, and spring than in summer. In the cooler half of the year, the counterclockwise winds that characterize a low may be quite strong, while in summer the lows of nontropical origin feature rather weak circulation. As long as a low remains connected with the belt of strongest upper-atmospheric winds, the weather generated by it usually is fairly predictable. However, if the low becomes detached and drifts around aimlessly, its behavior becomes much more unpredictable.

The Jet Stream

Within the broad belt of westerly winds that incessantly blow high in the atmosphere across the United States, there is at least one relatively narrow channel of winds having much stronger speeds. These cores of very strong winds, with speeds two or three times greater than that of the broad westerly current, make up the *jet stream*. The jet is a key element of the large-scale weather pattern over Texas because it often serves to move air masses into and out of the state and it also abets the formation of cyclones, or low-pressure centers, along fronts at Earth's surface. Actually, there are two types of jet streams that affect the weather over Texas: the polar jet and the subtropical jet. The *polar jet*, at an altitude of about 30,000 feet above mean sea level, is commonly found in the Texas atmosphere only in the colder half of the year. The rest of the time it is too far north to have any pronounced impact on conditions in Texas. The *subtropical jet*, on the other hand, is far more influential. It is a frequent feature of the upper-level wind flow over Texas in spring and autumn and, at times, in summer and winter.

Many of the winter snowstorms and springtime dust storms and windstorms that rage across parts of Texas are set in motion due largely to the presence of a strong jet high in the atmosphere over the southern Rockies. Indeed, the lee side of the Rockies, in an area that encompasses the Texas Panhandle, is a prime location for the development of these intense surface storms, a process also known as "cyclogenesis." With a core of jet-speed winds having velocities of 150–250 miles per hour (mph) at an elevation of 25,000–30,000 feet and surging northeastward out of the Rockies into the Great Plains, tons of lower-level Gulf moisture are pulled northwestward across Texas into this cyclogenetic area. The low-pressure center undergoes rapid intensification and moves slowly eastward across the Panhandle into Oklahoma, almost always spreading snow

in winter and early spring in the state's northern fringe and providing generous, widespread rains to other sections of Texas. A strong cold front trails from the low and sweeps steadily across Texas, ushering in cooler and much drier air.

The jet stream contributes to a marked change in the weather along the Texas coast during the cooler half of the year also. Sometimes, cold fronts that puncture the state sag and then stall either along the Texas coastline or a short distance offshore. When these fronts bog down, the rapid clearing that often takes place after frontal passage does not occur. Rather, low overcasts with intermittent light rain or drizzle coupled with dense fog may linger along the Texas coast and for many miles inland for several days. Skies eventually break up only when a "wave" or low-pressure center is spawned somewhere along the virtually stagnant front and moves northeastward out of the western Gulf. These surface storm centers are given birth as a result of the configuration of the belt of very strong winds racing northeastward across the Gulf at high altitudes. Before these low centers get fully organized and then move away from the Texas coastline toward the central Gulf coast, they likely will shed very appreciable rains on the Texas coastal plain. A large percentage of the 3-to-5-inch rains that typically occur in each month of winter in the Upper Coast stems from these "oceanic" cyclones.

Low pressure centers, or lows, may form elsewhere in Texas, also. Very often in spring and autumn, when a weak cool front pushes through part of the state and then stalls, a low will later form on the front in areas as diverse as the Permian Basin, the Red River Valley, or the piney woods of East Texas. When this happens, the front becomes "active" again and usually finishes its trek across the state. While the low is getting organized, however, thick layers of clouds usually supply generally appreciable rains.

## Clouds, Fog, Dew, and Frost

To the amateur weather watcher, the most obvious manifestation of the complex

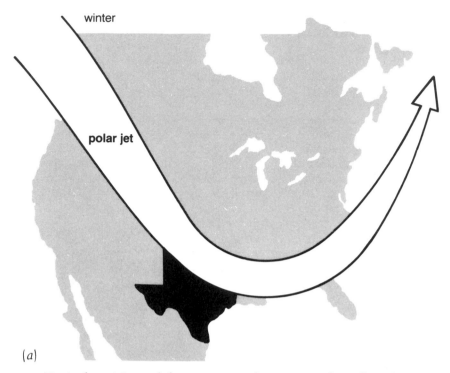

winter

polar jet

(a)

20. *Typical positions of the two types of jet stream that affect Texas weather during the cold and warm seasons. In winter the polar jet often extends far southward across North America, furnishing winds of 125–200 mph at the 30,000-foot level and propelling bitterly cold*

processes that continually go on high in the atmosphere is the appearance of various kinds of clouds. Indeed, clouds often tip off the professional meteorologist to significant weather changes taking place in the upper atmosphere that might not be discerned by using a thermometer or barometer. Of course, persons who just enjoy watching the sky do not have access to weather balloons or the data they supply to allow them to know in what ways the atmosphere is undergoing changes for better or for worse. But by being able to identify certain types of clouds, they can be better informed about impending weather changes. The altitude of clouds is also significant, for each cloud type is formed differently—in a way that is related to the type of weather that it portends. When one understands the difference between high, wispy cirrus clouds and a lower and more ominous sheet of cirrostratus, for example, one can make a fair guess as to whether

conditions will be fair or inclement and how soon a change can be expected.

Clouds consist of multitudes of water droplets, ice crystals, or a combination of the two. Sometimes, especially in winter, when the cloud cover is high and thin enough to allow the sun, or moon, to shine through it, certain optical effects are produced that allow one to ascertain if the clouds are made of water droplets or ice crystals. On several occasions every winter, a *halo*, or a narrow, bright ring around the sun or moon, is produced when rays of light from the sun or moon are refracted (or bent) by the cloud cover. The halo indicates that the cloud layer is high and made of ice crystals. On the other hand, if the cloud cover is composed of water droplets, a *corona*, or a bright ring made of various colors, may form to encompass the sun or moon. The diameter of the corona, which stems from the diffraction of light rays by spherically shaped water droplets making

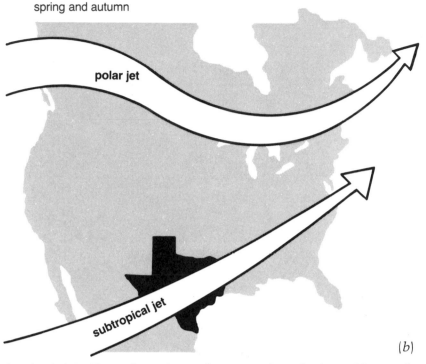

spring and autumn

polar jet

subtropical jet

(b)

*Arctic air into Texas. In spring and autumn the polar jet resides too far poleward to directly affect Texas' weather; instead, the subtropical jet, consisting of moist tropical air flowing over Texas at speeds of 40–80 mph at mid and high levels, becomes the main impetus for inclement conditions.*

up the cloud layer, is considerably smaller than that of the halo. Usually a corona is observed around the moon and not the sun, whose beams of light may be too powerful for the ring surrounding it to be seen.

A cloud will not form, even if the relative humidity reaches 100%, unless microscopic particles are present upon which the water droplets can cluster. The atmosphere, however, is never lacking in these tiny substances. Their numbers are enormous, for they consist of dust swept up from the land surface by the wind, sea-salt particles resulting from the evaporation of ocean spray, and multitudinous products of combustion (such as smoke). These materials may be found in every region of Texas on any day of the year, though the concentration of each type is usually greatest near its source region. They are referred to as condensation nuclei, and they are known to be hygroscopic in nature because they have an affinity for water. Indeed, a

salt particle is such an effective hygroscopic nucleus that it will attract moisture even before the relative humidity of the air reaches 100%. Anyone who has had to contend with the clogging of a saltshaker in damp weather can attest to salt's propensity for drawing water vapor to it.

Varieties of Clouds

Basically, clouds are formed in two ways: (a) by rising air currents and (b) by the condensation of water vapor in the air into a visible cloud shape. The first method leads to the development of *cumulus* (or *cumuliform*) clouds, so named because the term *cumulus* means "accumulated or piled up." Cumulus clouds, because they form in air that is buoyant, have significant vertical development and consist of elements that normally are individual heaps. The second method results in the development of *stratus* (or *stratiform*) clouds, for the word *stratus* stands for "layered or sheetlike."

(a)

21. The three most prominent types of high clouds seen in Texas: (a) cirrus, (b) cirrostratus, and (c) cirrocumulus. (Texas Department of Water Resources, a and b)

This type of cloud is found in air that is ascending very gently and uniformly as a result of large-scale convergence of air near Earth's surface. The vertical movement of air associated with cumuliform clouds is nearly always much more vigorous but far less uniform than that of stratiform clouds. Of the ten basic forms, or genera, of clouds that occur in the atmosphere, nine of them contain either the term stratus or the term cumulus.

The ten genera of clouds are classified according to height: (a) high, (b) middle, (c) low, and (d) clouds having great vertical extent. As a rule, high clouds consist entirely of ice crystals, whereas the middle and low varieties are composed primarily of water droplets. However, the upper segments of middle clouds may feature the coexistence of water droplets and ice crystals. Clouds with large vertical dimensions—such as the thunderhead—usually have lower portions made up solely of water droplets, middle segments consisting of both water

and ice, and upper sections composed entirely of ice. The average height of the bases of low clouds ranges from the surface of Earth to about 6,500 feet, middle clouds from 6,500 to 23,000 feet, and high clouds between 16,500 and 45,000 feet. The cumuliform clouds having much vertical extent, often referred to as "towering cumuli," are in a class by themselves, for their bases may be only a few thousand feet above ground level while their tops on a few occasions may extend as high as 70,000 feet.

Three main types of high clouds may be seen on many days of the year in Texas. The most familiar kind is *cirrus*, often observed in the "mare's tails" pattern, or as wisps that are thicker at one end than the other. Cirrus clouds, when unattached and arranged irregularly in the sky, usually hint of fair weather, but if they are systematically arranged in bands, a spell of inclement weather is in the offing. *Cirrostratus* clouds consist of thin sheets or layers of ice

(b)

(c)

(a)

(b)

(c)

*22. Texas' most common types of middle clouds are (a) altostratus, (b) altocumulus, and (c) nimbostratus. (Texas Department of Water Resources, a and c; Texas Department of Highways and Public Transportation, b)*

crystals that give the sky a milky appearance. This is the type of cloud that sometimes produces halos around the sun or moon and, in the process, suggests a deterioration in the weather that often leads to rain. As an arrangement sometimes labeled a "mackerel" or "buttermilk" sky, *cirrocumulus* clouds appear as thin, white patches of clouds having a rippled appearance. None of the three types of high clouds is capable of supplying rain, or any other form of precipitation, to Earth's surface. Yet they provide clues as to whether fair weather or a spell of rainy conditions is imminent. If cirrocumulus clouds thicken and merge together to form a solid layer or sheet, then wet weather is probably imminent. Cirriform clouds are seen in Texas in every season of the year, though with greater frequency in winter than in summer.

Middle clouds are less often in evidence in Texas than low or high clouds. Still, they are common in every season of the year, though their frequency of occurrence is less in summer than in winter. The most prevalent of the three types of middle clouds is the *altostratus* variety, a uniform, fairly dense sheet of gray or bluish color that often covers the entire sky. The composition of altostratus clouds is complex, since they usually are several thousand feet thick and transcend the freezing level. Altostratus clouds look somewhat like cirrostratus, but they are thicker, and one can distinguish the two by remembering that, though the sun or moon may shine wanly through altostratus, the sunlight or moonlight does not create a halo. Altostratus clouds frequently come after the formation of cirrostratus and are followed by widespread and fairly continuous precipitation. *Altocumulus* clouds occur as waves or patches of puffy or globular clouds and are white or gray in appearance. They consist of small liquid water droplets and closely resemble their cousins of a higher alti-

(a)

23. *The most prevalent kinds of low clouds seen in Texas include (a) cumulus, (b) stratus, and (c) stratocumulus. (Texas Department of Water Resources, b and c)*

tude—the cirrocumulus. Altocumulus can be differentiated from cirrocumulus by keeping in mind that the former sometimes cast shadows. If the amateur weather watcher has difficulty discerning middle and high clouds, he or she can take solace in the fact that even trained weather observers mistake altostratus for cirrostratus or altocumulus for cirrocumulus. A third species of middle cloud is the *nimbostratus*, a dense, shapeless, and gray-colored or dark cloud from which continuous, and sometimes heavy, rain often falls. It is made up of large water droplets and sometimes a mixture of falling raindrops and snowflakes. Virga, or rain that falls out of a cloud but never reaches the ground, may be seen with nimbostratus, which sometimes serves as a prism by diffracting sunlight into its many constituent colors. If winds are strong, *fractostratus*, or low scud clouds, often accompany nimbostratus clouds.

Texans see more low clouds than any other genera throughout the year. The most easily recognized and often observed cloud type is the *cumulus*, a puffy, white cloud whose top is usually dome-shaped with a cauliflower structure and whose bottom is nearly horizontal. Cumulus clouds are the product of rising currents of air from the layer of atmosphere near the ground, and so their shape is constantly undergoing changes. As a rule, if they do not grow together or to great heights, cumulus clouds are a sign of fair weather. If they sustain considerable vertical development, however, they may grow into *cumulonimbus* clouds, the most spectacular of all the cloud genera, whose summits invariably reach great altitudes. These mountainous clouds, known commonly as thunderheads, fit all three cloud classifications because they have bases only a few thousand feet above the surface (as much as 8,000 to

(b)

(c)

10,000 feet in the semiarid plains of western Texas) and tops that invariably extend as high as several tens of thousands of feet.

It is the height of the ceiling of the troposphere (the portion of Earth's atmosphere nearest the surface in which most weather events take place) that almost always determines how tall the thunderhead will grow. This is because the rising dome, when it reaches the upper limit of the troposphere, encounters warmer air that characterizes the stratosphere (the portion of the atmosphere above the troposphere). The top portion of the thunderhead then flattens out, leaving the classic anvil shape that is the signature of the full-fledged thunderhead. Indeed, the formation of the anvil top of the cumulonimbus indicates that the mature stage of growth has been attained by the cloud and it soon will begin to dissipate. Cumulonimbus clouds are the instigators of most local severe-storm phenomena that plague Texas almost the year around: high winds, hail, lightning, flash-flooding rains, and tornadoes. Cumuliform clouds—including the smaller cumulus and the larger cumulonimbus—are most prevalent in summer, though they are very common as well in spring and autumn. They are not scarce in winter, although at times a week or longer may pass without their developing anywhere in Texas.

The other varieties of low clouds are the stratus and stratocumulus. *Stratus* is a low, uniform, gray layer of cloud that resembles fog, but it does not rest on the ground. As the lowest of all cloud types, it is made up of widely dispersed water droplets. The bases of stratus clouds may be as low as a few hundred feet above the ground and as high as several thousand feet. They seldom produce anything more substantive than a fine light rain and often yield nothing more substantive than drizzle. They are distinguished from stratocumulus in that they are grayer and are more uniform at the base. *Stratocumulus* clouds, on the other hand, have a rounded appearance and nonuniform bases. They do not produce rain, but they may be transformed into nimbostratus, which are capable of generating good rains. Stratus clouds are very common in winter, particularly in the coastal plain. Stratocumulus clouds are found frequently in the warm season in the southeastern quadrant of Texas; they make up the low cloud cover that forms on many mornings in late spring, summer, and early autumn before dawn but then dissipate when the sun sufficiently warms up the layer of air near the ground.

The lover of sunshine will find the western tip of Texas to be the most appealing place, for it is the Trans Pecos that customarily sustains the least amount of cloud cover, regardless of the season of the year. Only one out of every six days is overcast in spring, summer, and autumn in El Paso (*see Appendix F-6*). Even in winter, cloudy skies typically occur on only two out of

every seven days, while half of the days in the month of January are clear. Autumn is even more cloud free than winter in the Trans Pecos, with an average of two out of every three days categorized as clear. By contrast, because of its proximity to the state's primary source of moisture, the coastal plain is traditionally the cloudiest region of all. A bit more than half of the days in winter and spring are marked by a substantial cloud cover not only in the Upper Coast but elsewhere throughout the eastern half of Texas as well. In fact, on a year-round basis, the amount of cloudiness in North Central and East Texas is only slightly less than that in South Central Texas. Elevations in much of these regions do not differ greatly, and with the prevailing wind much of the time being from the south, low-level moisture is usually spread fairly uniformly over all of these three regions. There are more cloud-free days in autumn than in any other season of the year in virtually all of Texas.

Another clue to the identity of the types of air masses and clouds that predominate in various regions of Texas is the amount of sunshine gauged every month. The average amount of sunshine, given as a percentage of the maximum amount possible in each month of the year, for locales in Texas where sunshine measurements are made routinely, is given in Appendix F-7. The far-western fringe of Texas, including El Paso, where an average of about 83% of the total amount of sunshine reaches the ground each year, is the sunniest place in Texas. By contrast, sunshine is least abundant in the southeastern quarter of the state, especially during winter, when less than half of all daylight hours are characteristically sunny.

Fog

A special variety of low cloud that plays a prominent role in human affairs is fog. As the aggregate of many millions of minute water droplets that break the path of light rays, fog may reduce visibilities to bothersome, if not perilous, levels. Hence, the impact of fog on society is mostly a negative one: a thick, "pea-soup" fog can so enshroud an airport as to put it temporarily

24. *The cumulonimbus, or thunderhead, with its classic anvil-shaped top.*

out of commission, it can make navigation on inland and coastal waterways difficult and treacherous, and it can slow highway motor traffic and sometimes contribute to serious auto accidents. Not to be forgotten, however, is that on occasions fog can be a blessing also: by retarding the rate at which air cools radiationally at night, a thick fog can spare an area from an untimely killing frost.

Fog simply is a dense cloud of visible water droplets, the bottom of which touches the ground. If it does not make contact with Earth's surface, the phenomenon technically is termed a stratus cloud. Much of the time fog is formed by the cooling of the humid layer of air next to the surface below its dew-point temperature. As long as the temperature of this air (and any body of air, for that matter) remains above its dew-point temperature, the air continues to be unsaturated. That is to say, it is capable of holding additional water vapor before it becomes overloaded and must give up some of its moisture in the form of visible water droplets. If additional water vapor is not supplied, then condensation of some of the vapor in the air will not occur until the temperature of the air drops to the dew-point level.

When air that is unsaturated cools, its relative humidity increases. When the temperature reaches the dew point, the air is saturated and the relative humidity becomes 100%. For instance, it is known that air having a temperature of 75°F (24°C) and a dew-point temperature of 55°F (13°C) has a relative humidity of 50%. It follows, then, that for the air to become saturated and a mist or fog to occur the temperature must drop to 55°F (13°C) (assuming no additional moisture is pumped in or is withdrawn during this time). The astute weather watcher can make a good estimate of the likelihood of fog at night by determining the dew-point temperature and comparing it with the projected drop in air temperature. If the two readings are forecasted to coincide, then fog is a good bet.

Radiational cooling is responsible for much of the fog that occurs in Texas. *Radiation fog* most often occurs at night in autumn, winter, and spring when skies are clear, the wind is nearly—but not entirely—calm, and the near-surface layer of air is rich in moisture. A sky devoid of clouds aids the formation of fog by allowing Earth's surface to cool itself quickly—a process accomplished by Earth emitting heat energy gained during the daytime into

*25. Fog is a phenomenon dependent upon a saturation of the air with moisture and an ample supply of condensation nuclei. (Texas Department of Highways and Public Transportation)*

space without interference from a cloud cover. A light breeze helps fog to form by mixing the coldest air next to the surface with warmer air immediately above it. Without this mixing, only dew is likely to form. On the other hand, if winds are blustery, too much mixing of the lowest layer of air takes place and, as a consequence, the shallow layer of cold air needed to cause fog is never allowed to form.

Radiation fog often develops during summer nights, only to burn off and dissipate not long after daybreak. The layer of fog, which may be as little as fifty feet thick or as much as several hundreds of feet thick, burns off from beneath, not from above.

Much of the sun's warmth penetrates the fog layer and warms the ground, thereby setting up thermal currents that punch up through the fog layer and mix enough dry air with the foggy air to eradicate it. On a localized basis, radiation fog is far more prevalent in valleys than on hillsides. Depressions serve as reservoirs for the drainage of the coldest and, hence, the heaviest air. Where the landscape is characterized by rolling hills and plains, it is not uncommon for the upland areas to be clear while nearby lowlands are enshrouded in fog. Valley fog usually is not thick and evaporates rapidly after daybreak.

A variety of fog that is very common in

the Texas coastal plain is *advection fog*. It is so called because its formation is dependent upon the horizontal movement of moist and relatively warm air from the Gulf of Mexico over a cooler land surface. It is prevalent during late autumn, winter, and spring all along the Texas coastline and sometimes up to 100 miles inland. As humid subtropical air drifts farther inland over a progressively colder land surface at night, it is chilled to its dew point, at which moment it becomes saturated and a thick fog develops. Quite often, a light drizzle will accompany advection fog. The fog grows thick enough to restrict visibilities to a small fraction of a mile; indeed, it is not uncommon for visibilities to be lowered to only a few feet.

In winter a highly localized phenomenon often occurs when cold polar or Arctic air invades an area and passes over the much warmer water surface of a lake or reservoir. This incoming cold air is warmed at its bottom by interacting with the water, while at the same time some of the water from the river or lake evaporates into the cold air. The few feet of cold air next to the water surface then becomes saturated and condensation of the vapor in the cold air leads to the formation of *steam fog*. Texans can see the rising tufts of steam fog on cold, crisp winter mornings as they drive near lakes or over rivers and streams.

In the upland plains and plateaus of western Texas, fog sometimes forms in winter when moist air from the coastal plain moves northwestward up over the higher terrain. As the moist subtropical air slopes upward, it expands and cools. If it cools to its dew-point level, the saturated air gives rise to *upslope fog*. This variety of fog is commonly seen in the Edwards Plateau, as well as farther north in the High and Low Rolling Plains. Sometimes it thickens enough to lower visibilities to one mile or less. Yet another type of fog that at times plagues large portions of Texas, usually in the coldest months of the year, is *rain*, or *frontal*, fog, which is the result of air being saturated by falling rain. The rain, usually triggered by a slow-moving cold front, falls into a mass of relatively dry air, thereby increasing the moisture content of the air

and lowering the temperature at the same time.

Few areas of Texas escape being socked in by dense fog at least a few times every year. The majority of incidences of "heavy" fog—dense enough to lower the visibility to one-quarter mile or less—take place in winter, while in summer fog is seldom so thick as to drop visibilities to such a small fraction of a mile (*see Appendix F-8*). The foggiest area of Texas is the stretch of coastal plain extending from the coastal bend near Corpus Christi up the coast to the mouth of the Sabine River. In winter fog beclouds the lower atmosphere at some time on more than half the days in such places as Houston, Beaumont–Port Arthur, and Victoria. In fact, heavy fog typically occurs on one out of every four days in these locales in winter. The Upper Coast also experiences the greatest frequency of occurrence of heavy fog in each of the other three seasons of the year. Very thick fog hampers activity in South Central Texas and in the Lower Valley on at least a few days each month in spring and in late autumn. In all of the state except the upper half of the coastal plain, heavy fog is a rarity in summer.

The most fog-free sector of Texas is the portion of the Trans Pecos west of the Guadalupe, Davis, and Chisos mountains. Low-level moisture from the Gulf of Mexico rarely reaches this westernmost periphery of the state in amounts sufficient to allow fog to form. Thick fog at El Paso in summer is about as scarce as a snow cover in the lower Rio Grande Valley in January. Even in winter, fog seldom thickens enough to plunge visibilities to less than one or a few miles.

## Dew and Frost

Whereas radiational cooling is a primary factor in the formation of much of the fog that hampers parts of Texas every year, it also leads to other forms of condensation at or near the surface of Earth. A fine clear night with little or no wind and an unmixed mass of cool air concentrated at or just above the ground often causes the formation of *dew*. Dew occurs when the temperature of the air next to the ground dips

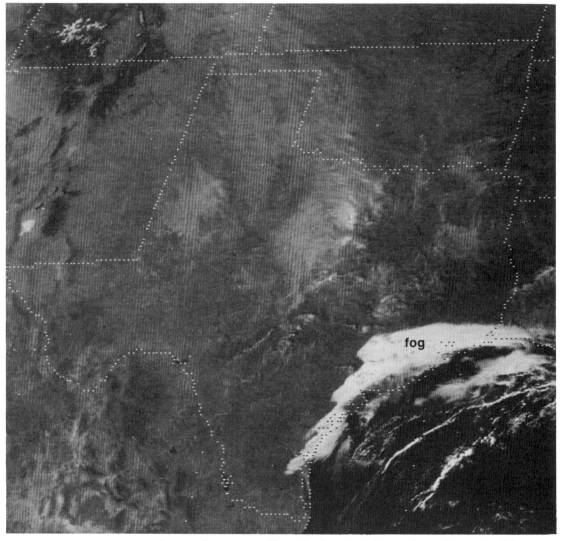

26. *A view from a satellite of advection fog enshrouding a sizable portion of Texas' coastal plain. (National Oceanic and Atmospheric Administration)*

to its dew point, at which instance water vapor in the air condenses to form water drops on such objects as chilled blades of grass. Dew usually makes little contribution to an area's moisture supply, although in the semiarid region of western Texas during long periods of rainless weather, it may be sufficient to contribute to the growth of plants that are able to absorb dew on their leaves. If the temperature at ground level drops to the freezing point, white frost, instead of dew, forms. This frozen type of condensation is known as *hoar frost* if it forms rather slowly during a crisp autumn, winter, or early spring night. If, on the other hand, a transparent, smooth coating of ice forms on the ground quickly, usually on a damp surface whose temperature plunges rapidly below freezing, then *glazed frost* is observed. Regardless of the rate at which they form, these two kinds of frost usually seriously affect the growth of grasses and other plants.

The appearance of frost attests to the

presence of cool, dry polar or Arctic air that
begins infiltrating Texas on a rather steady
basis in mid or late autumn. With each
subsequent intrusion of cool, dry air from
Canada, nighttime temperatures fall several degrees lower until near-freezing temperatures occur in the Panhandle, usually
in October, and elsewhere in northern
Texas a few weeks later. The first date of
frost is not always coincident with the first
freeze day. Indeed, frost often occurs with
reported minimum temperatures in the
middle or upper 30s, or as much as 4° to 6°F
above the freeze mark. This is because the
temperature reported by the NWS's cooperative observers is read from a thermometer
mounted in a weather instrument shelter at
a level of about five feet above the ground.
On a still, crisp night the temperature can
vary dramatically from the ground to as little as five feet or so above the ground. If
frost is seen covering the ground, you can
be certain that the temperature at ground
level reached the freeze mark.

# 3. Floods: When the Rain Pours

For more than six weeks during the summer of 1978, the center of Texas had languished in the painful throes of a severe drought. With the passing of each rainless day, cracks in the ground yawned a bit wider, while reservoir levels sank a bit lower. The eyes of longtime residents looked toward the Gulf of Mexico, for they were aware of the fact that, if the drought were ever to be extinguished in midsummer, relief most likely would have to come from the tropics. Sure enough, help seemed to be on the way when the season's first tropical storm—dubbed Amelia—was born not far off the lower Texas coastline one sizzling Saturday afternoon in late July. Unlike many of her predecessors, Amelia moved ashore with a near minimum of ferocity. She spilled only modest amounts of rainfall inland along the lower third of the Texas coastline, and for a short while it seemed that the rapidly diminishing storm would not subsist long enough to propel thirst-slaking rains farther north.

Drought-weary Texans were not to be disappointed, however. In fact, concern quickly shifted from getting enough rain to surviving a deluge when Amelia unleashed torrents of rain that spurred record flash floods in the Texas Hill Country in the predawn darkness of August 2. Within minutes tranquil rivers were transformed into veritable tidal waves as huge walls of water, moving at incredible speeds, tore through numerous communities situated within or near the rivers' flood plains. The flash floods were nothing short of catastrophic, resulting in tens of millions of dollars in property damages and the loss of twenty-seven lives. Downpours of more than 20 inches cascaded onto the headwaters of the Medina, Sabinal, and Guadalupe rivers, creating record flood crests and discharge rates that led to the engulfment of resort camps and picnic areas. Campers at Camp Bandina were alarmed to find their site inundated in a matter of seconds; eight of them drowned, while others, eventually rescued by helicopters, clung tenaciously to trees. Massive cypress trees lining the low-water banks of the rivers and ranging in size up to six feet in diameter were either yanked from the ground or snapped in two by the raging flood. Four unsuspecting persons spending the night in a cabin near the edge of the Medina River at Peaceful Valley Ranch were swept to their deaths. The Guadalupe River, rising at a rate of one foot per minute at Kerrville, reached the U.S. Highway 281 bridge, which stands fifty-nine feet above the normally serene river near the community of Spring Branch. The river flowed at a rate of 149 billion gallons per day, or more than twice the previous record flow established in 1959. (The normal flow rate is 104 million gallons per day.)

Although most tropical storms expire rapidly after moving inland, Amelia's remnants were far from finished. They drifted northward into western North Central Texas on the following day, dumping more

than 30 inches of rain on parts of Shackel-
ford County. A wall of water 20 feet high
formed on Little Hubbard Creek and crashed
through Albany, covering more than three-
fourths of the city and drowning six of its
residents. Many inhabitants of the city
spent the night on roofs of houses and on
top of oil derricks and in trees before they
were rescued the next day by helicopters
and boats. A nearby earthen dam developed
a fault 50 feet deep and 25 feet wide, and
water pouring out of the dam rushed into
Stephens County, destroying twelve homes
and damaging three hundred others. The
Clear Fork of the Brazos River swelled to
a width of 2 miles near Graham, and a
17-foot crest on the river washed ranch-
lands and submerged the U.S. Highway 183
bridge.

Few events in Texas' weather history attest as suasively to nature's proclivity for
carrying a blessing to excess than Amelia's
killer floods. Of all Earth's vast resources,
none is more vital to the survival of its
inhabitants than water. Moreover, a lack of
sufficient water has led to the extinction
of some civilizations, while an abundance
of water has contributed to the prosperity
of many others. With the growth rate of
Texas' population steadily ascending and
with the realization that the state's supply
of fresh water is virtually fixed, we confront the fact that the chances for continued prosperity—even our own survival—vary with the amount, distribution,
and judicious use of the available freshwater supply.

**Patterns of Precipitation**
The Water Cycle

The need to plan and implement means
of securing suitable-quality water for
Texas' future needs would not be so dire
if the water now held in the multitude of
reservoirs and rivers that pockmark and
dissect the state were not in continual transition. Great quantities of freshwater are
lost everyday when some of the water
stored in lakes and reservoirs is removed
into the atmosphere by evaporation, when
some of the contents of rivers and streams
pour into the Gulf of Mexico and mix with
saltwater to become unfit for human con-

sumption, and when an unsaturated atmosphere extracts moisture from the ground
through the process of evapotranspiration.
Yet, nature provides compensation for this
loss by drawing water from the Gulf of
Mexico and the Pacific Ocean and depositing it on land as rain. These exchanges of
water among the land surface, atmosphere,
and the adjacent ocean make up what is
called the hydrologic, or water, cycle. As
illustrated in fig. 27, some of the water
vapor supplied to the atmosphere by the
ocean is carried inland where it combines
with smaller quantities of land-evaporated
water vapor to form precipitation. Some of
the precipitation that reaches the ground
soaks into it and replenishes the supply of
groundwater, while some is returned to the
ocean by runoff from rivers and streams.
The remainder of the precipitation is given
back to the atmosphere by evaporation.
The hydrologic cycle is seldom ever completed locally; rather, moisture evaporated
from coastal waters usually is precipitated
many miles away and far inland.

The circulation of the atmosphere furnishes an almost continual exchange of air
masses from the land to the sea and vice
versa that facilitates the operation of the
hydrologic cycle. Humid air from the Gulf
of Mexico and from the eastern Pacific
Ocean near Mexico is channeled into Texas,
where it condenses to form clouds and rain
droplets, some of which ultimately fall to
the surface as rain. Much drier continental
air masses then envelop the state and absorb much of the moisture evaporated from
the land. Thus, not only do rivers and other
waterways (both surface and subterranean)
transport land-precipitated moisture back
to the Gulf of Mexico, but so do the dry polar
air masses that penetrate the state in all but
the hottest and driest periods of the year.

The displacement of atmospheric moisture is of fundamental importance to
weather for a variety of reasons: (a) it often
condenses to form clouds and precipitation
and, occasionally, fog, dew, and frost; (b) its
concentration in the air influences the rate
of evaporation, a process of monumental
significance to animal and plant behavior;
(c) it absorbs radiation from both the sun
and the land surface; and (d) its transforma-
tion from one state to another (vapor to liq-

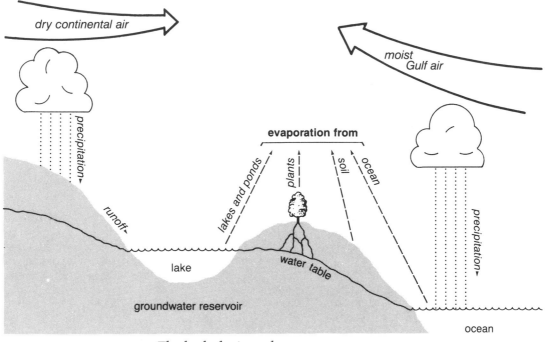

*27. The hydrologic cycle.*

uid, for example) releases heat energy that contributes to the vertical motion of air in the lower atmosphere. None of these attributes or tendencies, obviously, is more important than the others. In terms of supplying vital amounts of water to Earth's surface, however, the process of condensation is most crucial to the formation of precipitation. Water vapor will not condense into droplets until it is cooled to its dew-point temperature or unless more water vapor is added.

Few clouds that form ever yield precipitation. Even low clouds, such as the billowy cumulus cloud and the layered stratus cloud, often fail to shed any meaningful precipitation. This may be due to the overabundance of condensation nuclei, where too many particles compete for the available supply of cloud moisture. The result is too many water droplets too small in size (and weight) to fall out of the cloud as rain. It is upon this tendency of nature toward excess that several attempts to increase rain through cloud seeding in recent years have been predicated. While nearly a half-dozen different rain-increase programs have

been performed in parts of Texas since legislation was enacted in 1967 to regulate all weather-modification activities, none of them has been evaluated to the degree satisfactory to atmospheric scientists.

The Distribution of Rainfall

All but one of the attempts to increase rainfall in Texas since 1970 have centered in the central and western sectors of the state, for it is in these regions that rainfall has often been too scanty for the maintenance of a sufficient supply of water for crops, livestock, and other water-dependent enterprises. An examination of fig. 28 reveals that more than half of Texas, in an "average" year, collects less than 30 inches of precipitation, virtually all of which occurs as rainfall. In fact the western third of the state normally receives less than 20 inches in a year. By contrast, the eastern third of Texas is ordinarily endowed with at least 35 inches of rain over a twelve-month period. So diverse are the climatic regimes that typify the western and eastern ends of the state that average annual precipitation varies from a bit less than 8 inches at El Paso at the extreme western tip to more than 56 inches in the lower Sabine River Valley of extreme East Texas. On the other hand, yearly rainfall totals vary little from north to south across Texas, as indicated in the figure by the lines connecting points having equal rainfall that roughly approximate lines of longitude. Average annual precipitation in the northern periphery of the Texas Panhandle amounts to about 20 inches, the same sum that characterizes the Lower Valley region of extreme southern Texas. As a general rule, precipitation totals, on an annual basis, decrease about one inch for each fifteen miles from east to west across Texas.

Accordingly, the Trans Pecos is the driest region of the state with an average region-wide precipitation total for the year of only 11.65 inches. On the other hand, the Upper Coast and East Texas are traditionally the wettest with average annual totals of 45.93 inches and 44.72 inches, respectively. (*See Appendix C-1* for precipitation data.)

Two areas of Texas are characterized by sharp gradients in average annual rainfall. One is the central mountainous section of the Trans Pecos, with an average yearly rainfall of 16–18 inches, or 4 to 6 inches more than that typifying points 50 miles or more to the east as well as to the west. This rainfall "oasis" is due to the influence

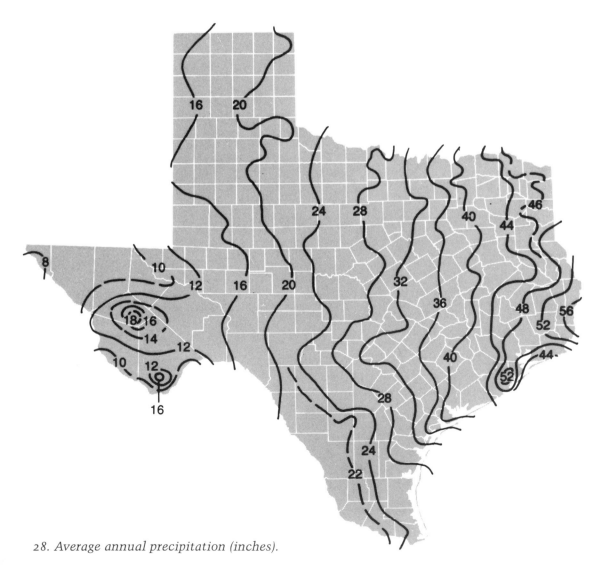

28. *Average annual precipitation (inches).*

of the mountainous topography, which forces moisture-laden winds from the Gulf of Mexico—and even the Pacific Ocean—to be lifted great distances. This ascension of humid air to levels of lesser pressure results in cooling of the air to its saturation point, a process that is conducive to the formation of rain-bearing clouds. Any visitor to the Davis Mountains who beholds the abundance of evergreen vegetation can appreciate the role of the terrain in "squeezing out" quantities of moisture from the atmosphere that lower elevations elsewhere in the Trans Pecos do not receive. The other region where rainfall does not decrease fairly uniformly in a longitudinal direction is along the Balcones Escarpment, which is the juncture of the Edwards Plateau with South Central Texas. Average annual rainfall at Boerne (in Kendall County) is virtually as abundant as that in Austin, a city that is more than 60 miles to the east. This anomaly in the rainfall pattern is believed to be a result of the influence of the area's topography. The Balcones Escarpment features sharp rises in elevation of 1,000 feet or more over distances of only a few miles. Warm, moisture-laden air from the Gulf being pumped inland through central Texas is lifted almost abruptly by this geographic feature. If the humid air is channeled into the region in sufficient quantities—such as when remnants of tropical storms drift inland—the additional amount of rain extracted by the sudden uplift of the wind currents can be sizable.

It is rare that Texas' driest point in any year is situated in some region other than the westernmost Trans Pecos. In most years some locales in this arid region receive less than 8 inches of rainfall. Indeed, in several of the drought-plagued years of the 1950s, some communities there recorded less than 3 inches of rain (see Appendix C-4). At the height of the extreme drought in 1956, Presidio collected only 1.64 inches of rainfall during the whole year. Nearly all the scantiest annual rainfall totals measured in Texas since 1950 were registered somewhere in the Trans Pecos, and all but a few occurred in one of the seven years of intense drought that marked the 1950s in Texas. On the other hand, the most bountiful amounts of rainfall in any year almost invariably are received in either East Texas or the Upper Coast. Yearly sums of 60 inches or more are not uncommon in at least one locale in either of these regions every year. Indeed, in five of the years between 1951 and 1980, one or more locales collected 80 inches of rain or more. Due largely to several weather disturbances of tropical origin, at least one Upper Coast city near the coastline received 100 inches or more of rain in the years 1973 and 1979. This includes the total for 1979 of 106.44 inches of rain gauged at Freeport.

Rainfall is seldom, if ever, spread uniformly throughout the year. Some periods furnish the bulk of the year's rainfall, while others constitute the "dry season." Spring is the wettest season of the year in most of Texas, with May normally a bit wetter than April. Only in the western third of the state does this rule not apply, for in the High Plains and Trans Pecos, rainfall is most plentiful in summer. Much of the rain received in those regions is generated by scattered and mostly semiorganized convective clouds (also regarded as thunderheads), which are numerous on many days from June through September. In fact, rainfall produced during the year's four warmest months in these two regions of highest elevation constitutes about two-thirds of the total precipitation that occurs in the "typical" year. Elsewhere in Texas, at such points as Dallas–Fort Worth, San Antonio, and Houston, rainfall reaches a secondary peak at the end of summer or at the onset of autumn; September rainfall is almost as abundant as that in April, for instance. This pattern reflects the influence of tropical weather systems—most notably hurricanes and tropical storms—that migrate out of the Gulf of Mexico at that time of the year. In fact, in some years one or more major tropical disturbances may supply far more rainfall in August or September—or even October—than is collected earlier in the year in spring from nontropical weather events. Average monthly rainfall statistics for selected locales in Texas undeniably are biased to some degree by the very substantial rains that have fallen in conjunction with tropical disturbances on several occasions since 1951 (see Appendix C-2).

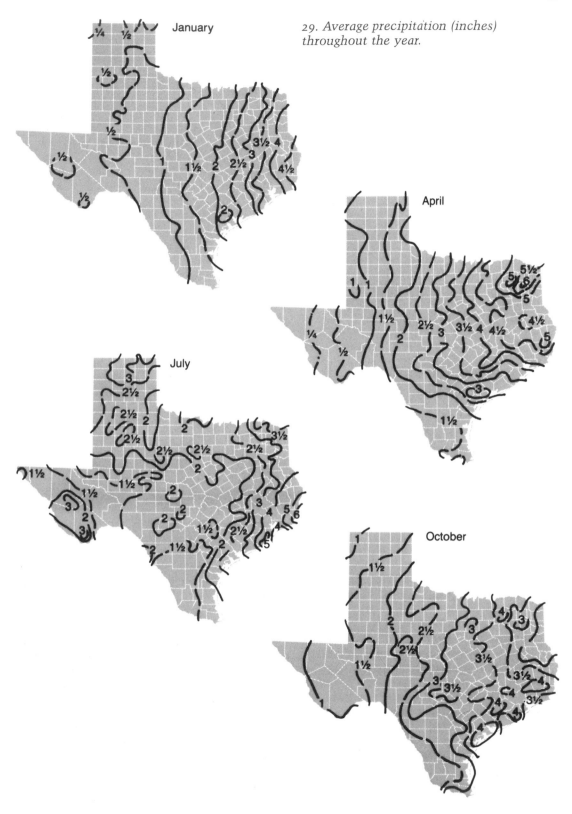

29. *Average precipitation (inches) throughout the year.*

Except in western Texas, summer rainfall is almost as lean as precipitation that comes in winter. Indeed, in East Texas, at such points as Lufkin, rainfall in July and August is less than that observed in December and January. Again, whether a summer in Texas is unusually wet or dry is often dictated by the degree of activity, or inactivity, of tropical weather systems in the Gulf of Mexico. Elsewhere, however, the three summer months ordinarily are a bit wetter than their winter counterparts. In Texas' southern quadrant, at such locales as Houston, San Antonio, and Corpus Christi, average monthly rainfall "bottoms out" not during winter but at the onset of spring (March).

It is not surprising, with the extreme eastern section of Texas collecting the greatest amounts of rainfall in nearly every year, that the same area also features the largest number of rainy days. A "rain day" consists of any 24-hour period during which enough precipitation falls to be measured, with the minimum amount of measurable precipitation being 0.01 inch. About 150 days bring precipitation totaling 0.01 inch or more in the Sabine River Valley, while in the Trans Pecos the number of rain days is only 40–50 in a typical year. Since a rainfall amount of only one or a few hundredths of an inch is often inconsequential, days when one-tenth (0.10) of an inch of rain occurred are more noteworthy. Even with a heavier threshold value for a rain day, it is seen, in fig. 30, that two to three times as many rain days occur in East Texas as in the western fringe of the state in every season of the year. Days that furnish one-half (0.50) inch of rain are most numerous in the Upper Coast and East Texas as well, though the number of such days is appreciable in western sections during the summer month of July (see fig. 31). Substantial rains of 1.00 inch or more are very occasional in the western half of Texas in midsummer, and they are infrequent even in eastern sections (see fig. 32).

## When Rains Become Excessive

Of all the elements composing the hydrologic cycle, meteorologists are most interested in precipitation. After all, precipitation is the cause of a host of effects, both beneficial and detrimental. On the one hand, rainfall replenishes water supplies not only for human consumption but also for industrial use, it spurs plant growth, it cleanses the atmosphere by removing pollutants, and it furnishes water in amounts needed for navigation and recreation. Often, however, this vital commodity comes in quantities and in time intervals too exaggerated for the land surface to accommodate. Excessive rainfall leads to flooding, which in turn brings about massive economic loss. The injurious impact of too much water dumped too quickly on a land surface is felt especially in urban areas where development has taken place in floodplains.

### Runoff

Because precipitation in Texas is almost always some form of rainfall, the impact on the flow of water in rivers and streams is immediate. Only in the Panhandle in winter is the snowpack ever substantial enough to provide water in a delayed storage mode. Runoff, or that portion of precipitation that returns to the Gulf of Mexico or reservoirs within Texas either by flowing over the land surface or through the soil and water table, depends largely upon rainfall amount and intensity (i.e., the amount of rain that falls within a certain amount of time). Peak runoff on most rivers in Texas occurs in late spring or early summer, coincident with or just subsequent to the relatively heavy rains that usually take place in April, May, and early June. Some rivers in the southern section of the state—such as the Nueces—sustain maximum runoff in autumn due to the influence of tropical cyclones. In the west—on the Pecos River, for example—peak runoff comes during or soon after relatively heavy summer rains and again in autumn from substantial rainfall induced by tropical cyclones. Most often, if rainfall is of low intensity, nearly all of it will seep into the soil and thus take a great deal of time in ultimately reaching a river or stream. On the other hand, if a large amount of rain falls in a matter of minutes (or at most a few hours), that intensity may exceed the soil's infiltration capacity. That is to say, the soil cannot absorb all the water it is

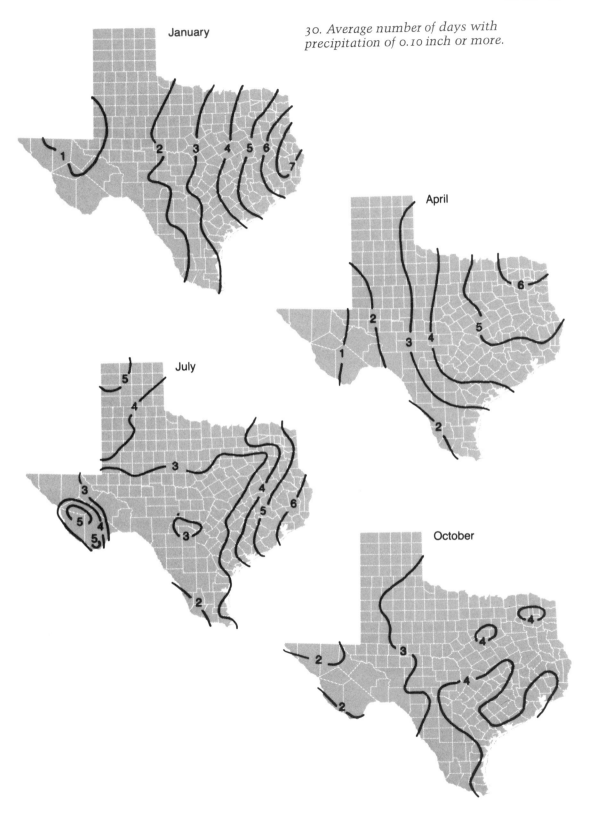

30. *Average number of days with precipitation of 0.10 inch or more.*

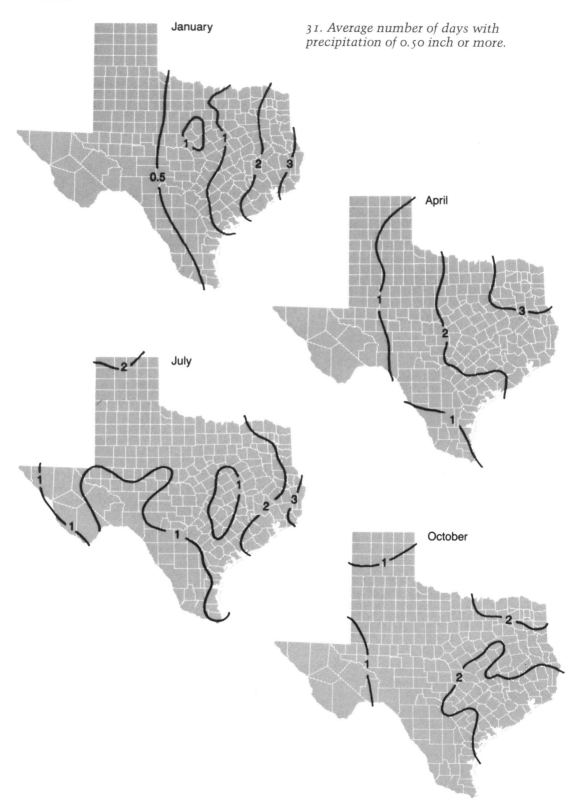

January

April

July

October

31. Average number of days with
precipitation of 0.50 inch or more.

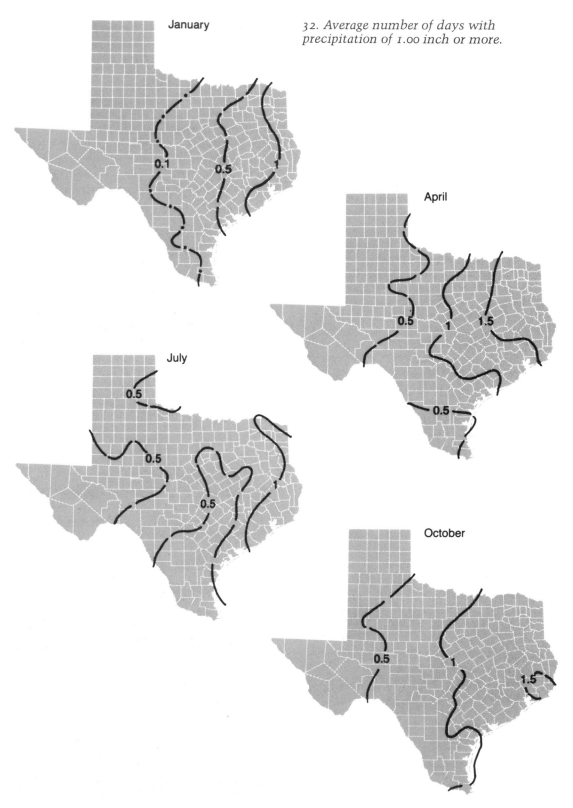

32. *Average number of days with precipitation of 1.00 inch or more.*

receiving, so some portion of the rainfall becomes residual water that flows quickly to a nearby river or stream. If this excess water is of sufficiently large quantities, flooding results. How much rain infiltrates the ground and how much "runs off" into a nearby conduit also depends upon the type of soil on which the rain falls and how much moisture that soil already contains. If the soil is largely or entirely sand, almost any reasonable amount of rain can be expected to be absorbed with little or no sur-

face runoff. Clay soils, however, have very low infiltration capacities, and even rains of low intensity are likely to lead to some surface runoff.

Soil having a high humus content and covered by a dense layer of grass enhances the absorption of rainwater by acting as a sponge. The presence of vegetation will also determine the amount of rain that runs off. Plants improve soil structure and their roots provide a means of channeling excess soil water to greater depths within

*33. Normal distribution of runoff, shown as a percentage of the annual amount, for various rivers at selected points in Texas. (Texas Department of Water Resources)*

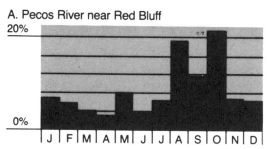

A. Pecos River near Red Bluff

B. Red River at Denison Dam

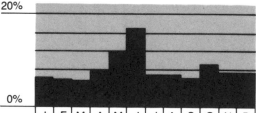

C. Sabine River near Bon Wier

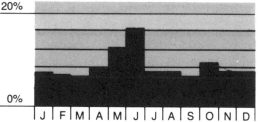

D. Brazos River near Bryan

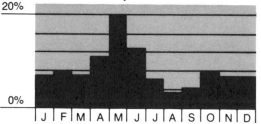

E. Nueces River near Mathis

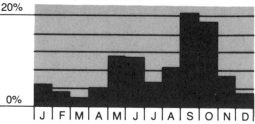

the ground. They also retard the horizontal movement of water and thus help hold down the rate of soil erosion. They have the ability to retain some rainfall on their external structures and to slow the rate of fall of raindrops. One other consideration in determining the likelihood of flooding is the duration of rainfall. The ability of the ground to absorb rainwater decreases as the duration of a rainy spell increases. Eventually, if the rain continues long enough, the ground's infiltration capacity becomes so small that even low-intensity rains lead to substantial runoff. This explains why, when a hurricane or other major tropical disturbance surges inland across the Texas coastline, rainfall that lasts for many hours may soak the ground to the extent that the soil's infiltration capacity reaches zero. At that instant, since the ground can take no more water, every drop of rain that falls becomes runoff—and potentially an element of a flood.

## Types of Floods

Two types of storm events are the initiating agents for the kinds of floods witnessed in Texas. One is the general, prolonged rain that through sheer quantity of water leads to widespread flooding throughout a watershed. The other, which is common particularly in the warmer half of the year, is the very intense thunderstorm that may crop up in a matter of minutes and cover only a few square miles of land surface. At times these huge thunderstorms generate rains of 2–3 inches in less than half an hour, and this enormous amount of water is deposited on such a small area of land surface in so short a period of time that most of the water cannot be absorbed by the ground. Consequently, a flash flood takes place when this large volume of rainwater quickly leaves the watershed of a river or stream as runoff. If that same quantity of rainfall had been distributed rather uniformly over the whole watershed, little if any runoff would have occurred. Yet, the concentration of such a heavy downpour in a localized area results in the rapid peaking of streamflow that is followed by a rapid recession of the flow.

Destructive floods occur somewhere in Texas every year, and on several occasions in many years lives have been lost and damage to property has run into many thousands, if not millions, of dollars. Sizable floods have occurred in every month, but as a general rule, spring and early autumn are the worst seasons for floods (*see Appendix E-3*). Runoff into rivers and streams in most of Texas reaches a peak in spring in a typical year, then declines during summer only to ascend again with the onset of autumn. Spring is frequently the season of floods because it marks the end of the period of soil-moisture accumulation. With the beginning of springtime rains, the soil, depleted of much of its moisture by the relatively dry winter, has its moisture supply replenished. Frequent, heavy rains— a common trait of a Texas spring—eventually soak the ground, so any surplus rainfall runs off to swell rivers and streams to near or above flood stage. The secondary maximum of runoff in late summer and early autumn is a reflection of the substantial rains generated in many years by one or more major tropical weather disturbances or by slow-moving cold fronts that permeate the entire state.

Since heavy thunderstorms develop often in every sector of Texas in late spring, summer, and early autumn, flash floods are a phenomenon to be expected anywhere in the state at that time of year. They are common in the mountainous Trans Pecos, particularly in summer, when a sudden deluge transforms a normally quiet stream or dry gulch into a torrent of water that sweeps away everything in its path. Flash floods are common in the High and Low Rolling Plains and the Edwards Plateau, for ground and tree cover is sparse in much of these regions and the slopes of streams and creeks are fairly steep. However, no portion of Texas—and, for that matter, the entire United States—is as prone to be afflicted by flash floods as the segment of central Texas in the vicinity of the Balcones Escarpment. This topographic anomaly at the juncture of the Edwards Plateau with the expansive coastal plain is cut in hundreds of places by rivers, streams, creeks, and arroyos that fill rapidly when appreciable rains fall upstream on their watersheds. Often, these narrow waterways overflow in a matter of minutes, and the water that rages along carries a deadly cargo of smashed structures, uprooted trees, boulders, mud, and other debris. The devastating impact of flash floods generated in August 1978 by Tropical Storm Amelia on scores of communities in the Texas Hill Country serves as a solemn reminder of how vulnerable the region is to the sudden eruption of flash floods.

Many of the floods that occur in the eastern half of Texas are more extensive and are slower to materialize than the flash floods that often plague areas farther west. In North and South Central Texas, as well as in East Texas and the Upper Coast, the terrain is characterized in many places by broad flat valleys with considerable timber and brushland. The natural drainage channels in these areas have gentle slopes and limited capacity, and they follow meandering courses from headwater areas all the way to their mouths at the Gulf of Mexico.

Runoff is usually slow, but a broad, flat-crested, slow-moving flood is set in motion when substantial amounts of rain occur over periods lasting several days. Occasionally, bands of scattered convective storms or a line of frontal thunderstorms contribute enough rainfall to swell a river or stream out of its banks, but, most often, floods in eastern Texas stem from long-continued, warm, rainy weather. These persistently appreciable rains last a few or several days and are associated with a slow-moving upper-atmospheric storm system or a surface front that has stalled. In autumn, they may be the product of the dying residue of a hurricane or a tropical storm. As long as relatively warm, very moist Gulf air is fed into the large-scale weather system responsible for the inclement weather, the rain will continue and waterways will swell further. In some years, in the coastal plain, the lower basin

34. *Flash floods result when large volumes of rainwater are deposited in such short periods of time that the ground cannot absorb the water. (Texas Department of Public Safety)*

35. *Flooding often is aggravated along rivers and streams when debris piles up under bridges, thereby retarding the flow of water downstream. (Texas Department of Highways and Public Transportation)*

regions of some of Texas' largest rivers will remain inundated for many days and, sometimes, even several weeks.

Great Floods in Texas' History

The ferocity with which a rolling wall of water induced by flash-flooding rains can kill and destroy was aptly demonstrated not only by Amelia's deluge but also by the savage flood that tore through the town of Sanderson west of the Pecos River on June 11, 1965. Nightlong rains, including up to 8 inches in two hours, filled two creeks that converge on the western edge of the small Trans Pecos community of 2,300. One of the two—Three-Mile Draw—dumped such a torrent of water into the other—Sanderson Creek—that a wall of water as high as 15 feet was sent crashing through part of the town about daybreak. Houses, automobiles, and business structures were sent tumbling along like toys in a drainage canal. Twenty-six persons were drowned by the sudden, awesome gush of water. Another 450 people were left homeless.

Damage worth $2.5 million was done to homes, businesses, and automobiles. The ravaging flood uncovered graves and washed away many headstones and markers, some of which were found as far as four miles away from their original location. The flow of water at the time of the flood peak on Sanderson Creek, which traditionally is dry on all but a few days each year, was measured as 76,400 cubic feet per second.

Other parts of Texas, from the canyons of the Big Bend to the foothills of the Caprock and the Edwards Plateau, have been victimized repeatedly by raging floods in recent decades. No area, however, has been assaulted by as many catastrophic flash floods as the Balcones Escarpment, the sharp gradation in terrain that marks the eastern limit of the Edwards Plateau. Symbolic of the suddenness with which flash floods surge along the myriads of creeks and streams that slice through the Balcones Escarpment is the wall of water that swept through a residential section of Austin on the night of May 24–25, 1981. A torrential downpour amounting to about 10 inches over a four-hour period and falling on ground already saturated from previous heavy rains instigated a severe flash flood that roared along Shoal Creek and several other neighboring streams. Many of the thirteen drowning victims were at low-water crossings or were trying to drive through rapidly flowing water at the time tragedy befell them. Creeks rose so quickly that many residents of west Austin had to seek refuge in second-story attics. Residential and commercial property damage amounted to about $40 million.

Because of their proximity to the Balcones Escarpment, Austin and neighboring communities have been subjected to numerous other devastating floods. A cloudburst dumped almost 5 inches of rain on the capital city during a four-hour period early one night in October 1960; a resulting flash flood caused damages worth $2.5 million. On the same night, rains of as much as 10 inches elsewhere in central Texas sent rivers and streams raging out of their banks, and eleven persons were swept away. Similar flash floods erupted in central Texas a few days before Thanksgiving in 1974, claiming the lives of thirteen people, ten of whom resided in Travis County. Six months later, in May 1975, a barrage of vicious thunderstorms that battered Austin with hail and high winds also produced slashing rains that rapidly filled creeks and ravines in the area, drowning four people. West of Austin, in and near the Pedernales River basin, floods over the past several decades have been exceptionally harmful. The Pedernales swelled to its highest level ever when rains of 23–26 inches were spilled onto its watershed on September

*36. The broad valleys and flatlands of eastern and southern Texas, which serve as temporary storage areas for large amounts of runoff, sometimes remain inundated for weeks, especially in spring. (Texas Department of Public Safety)*

8–10, 1952. The resulting flood cost five lives, destroyed 17 homes, and damaged 454 others, with the total property loss amounting to "many millions" of dollars.

Other locales on the coastal side of the Balcones Escarpment have also been devastated by flash floods. Few in Texas' weather history were as pernicious or as costly as the ravaging floodwaters that covered much of the Texas coastal plain on May 11–12, 1972. The highest discharge rate in the history of Purgatory Creek forced the San Marcos River to overflow and caused $400,000 in damage to property in San Marcos. Rainfall of one foot or more in barely more than a two-hour period produced floods that killed seventeen people in New Braunfels and exacted a toll in property damage of $17.5 million. Rising water forced more than three thousand residents of the city to forsake their homes in the middle of the night; some had to be evacuated by helicopter. A peak discharge of 2,630 cubic feet per second was gauged on Trough Creek near New Braunfels, which is the equivalent of a torrent of water produced by a rainfall rate of 8½ inches per hour. No run-off rate of that magnitude for a watershed so small (less than one-half square mile) had ever been recorded anywhere in Texas up to that time. Rainwater poured into the Guadalupe River at New Braunfels with such speed that the usually sedate river rose from a level of 3 feet to 31 feet in just two hours.

Conceivably the most protracted spell of flood-producing rains since the Great Depression of the early 1930s ravaged the northeastern corner of Texas during the last week in April 1966. Day upon day of torrential downpours caused rivers and streams to rampage in the area between Fort Worth and Texarkana, and the tempestuous waters swept away bridges, washed out dams, and caused damage worth $27 million to dwellings, businesses,

*37. A house in Sanderson after the flash flood of June 1965. (Texas Department of Public Safety)*

*38. San Marcos during the May 1972 flood. (United Press Photo)*

farms, and roadways. Rains of 12–20 inches, including nearly 23 inches at Gladewater in only 2½ days, were responsible for thirty-three drownings, including fourteen in Dallas and Tarrant counties. Dams in the vicinity of Longview that had withstood floods for forty years broke under the force of rapidly rising water. The Dallas–Fort Worth area experienced some of the wildest flash floods imaginable during this period. Rainfall with an intensity as high as 11 inches per hour filled creeks and streams to overflowing almost instan-

taneously in the densely populated metropolis. Most of the area's fourteen drownings occurred on April 28 when the victims were trapped in automobiles that were carried away by floodwaters swirling across streets and roads.

The deep canyons that beautify the area within and upstream of the Big Bend and the Caprock Escarpment farther north in the High Plains have been the scenes of many swiftly flowing flash floods. The great flood generated by the remains of Hurricane Alice in June 1954 (discussed in more detail in Chapter 4) stemmed from cataracts that amounted to more than 27 inches in a 48-hour period in the vicinities of Langtry, Sheffield, and Ozona. The rapid release of such huge quantities of rain sent an 86-foot wall of water through the Pecos River canyon that obliterated a highway bridge near Comstock. The torrents of rain caused the Rio Grande to rise 50–60 feet (or 30–40 feet above flood stage) at Eagle Pass and Laredo. Another graphic illustration of canyon-filling floods stemmed from a 10-inch rain in less than two hours near Canyon in the High Plains on May 26, 1978, that sent a wall of water through Palo Duro Canyon, drowning four people and injuring fifteen others.

Floods, not necessarily of the flash variety, have been almost commonplace in Texas' Upper Coast region over the years. Perhaps the most deleterious onslaught of heavy rains not attributable to a tropical weather disturbance occurred on June 12–13, 1973, when a massive rainstorm dumped 10–15 inches of rain centered in the vicinities of Houston, Liberty, and Conroe. Ten persons were drowned, and an astounding $50 million in crop and property damage was incurred. Almost a repeat performance took place three years later, when more than 13 inches of rain struck Houston, causing eight deaths and property losses of $25 million.

Texas' second-largest metropolitan area has not been immune to the impetuosity of flash floods. Aside from the calamitous flood of April 1966 mentioned earlier, likely the most ruinous flood in modern history to afflict the Dallas–Fort Worth area occurred on May 16–17, 1949, when a 12-inch rain forced the Clear Fork of the Trinity River out of its banks. Ten square miles of the city of Fort Worth were flooded when levees constructed to contain the floodwaters broke. The flood's toll consisted of 10 deaths, 13,200 persons left homeless, and 3,000 residences and 250 businesses damaged or destroyed, with damages amounting to about $6 million. About half as much damage was wrought by another flood that occurred on the Trinity River, its tributaries, and other nearby creeks and streams on September 21–23, 1964. Considerable residential damage was inflicted on the Richland Hills suburb of Fort Worth along the Big Fossil Creek, while expensive homes in North Dallas were heavily damaged by flooding White Rock Creek.

Floods on numerous occasions prior to World War II have been almost as wrackful. Appendix E-3 includes a tabulation of the most significant floods, not caused by tropical cyclones, that ravaged parts of Texas during that time. The most calamitous flood originated from a mammoth rainstorm that dumped more than 20 inches of rain on a large portion of central Texas on September 8–10, 1921. That weather event probably is best remembered for the record amount of rainfall that cascaded down on the community of Thrall: an incredible 36.4 inches in one eighteen-hour period, a record that still stands as the greatest high-intensity rainfall in U.S. weather history. Floodwaters from this gargantuan rainstorm swept 215 people to their deaths, caused $19 million worth of damage to property and crops, and left water as much as 9 feet deep in the downtown section of San Antonio. Copious rains of a similar magnitude produced a highly disastrous flood on the Brazos River in the early summer of 1899. Rainfall totals of as much as 33 inches spread out over three days cost at least 30 lives and property damage amounting to $9 million. Fourteen years later, another flood occurred on the Brazos, nearly quadrupling the number of drowning victims; it took 117 lives and caused damage of more than $8.5 million. In April 1900 swollen creeks and streams fed the Colorado River to the extent that McDonald Dam at Austin gave way suddenly, and the wall of water that ensued downstream took 23 persons to their deaths. A four-day rain that measured more than 25 inches at San Angelo in September 1936 forced the Concho River to flood the city's business district and about five hundred homes.

## Predicting Flash Floods

Flash floods are the most dangerous type of flood because they are the most difficult to predict. This is due mainly to the fact that forecasting the occurrence of thunderstorms remains a formidable task. By contrast, floods that result from widespread rains usually are easier to anticipate. Still, this type of flood also presents a substantial hazard because of the considerably large volume of water involved. When meteorologists develop and broadcast a prediction of heavy rains, their job is not finished. Rather, they are faced with the challenge of predicting how much of the total precipitation will fall within the watersheds of rivers and streams, and when and for how long these heavy rains will occur. They are charged with supplying flood information to the operators of reservoirs, owners of farms, generators of electrical power, and various industrial and domestic interests on the rivers' flood plains. Consequently, to alert the citizenry of any excessive accumulation of water that poses a hazard, the National Oceanic and Atmospheric Administration (NOAA), of the U.S. Department of Commerce, maintains a careful, continuous watch on the nation's rivers, streams, and other waterways. By collecting and constantly analyzing a horde of special river and rainfall data, NOAA's NWS is equipped to provide river forecasts and flood advisories. Like the other warning services offered by the NWS—those pertaining to tornadoes, severe thunderstorms, hurricanes, and strong winds—flood warnings provide the public with time to evacuate flood-prone areas, to move property and livestock to higher ground, and to take what other necessary emergency procedures are warranted. The river forecast and flood-warning service saves millions of dollars in flood losses annually and, most significant of all, countless lives as well.

Once meteorologists at the NWS's River Forecast Center in Fort Worth ascertain the possibility of flash-flooding rains, advisories are issued to the public through radio, television, NOAA Weather Radio, and local emergency organizations. (Keep in mind that the news media, by law, are not obligated to supply that type of information; however, most do broadcast weather warnings as a public service.) A *flash flood watch*, usually issued at least six to twelve hours in advance, means that the populace should remain informed of ongoing weather events (by staying tuned to the news media) and should be ready for immediate action if a *flash flood warning* is released. The flash flood watch does not mean that a flood is in progress; rather, it suggests that weather conditions exist, or are expected to develop, that have the potential for fomenting floods. When the trained weather observer detects, by radar, satellite, or field-observer reports, that a flood has materialized, then a flash flood warning is given out. The warning admonishes those persons living in an area subject to flooding to take prompt action. It cannot be overstated that when you realize a flood is imminent you should act quickly to save yourself. Time to take the necessary precautions may consist of only a few minutes, if not seconds.

Flood-alert systems preserve lives and property only when there is an effective response from the community. All residents of a community with the potential of being affected by a nearby river or stream should understand what a forecast river height means in terms of their proximity to that water body. They ought to know where their property is situated with reference to the flood level of a nearby river. Equally important is the need for them to know where they can go to be safe from rising water. Most communities can either provide flood-mapping information or steer you to its source.

# 4. Hurricanes: Raging Tempests from the Gulf

It was not that the storm struck without ample warning. Indeed, its early detection and subsequent patternistic behavior led to the greatest mass movement of humanity in peacetime. As a result, towns all along the Texas coastline from Corpus Christi to Port Arthur were abandoned; only at Galveston, which was protected by a 17-foot seawall, did residents stay put in any great numbers. As with all approaching marine storms, shreds of high-level cirrus clouds signaled the impending change for the worse as did ocean swells that grew more pronounced with the passage of time. The dials on wall barometers slid unfalteringly, while outside the winds gradually but steadily intensified. Radios bleated out the news that an awesome cyclone, born six days earlier in the western Caribbean, was bearing down on the Texas coastline's midsection. During its slow, deliberate trek across the Gulf of Mexico, the colossal storm seldom deviated from a straight-line course, thus making it one of the most predictable hurricanes in Texas history. Hurricane Carla, with an eye measured at 30 miles in diameter and winds clocked at over an astonishing 150 mph, slammed ashore near Port Lavaca just after noon on September 11, 1961.

Prior to Carla's emergence from the eastern horizon, there was nothing to do but shore up and ship out. More than a quarter of a million Texas coastal residents did just that. Some stayed behind foolhardily to brave the elements, and thirty-four of them did not survive. Unquestionably, the evacuation of virtually all the inhabitants of many small, unprotected coastal towns helped ensure a rather modest death toll for such a violent storm as Carla. However, as one of the most intense hurricanes to strike Texas in recorded history, Carla wrought damage and destruction worth $408 million. She did not perish without a protracted struggle, meting out torrents of rains that flooded more than 1.5 million acres of land within Texas. In fact, Carla supplied heavy rains far inland in central Texas and could still be identified several days later while generating rain in North Dakota.

When the hundreds of thousands of evacuated coastal residents returned home a short while after Carla unleashed her fury, they saw the unmistakable signature of a hurricane: buckled utility poles hanging like corpses from taut power lines, twisted street signs pointing every direction but the correct one, mangled television antennae strewn on roofs and in yards, and boats of all sizes inverted in canals. Dead cattle were scattered about, leaving a sickening stench, and hundreds of rattlesnakes, swept out of their habitat by surging water, wriggled and hissed in yards and on streets. Some livestock, crazed by the storm ordeal, attacked people and had to be shot, and alligators were seen roaming about in some neighborhoods. Shrimp boats and pleasure craft were swept far ashore by the storm surge, with even heavy steel boats as much as 70 feet in length displaced up to 500 feet from the shoreline.

Carla practically destroyed the fishing resort of Port O'Connor. Nothing but flattened debris was left of the downtown section, and nearly all the homes on the street fronting the bay were destroyed or so badly broken as to be irreparable. Three brothers, who donned life preservers and endured the raging storm in a house specifically designed to withstand hurricanes, watched a wall barometer drop steadily as the eye of the storm moved ever closer around four o'clock on the afternoon of September 11. One of them described the scene when the eye enveloped Port O'Connor: "Everything got deathly still. The sun peeped out for a few minutes, and the wind and rain died altogether. There were even some birds flying around."

Port Lavaca also bore the brunt of Carla, but, because of a total evacuation hours before she struck, the city sustained no fatalities or serious injuries. Tides there were measured at 18.5 feet above normal, and winds estimated at 175 mph pummeled the city. The old causeway heading out of the city toward Point Comfort was demolished, while a second, modern causeway nearby was severely damaged with huge concrete slabs separated and lifted partially atop other broken pieces of the conduit. A heavy steel air force patrol boat some 70 feet in length was driven 50 yards onto land, and many huge planks and heavy chunks of driftwood littered the harbor. Meanwhile, at the height of the storm siege in Palacios, seawater covered nearly all the town at a depth estimated at 13 feet. Only the hospital, situated at the highest point in town and used as a haven for some two hundred residents who chose to ride out the storm, avoided getting wet, although water crept to within a few feet of the front door.

While very few tropical weather disturbances ever rip into Texas' coastal flank with a ferocity equal to that of the unforgettable Carla, most of the scenes described are duplicated somewhere along the Texas coast every few years. Texans who reside near the coastline are faced with the prospect that the forthcoming summer or autumn likely will provide one or more maritime storms that will prove to be destructive—if not catastrophic—for some sector of the state's coastal region. When oppressive heat and humidity become fixtures of the daily weather pattern in southeastern Texas, the Gulf of Mexico begins fostering numerous disturbances of varying sizes and intensities. Occasionally, some of these evolve into organized "waves" and furnish a large percentage of the rainfall that typically soaks the Texas coastal plain during the months of peak warmth. Unfortunately, one or more of these concentrated masses of convection sustain explosive growth and are transformed into identifiable storm centers with a definite cyclonic circulation pattern that may attain the status of a tropical storm or hurricane. In other instances, a similar sequence of events much farther east in the Atlantic Ocean or Caribbean Sea leads to a tropical storm or hurricane that migrates westward into the Gulf of Mexico, where warm sea-surface temperatures and a compatible circulation pattern in the atmosphere sustain the storm or nourish it further.

## The Birth of a Sea Giant

Easily recognized by astronauts and weather satellites orbiting far above the outer limit of Earth's atmosphere, the hurricane hovers over the sea as a gigantic whirlwind of awe-inspiring violence. Because of its intensity, size, and duration, the hurricane is the most destructive weather phenomenon known. Although the origin of the term is not known for certain, "hurricane" likely stems from the Spanish word *huracán*, which itself was a derivative of either Hunraken (the Mayan storm god) or Hurakan (the Quiche god of thunder and lightning).

Unique in both ferocity and structure, the hurricane is a highly advanced stage of a "tropical cyclone," which is an all-encompassing term used to describe all cyclonic circulations originating over tropical waters. The immense swirling mass designated a hurricane was in its infancy merely an "easterly wave," which then evolved into a "tropical depression," and, later, a "tropical storm." These antecedent forms of the hurricane, described in fig. 39, are differentiated by pressure pattern and

| symbol | stage | range of wind speeds | |
|---|---|---|---|
| | | mph | knots |
| | easterly wave | | |
| L | tropical depression | less than 39 | less than 34 |
| 6 | tropical storm | 39–73 | 34–63 |
| 6 | hurricane | 74 or more | 64 or more |
| | *major* | *101–135* | *88–117* |
| | *extreme* | *136 or more* | *118 or more* |

*39. Classification of the various stages of a tropical cyclone.*

the measure of highest sustained wind speeds. Easterly waves are very common events in the Caribbean and the Gulf during summer and early autumn; only a smattering of them ever develop into depressions or tropical storms and fewer still ever mature into hurricanes.

Why is the onset of summer nearly always synonymous with the beginning of tropical cyclone activity in the Atlantic Ocean? The position of the semipermanent subtropical ridge (or high-pressure cell) is a key factor in the instigation of hurricanes and lesser tropical disturbances from early summer (June) through mid-autumn (October). As the sun ascends to its highest point in the sky and the length of daylight reaches a maximum late in June, the large zone of high pressure perennially migrates northward in the Atlantic and assumes a surface position near the Bermuda Islands. The flow of air around this dome of high pressure is anticylonic, or clockwise; to the south of the center of the high, the surface wind flow is mainly from east to west (*see fig. 40*). These "easterly tradewinds" sometimes extend as high as 40,000 feet or more during summer and cover vast areas of the southern half of the North Atlantic Ocean.

During summer, the Bermuda high-pressure system experiences alternating stages of enervation and strengthening, and the center of the massive high may drift away momentarily from its normal position in proximity to the Bermuda Islands. Sometimes—especially in early summer and again in early autumn—when the "Bermuda High" is relatively weak and displaced farther south in the Atlantic from its usual location, a trough of low pressure may impinge on the tropics. If the southern end of this trough gets trapped by the prevailing easterlies of the tropics, an "easterly wave" may form. This wave propagates westward and organizes the circulation pattern near the sea surface into areas of converging and diverging airflow. Where air converges, the depth of the moist layer of air near the ocean surface expands, and this circumstance abets the formation of large cumulonimbus clouds (thunderheads), which may grow to heights of 40,000 feet or more above the ocean.

From June to October hardly a day goes by that one or more easterly waves are not recognizable in some part of the Caribbean Sea. Many of them travel westward without undergoing any intensification. Occasionally, an easterly wave is further organized when forces high in the atmosphere intensify the circulation within the wave. In a matter of hours a vortex (much like a whirlpool) forms within the wave, and further maturation of the circulation pattern leads to the development of a tropical depression, then a tropical storm, and possibly a hurricane.

The embryo of a tropical storm or hurricane may also originate in another way. Between the large high-pressure cell centered in the Atlantic near Bermuda and its counterpart in the Southern Hemisphere is a low-pressure band encircling Earth in the vicinity of the Equator that is known as the "equatorial trough," or "intertropical convergence zone" (ITCZ). This lateral band migrates northward, following the sun, during spring and early summer ultimately to a point about 12°N latitude in August (*see fig. 42*). At times the ITCZ is observed to be 100 miles wide and is marked by huge clusters of thunderstorms that soar to heights of 35,000–50,000 feet over the open seas. During summer, with the ITCZ well north of the Equator, an eddy within the ITCZ may be spun out northward into the prevailing easterlies, where it undergoes intensification and grows to become a depression.

It is believed that the key to the sustenance and intensification of a tropical cyclone is the interaction of low-level and high-level winds within and near the cyclone (*see fig. 43*). The storm's circulation pattern is maintained—and perhaps even enhanced—when more air is pumped out the top than is brought in at the bottom near the ocean surface. On the other hand, if more air converges on the storm than is discharged, the system fills up and

the storm dissipates. It is important to remember, too, that the inflowing air must carry a sufficient amount of heat and moisture to fuel the vast atmospheric power cell.

Still, it is not known for certain that such ingredients as low-level inflow and high-level discharge of energy actually cause hurricanes. Rather, they may be merely symptoms of other, more subtle influences that contribute to hurricane growth and intensification. Unquestionably, the giant heat engine is both unpredictable and inefficient. The relatively infrequent occurrence of a hurricane suggests that a multitude of peculiar conditions must be met before it is generated.

## Looking Back:
## The Most Memorable Hurricanes
Intruders from the Gulf of Mexico

Whereas Carla is regarded by most veteran weather observers as the most catastrophic hurricane to hit Texas since the start of the twentieth century, its death toll pales in comparison with that exacted by the "West India Hurricane" that slammed into Galveston Island on September 8, 1900. Because it took an estimated 5,000–8,000 lives, that awesome storm is still recognized as the single worst weather disaster in U.S. history. Secondarily, it also cost Texans a loss in property valued at $30–$40 million, an astounding price when thought of in terms of the value of a dollar four or five generations ago. A wind ve-

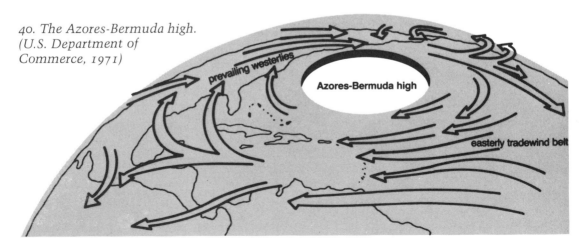

40. The Azores-Bermuda high. (U.S. Department of Commerce, 1971)

*41. A typical late-summer day in the Gulf of Mexico and Caribbean Sea marked by one or several easterly waves. (National Oceanic and Atmospheric Administration)*

locity in excess of 100 mph was observed on Galveston Island as the eye of the storm came ashore; likely a higher wind speed would have been gauged had the wind-measuring device not been blown away. Much of the city of Galveston was inundated by a storm surge that brought water as much as 8–15 feet deep into the streets. The entire southern, eastern, and western sectors of the city, from two to five blocks inland, were swept clean—not one building was left standing. Most of those who died, including entire families, were drowned. Streets were filled with debris and, three weeks after the hurricane hit, bodies were

still being removed from the drift at a rate of twenty to thirty per day. The stupendous loss of life was due not to inadequate warning of the impending disaster. Indeed, storm warnings had been hoisted one to two days in advance of the hurricane's onslaught. By heeding warnings as little as six hours prior to the onset of dangerous conditions, some 12,000 island and coast-line inhabitants saved themselves. Of those who chose to stay, fewer than one hundred—or no more than 15% of the total remaining on the island—survived.

Hapless Galveston Island was lashed a second time by a vicious hurricane just fif-

intertropical convergence zone

*42. A global view of cloud patterns over the Western Hemisphere on a typical late-summer day; the intertropical convergence zone (ITCZ) is seen girding Earth's mid-section. (National Oceanic and Atmospheric Administration)*

teen years after the great hurricane of 1900. Due largely to the seawall constructed on the island a few years earlier, the death toll from the hurricane that smashed into the island before dawn on August 17, 1915, totaled 275, a striking reduction from the thousands of lives lost in 1900. The seawall's value was evidenced by the fact that only 10% of the homes on the island not

protected by the seawall were left standing by the intense cyclone, whose sustained winds topped out at 120 mph near the time of landfall. Still, damage and destruction along Galveston Bay and in adjacent coastal areas were massive: violent gusts of wind combined with an enormous storm surge and flash-flooding rains to cause property losses of $56 million in Texas, with more

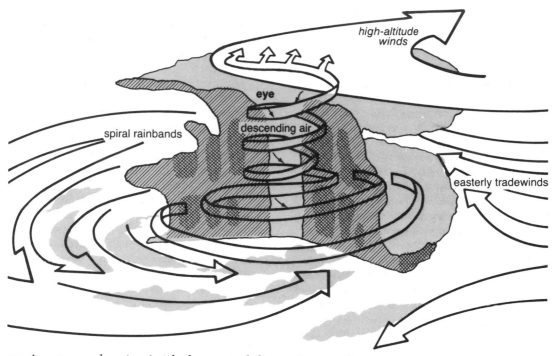

*43. A cutaway drawing (with the vertical dimension greatly exaggerated) of a typical Texas hurricane; low-level easterly tradewinds feed moisture and heat energy to the storm, while high-level winds vent the system by helping to draw moist air up through the storm's "chimney." (U.S. Department of Commerce, 1971)*

than one-tenth of that total sustained at Galveston. Winds of 60 mph were felt as far inland as San Antonio, where one-fourth of the cotton crop was destroyed. Storm rainfall totals, which ranged from 10 to 19 inches in the eastern sector of Texas, were probably underestimates of the actual volume of water that fell. This is due to the fact that much of the rain, which was driven by very strong winds, fell in almost horizontal sheets and hence was hardly detectable by rain gauges.

Like Galveston, the little community of Indianola, once situated on the coastline in Calhoun County, bore the brunt of more than one of Texas' all-time great hurricanes. Although details are sketchy on the hurricanes that hit the town of 6,000 residents in the latter portion of the nineteenth century, it is known that the storm surge associated with the hurricane on September 16, 1875, carried away three-fourths of

the town and killed 176 of its inhabitants. Eleven years later, on August 20, 1886, another great cyclone sent a storm surge that carried away or left uninhabitable every building in the town. Survivors abandoned hope of restoring trade and, because they feared yet another weather calamity, moved away to Port Lavaca, Victoria, and Cuero. Indianola was never rebuilt; today the site is marked by a park and an impressive statue of French explorer René Robert Cavelier, Sieur de la Salle.

In terms of fatalities, second only to the West India Hurricane that devastated Galveston in 1900 was the great cyclone that slammed into Corpus Christi on September 14, 1919. The coastal bend city, which had been hit by another severe hurricane just three years earlier, caught the full force of the hurricane, whose eye crossed the coastline just south of town. A storm surge combined with winds measured as high as 110 mph erased 284 lives and caused damage and destruction amounting to more than $20 million. The memorable killer hurricane forced tides up to 16 feet above normal as far away as Galveston.

No hurricane generated as much rainfall or as many tornadoes as Beulah. As the third-largest hurricane in Texas weather history, Beulah dumped torrential rains of 10 inches or more on a vast area of Southern Texas from San Antonio southwestward to the Rio Grande and southeastward to the Gulf of Mexico. The mammoth flooding that resulted on many streams and rivers in this part of the state was the major contributing factor to an astounding $150 million in damage wrought on property and crops. To make matters worse, before and subsequent to Beulah's landfall around daybreak on September 20, 1967, more than one hundred tornadoes were spawned by the sprawling storm. As a rule, hurricane-produced tornadoes are much smaller and remain on the ground for shorter periods of time than those that develop at other times of the year in interior portions of Texas. This may explain why, out of more than one hundred tornadoes sighted during Beulah's lifetime, only a few caused significant damage and only two resulted in deaths.

In the 36 hours following Beulah's landfall, the slowly disintegrating severe storm traced out a path rather unique to most tropical cyclones entering Texas. After drifting northwestward during the first 24 hours of its existence over land, it inexplicably turned and moved southwestward toward the Rio Grande south of Laredo and into Mexico before vanishing on September 23 (*see fig. 45*). That course fostered rains of rare magnitude in South Central and Southern Texas, including a total over five days

*44. Mammoth flooding produced by Hurricane Beulah in September 1967. (Texas Department of Public Safety)*

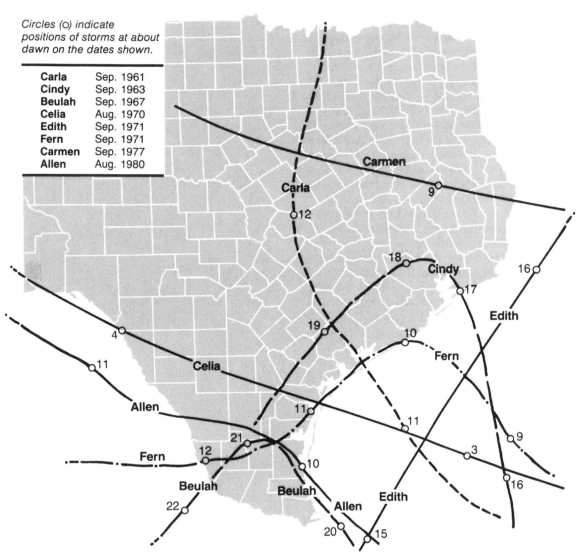

*Circles (o) indicate positions of storms at about dawn on the dates shown.*

| | |
|---|---|
| **Carla** | Sep. 1961 |
| **Cindy** | Sep. 1963 |
| **Beulah** | Sep. 1967 |
| **Celia** | Aug. 1970 |
| **Edith** | Sep. 1971 |
| **Fern** | Sep. 1971 |
| **Carmen** | Sep. 1977 |
| **Allen** | Aug. 1980 |

*45. Tracks of Atlantic hurricanes that affected Texas during 1960–1982. (National Weather Service)*

of 27.38 inches at Pettus in Bee County. Numerous locales between Corpus Christi and the Rio Grande collected more rainfall in less than four days than they normally receive in a whole year. No river or stream south of San Antonio was spared from major flooding; indeed, many of them sustained the highest flood levels in recorded history. The San Antonio River established an all-time high-water mark when it crested at 18.4 feet above its flood stage of 35.0 feet. Many locales in the area south of a line connecting Laredo, San Antonio, and

Matagorda were isolated for more than a week because of flooded roads. The effects of raging floodwaters were especially hard to forget for residents of Ganado, where oily residues carried from nearby oil fields by the rising water were deposited on most buildings in that community.

While Hurricane Celia was the costliest hurricane to strike the Texas coastline in terms of damage to property, virtually all the damage stemmed not from flash-flooding rains or storm surge but from violent winds. Celia was not much of a rainmaker: heavy rains of up to 6–8 inches were confined to a three-county area encompassing Corpus Christi, and the band of moderately heavy rains (with storm totals of little more than 2 inches) was unusually narrow—merely 50 miles either side of the hurricane center. What distinguishes Celia as one of the great hurricanes of all time is the nearly one-half billion dollars ($453,773,000) in damage to property and crops in southern Texas from the coastal bend section to the Rio Grande. Such heavy losses were due to the fact that Celia unleashed her greatest fury on the major metropolitan area of Corpus Christi, where a peak wind gust of 161 mph was gauged at mid-afternoon on August 3, 1970, or within one hour after the storm's eye crossed the coastline midway between Corpus Christi and Aransas Pass. Towns like Pearsall and Jourdanton, located within 40 miles of the hurricane eye, received no rainfall at all. The highest storm surge, measured at Port Aransas, was only 9.2 feet. Another bizarre aspect of Celia was the location of the storm's strongest winds. Unlike most hurricanes, which feature highest winds in the right-front quadrant, Celia manifested greatest wind speeds in the left-rear quadrant. Furthermore, these strongest winds were observed in streaks spaced about 1.5 miles apart, with almost no damage resulting between them. Hurricane-force winds were felt as far inland as Del Rio, while wind gusts of more than 50 mph occurred in the Big Bend area.

Occasionally, a tropical cyclone so weak as to be little more than a mild bother as it trudges ashore develops into a catastrophic gusher once it bogs down in interior Texas. Tropical Storm Amelia, a benign system that cruised inland between Brownsville and Port Isabel on July 30, 1978, and furnished modest amounts of rain and minimal wind damage, was a classic case of a weakening storm that waited to unleash its fury some time after it invaded the coastal plain of Texas. No one remembers Amelia for its modest storm surge, temperate rainfall, or fresh gales at the time it moved

*46. Violent winds were largely responsible for the damage by Hurricane Celia in August 1970. (National Oceanic and Atmospheric Administration)*

ashore. Rather, its persistence and its ferocity in triggering some of the worst flooding of the twentieth century in the Texas Hill Country and farther north in the Low Rolling Plains will forever be its distinguishing traits. Just hours after it seemed to perish some 200 miles from its primary source of energy—the Gulf of Mexico—the storm revived while nestled in the rolling hills of South Central Texas—spilling torrents of rain along the Balcones Escarpment; soaking the watersheds of the Guadalupe, Medina, and Sabinal rivers almost instantaneously; and filling those rivers and numerous other streams and creeks in the area to overflowing. Before dawn broke on August 2, the raging floodwaters swept away 25 unsuspecting riverfront residents and campers. Even after taking that death toll, as well as injuring 150 other people and causing $50 million in damages in flood-ravaged Bandera, Kerr, and Medina counties, Amelia was far from finished. Again the storm flirted with expiration as it drifted northward across the Edwards Plateau, only to be reinvigorated near Abilene. Copious rains—amounting to 20 inches in some locales—forced rivers and streams to burst out of their banks and reservoirs to fill quickly and overflow. At long last, the irrepressible storm waned and died over North Central Texas on August 12 but not before it had bestowed welcome rains of 5 inches on a drought-plagued stretch of terrain south of the Red River.

During the time that it moved onshore and drifted westward through Texas' Upper Coast in mid-September 1963, Hurricane Cindy produced torrents of rain that led to flooding on a scale similar to that instigated by Amelia. Surprisingly, however, a hurricane that cost Texas residents a loss in property of $12.1 million was not responsible for a single death or serious injury. After it made landfall at High Island on the morning of September 17, 1963, Cindy stalled for nearly 24 hours and sent incessant torrents of rainfall that totaled 15–20 inches over a 72-hour period in Texas' three easternmost counties.

Hurricane Audrey not only demonstrated that severe tropical cyclones can occur very early in the season but also lent credence to the fact that the storm tide may be the most destructive feature of some hurricanes. In spite of warnings issued by the U.S. Weather Bureau more than twelve hours in advance of the storm's landfall on June 27, 1957, few people evacuated the coastline from High Island to Morgan City, Louisiana. The hurricane-induced sea surge sent floodwaters streaming inland, and 390 coastal residents (only 9 of whom lived in Texas) died from drowning at the outset or after falling from trees infested with snakes also seeking refuge from the rising water. Tens of thousands of other coastal inhabitants were saved, however, as a result of a carefully orchestrated mass evacuation from low-lying areas to higher ground in areas like the Bolivar Peninsula and Sabine Pass.

The greatest floods in recorded history in the mid–Rio Grande Valley stemmed from Hurricane Alice, an exceptionally early and prolific rainmaker that entered Texas by way of Mexico. After Alice made landfall in northeastern Mexico south of Brownsville on June 25, 1954, the slowly collapsing storm meted out driving rains over a 48-hour period that flooded much of the lower Pecos River Valley. Flash floods in normally dry arroyos washed out railway and highway bridges, including the one over the Pecos River upstream from Del Rio. A Southern Pacific train with about 265 passengers, along with about 200 automobiles, was stranded at Langtry; some of the travelers had to be evacuated by helicopter. More than 27 inches of rain fell in 48 hours at Pandale, while rains of 12 to 21 inches were common in lesser time in Comstock, Sonora, and Sheffield. Flood damage was particularly heavy at Laredo, although practically all deaths and extreme property damage attributed to Alice occurred on the Mexican side of the river.

Few major tropical cyclones materialize after mid-October, when the vast high-pressure cell in the Atlantic is well on its way toward the Equator and the atmosphere and the ocean surface are undergoing a coincident cooling. Hurricane Jeanne, the next-to-last entry in summer of 1980, proved to be an exception when it formed in the eastern Gulf of Mexico and threatened the lower Texas coastline in mid-November. Indeed, Jeanne came within only a few hours of being the only hurricane to strike Texas during the month of November in at least the last 110 years. Undoubtedly, the modest hurricane's belatedness was due to a continuation of uncommon warmth that bathed Texas and all of the Gulf of Mexico to an extreme, record-setting degree for practically all summer of 1980. Jeanne posed a serious threat to the lower Texas coast until it moved to within a few hundred miles of Brownsville, at which time it stalled and then retreated into the central Gulf in the face of a potent Arctic front that barreled through Texas and had earlier instigated a record snowstorm in the High Plains.

## Survivors from the Eastern North Pacific

While virtually all the concern shared by Texans regarding tropical disturbances centers each year on the Gulf of Mexico, the warm waters of the eastern North Pacific offshore western Mexico foster as many—if not more—tropical cyclones than the Atlantic and all its appendages. Most of the tropical storms and hurricanes spawned in the eastern Pacific during the period of June–November drift northwestward through open water without ever impinging on Central America. Occasionally, however, a major tropical system will deviate from this common trajectory and hit the western coastline of Mexico (see fig. 47). These exceptions may be intense enough or thrive long enough to whip high-level clouds across mainland Mexico into the western and southern extremities of Texas. On rare occasions (say, once every 3–5 years), a cyclone may remain intact during its transcontinental trek and provide parts of Texas with meaningful rain. One such hardy storm attacked Texas on its western flank with a vengeance in September 1978. The rare phenomenon—remnants of Tropical Storm Paul, which slammed into western Mexico on September 26—triggered historic flood crests on the Rio Grande above the Big Bend all the way to Falcon Reservoir. Rains totaling 5–15 inches—equal to what the region typically receives over a whole year—soaked the Trans Pecos region of Texas from the Guadalupe Mountains to the Davis Mountains for one whole

week late that month and combined with equally heavy rains in northern Mexico to swell the Rio Grande at Presidio to record-setting levels (see fig. 48). The Rio Grande swelled to more than 25 feet (or 12 feet above flood stage) on three separate occasions in late September, inundating 7,000 acres of rich but unplanted farmland around Presidio and wiping out the railroad bridge connecting the United States and Mexico. Ordinarily no more than a trickle, the Rio Grande swelled to a width of two miles at the tiny community of Candelaria, which was isolated by road washouts for five days.

Likely the most bounteous rainfall produced by an eastern North Pacific hurricane in Texas weather history occurred near the Red River in October 1981. Less than twenty-four hours after Hurricane Norma crashed into the western Mexican coastline near Mazatlán, a barrage of cloudbursts feeding off the copious moisture supplied by the dying hurricane over the Chihuahuan desert dished out rains that, in some spots, were the equivalent of a whole year's worth of normal precipitation. Three-day rainfall amounts totaled more than 25 inches at Gainesville, Breckenridge, and Bridgeport, while much of the remainder of western North Central Texas collected 5–10 inches of rain in the same period. Tarrant County sustained more than $50 million in damage done to property, and a deluge of more than 21 inches in less than twenty-four hours led to property losses of over $20 million in nearby Stephens County. In Wise County, Lake Bridgeport had been 30 feet below normal before Norma made landfall, but in only a few days the torrential downpours generated by the dissipating cyclone raised the level of the reservoir to normal. Grayson County was hit by a "hundred-year flood" on October 12, only to be struck four days later by a second flood of even greater magnitude. In Gainesville, Elm Creek poured out of its banks and flooded the city zoo, drowning many animals. However, one resident of the zoo—an elephant named Gerry—survived even though it was swept away from its habitat by the raging floodwaters. Apparently, the huge animal, who was rescued from a tree in which it had become lodged,

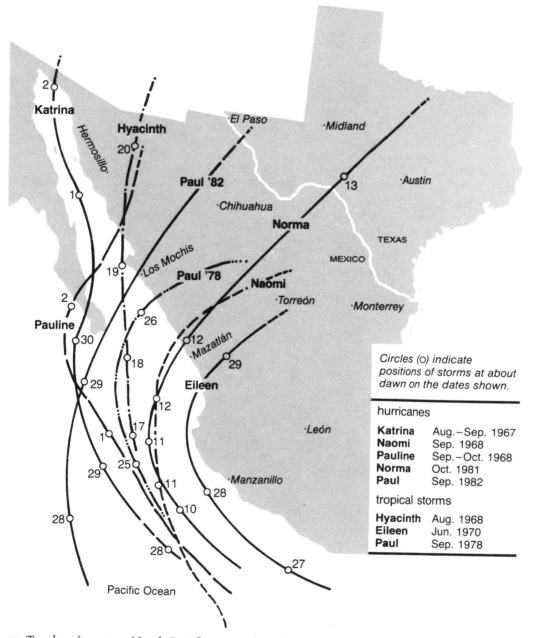

Circles (O) indicate
positions of storms at about
dawn on the dates shown.

| hurricanes | |
|---|---|
| **Katrina** | Aug.–Sep. 1967 |
| **Naomi** | Sep. 1968 |
| **Pauline** | Sep.–Oct. 1968 |
| **Norma** | Oct. 1981 |
| **Paul** | Sep. 1982 |

| tropical storms | |
|---|---|
| **Hyacinth** | Aug. 1968 |
| **Eileen** | Jun. 1970 |
| **Paul** | Sep. 1978 |

47. *Tracks of eastern North Pacific tropical cyclones that had a significant impact on Texas' weather during 1966–1982. (National Weather
Service)*

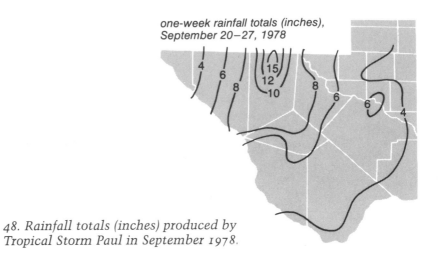

one-week rainfall totals (inches),
September 20–27, 1978

*48. Rainfall totals (inches) produced by
Tropical Storm Paul in September 1978.*

*49. The Presidio Valley, upstream from Big Bend National Park, after
Tropical Storm Paul.*

escaped death by holding its trunk above the floodwaters for more than a dozen hours.

## Profile of a Typical Texas Hurricane

Most hurricanes that strike the Texas coast are at least 100 miles in diameter. While, by definition, sustained wind speeds near the center of hurricanes exceed 73 mph, gale-force winds (with speeds of more than 40 mph) typically extend 150–250 miles in all directions from the center. Seen from above with the aid of weather satellites, the cyclonic spiral of the storm mass is distinctive (*see fig. 50*). Observed from land with weather radar, the storm is identified by heavy cloud bands that often produce torrential rains. Usually these bands of billowing cumuliform clouds are separated by areas of light rain or no rain at all. Wisps of cirrus clouds are invariably observed on the storm's periphery.

The best known feature of any hurricane is probably the "eye" of the storm (*see fig. 51*). No other phenomenon displays a calm core as does the hurricane. Most hurricanes that either form in the Gulf of Mexico or migrate into the Gulf from the Caribbean Sea possess eyes with diameters on the average of about 15 miles, although diameters of 25 to 30 miles are not rare. From the innermost wall of booming thunderstorms that rings the eye, winds drop off drastically to speeds of 15 mph or less near the center of the eye. As the opposite side of the wall approaches, however, winds increase just as dramatically but emanate from the other direction. People beleaguered by a hurricane and ill-advised as to the nature of the storm's eye are deceived into thinking the storm has passed at the time the eye moves overhead, for at this point the rainsqualls end, the clouds separate to allow bursts of sunlight, and an eerie stillness sets in. They begin moving about, only to discover to their horror that the storm is merely half-finished. It is common to find flocks of birds flying within the eye of a hurricane. These fowl get entrapped in the center of the storm while it is still in its infancy and often cannot escape until the storm runs aground or eventually blows itself out in colder waters thousands of miles from the birds' habitat.

The circulation of air around the center of a hurricane resembles the flow of fluid in a whirlpool. Winds on the rim of the storm follow a wide pattern and accelerate only as they approach the center of the vortex. On the storm's periphery wind speeds may amount to only 30–40 mph. It is within the innermost band of clouds encompassing the eye that wind velocities are usually the greatest—in many instances as much as 100 mph, though, as stated earlier, a few hurricanes that have pummeled the Texas coast in recent decades have mustered sustained wind speeds of nearly twice that much. Concomitant with this acceleration of wind speed from the edge of the hurricane to its center is the declination of atmospheric pressure. In the absence of storms, the barometric pressure at sea level in the tropics is generally close to 30 inches of mercury, or approximately the equivalent of 1,015 millibars (mb). Pressure may drop, however, as much as one inch of mercury per hour in some hurricanes. Most of the more intense hurricanes that affect Texas have lowest pressures at or near the center of the storm of 27.50 to 28.50 inches; in the metric system these limits would be 931 mb and 965 mb, respectively, since a bar (1,000 mb) is a unit of pressure equal to 29.53 inches of mercury.

The forces that control the movement of a hurricane also steer the storm to its demise. A tropical cyclone survives—and even thrives—as long as the system has access to heat and, hence, energy, from relatively warm water. Indeed, ample evidence has been uncovered suggesting that hurricane intensification is directly linked to sea-surface temperatures. Most cyclones will not form over water having a temperature cooler than 78°–80°F (26°–27°C). Decreases in water temperature also have a direct bearing on the rate at which a cyclone loses its vigor. Eventually, however, the steering currents responsible for taking the storm on a generally westerly or northerly track drive it onshore or over colder water beyond the tropics. It is not uncommon for steering winds to push a tropical storm or hurricane across cooler water

*50. As one of the Atlantic Ocean's most dangerous storms ever, Hurricane Allen—with its proverbial "eye" easily recognizable in this satellite photograph and its storm mass covering much of the Gulf of Mexico—bears down on the lower Texas coastline just 24 hours prior to its landfall on South Padre Island on August 10, 1980. (National Oceanic and Atmospheric Administration)*

within the Gulf of Mexico and significantly weaken it. It is rare, though, for a storm to meet water in the Gulf cold enough to cause it to die. Most often that happens to those storms that trek northward in the open Atlantic Ocean to latitudes of 40° or beyond.

No two hurricanes follow precisely the same track. The paths carved out are as unique as the storms themselves. Predict-

ing the movement of hurricanes is made difficult because usually it is not known to what extent the steering winds control the rate and direction of movement of the storm mass. Inadequate data on the upper atmosphere out over the sea compound the prediction problem. As a general rule, tropical storms and hurricanes that strike the Texas coastline move at speeds less than 30 mph at the time of landfall.

Texas coastline

intense thunderstorm cells

eye

*51. A close-up of the ominous "eye" of Hurricane Allen as the un-
forgettable giant steams toward Texas; at this point in its life early on
August 8, 1980, Allen was generating sustained winds near the center
of 185 mph. (National Oceanic and Atmospheric Administration)*

Within a few hours after landfall in
Texas, the typical hurricane is sapped of
much of its vigor because, once over land,
the storm's primary source of energy—the
relatively warm waters of the western Gulf
of Mexico—is obviously lost. Furthermore,
the circulation of the hurricane is ham-
pered by the additional, deleterious effects
of frictional drag. Although winds inevitably diminish, torrential rains may persist
for many hours. Most often, a Texas hurricane is transformed into a broad, major

extratropical weather feature that affects
the state's weather and, subsequently, that
of much of the nation to the north and east
for days following landfall. For instance,
the perishing remnant may link up with an
approaching cold front and induce very significant rainfall in north and west Texas.
Sometimes that rainfall is needed badly, for
the searing heat of a Texas summer fo-
ments a drought of such magnitude that
only a rainmaking system as massive as a
disintegrating hurricane can break it.

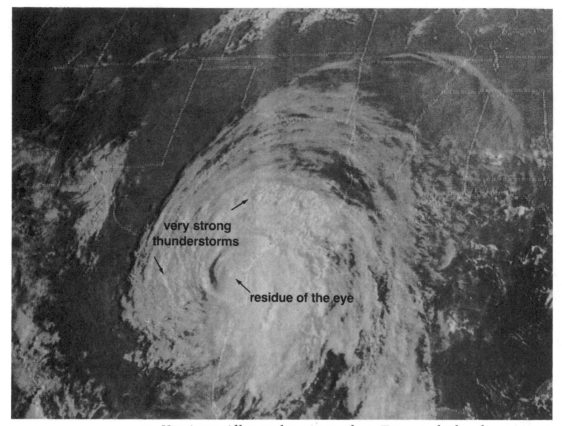

52. *Hurricane Allen onshore in southern Texas at daybreak on August 10, 1980; the gigantic storm blanketed two-thirds of Texas with clouds, and even though it slowly disintegrated as a well-organized cyclone once it moved inland, Allen continued to pound the southern quarter of the state with gale-force winds and slashing rains at the time of this satellite photograph. (National Oceanic and Atmospheric Administration)*

### The Hurricane's Legacy

More so than any other kind of natural weather disaster, the hurricane invariably engenders extensive damage when it crosses the coastline. No natural calamity that ever befell this nation can match the West India Hurricane of 1900 that cost 5,000–8,000 lives along the upper Texas coastline. Fortunately, due largely to vastly improved detection and warning systems, the death toll from tropical cyclones has lessened markedly since the beginning of the twentieth century. In contrast to the monumental hurricane that blasted Galveston in 1900, Hurricane Carla—an extremely strong cyclone regarded by most weather experts as even more severe than the 1900 storm—took no more than 40 lives in both Texas and Louisiana; associated with Carla was likely the largest mass exodus in the nation's history. With sophisticated radar installations strategically positioned near the coastline and orbital satellites constantly monitoring conditions in the Gulf of Mexico, along with an extensive local storm alerting system tied to television and radio, it is virtually unimaginable that a hurricane or tropical storm could arrive unnoticed and unannounced. On the other hand, with the value of oceanfront property continually rising and the coastal popu-

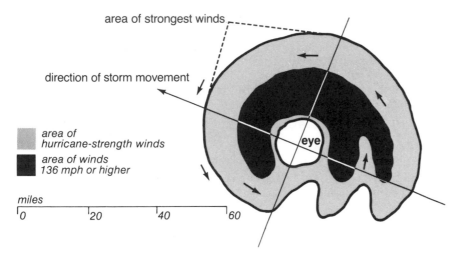

area of strongest winds

direction of storm movement

area of
hurricane-strength winds

area of winds
136 mph or higher

miles

| 0 | 20 | 40 | 60 |

eye

*53. A schematic diagram of the distribution of winds around the eye of an extreme hurricane like Hurricane Carla (1961). (Henry et al.,* Hurricanes on the Texas Coast*)*

lation constantly increasing, the toll in damage and destruction to property climbs still higher with each successive hurricane that comes on the scene. While the $30–40 million worth of property ruined by the Galveston hurricane of 1900 is regarded even to this day as an exceedingly costly toll, it pales in comparison with the $408 million in damage caused by Carla 61 years later.

A Ruinous Wind

The damage from hurricanes and other intense tropical cyclones stems from very strong winds, a devastating storm surge, or flooding from excessively heavy, attendant rains that frequently persist for several days after the storm makes landfall. With most cyclones that affect the Texas coast, winds within the system are not distributed uniformly around the eye of the storm. As illustrated by fig. 53, highest sustained wind speeds usually are observed in the area within 5–20 miles of the eye. A major determinant of wind damage is the angle at which the cyclone strikes the coastline. Though such an occurrence is relatively infrequent, a northeastward-moving cyclone may pass along and parallel to the coastline, thereby exposing a greater land area of the coastal plain to hurricane winds than that affected by an approaching storm that

traverses the coastline at an angle perpendicular to it.

Not many structures are built to withstand the kind of force generated by most hurricanes without some damage being sustained. Slight structural damage, including the removal of shingles from the roofs of homes, is common with wind speeds in the 40s. More appreciable damage to structures, along with the breakage of trees, results from winds in the 50–60 mph category. At this time a hazard may develop when overhead power and telephone lines are broken. Equally serious may be blowing debris from wrecked buildings that causes extensive damage to other structures and, especially, serious injuries to people in unsheltered areas.

An Engulfing Storm Surge

By far the greatest amount of destruction and number of deaths stem from the hurricane's storm surge, which is an abnormal rise in the level of the sea. The size of the storm surge depends largely on offshore topography, with the highest surge associated with shallow continental shelf regions like that off the Texas coast. The persistent pounding of a storm surge that may reach heights of 20 feet or more, combined with waves superimposed on the elevated mound of water, can destroy almost any structure

within its path (*see fig. 54*). The majority of
people who remained at home near the
Texas coastline at the time of the 1900
Galveston hurricane and also during the
two hurricanes that hit Indianola in 1875
and 1886 were buried by the storm surge,
and many other residents who waited too
long to abandon their homes were swept
away by the storm surge and never found.

The state of astronomical tides that regularly occur along the coast can enhance or
lessen the impact of a surge. Just as the
profile of the shoreline and the ocean bottom near the shore are of crucial importance, so too is the presence of estuaries,
inlets, and offshore islands. Even the amount
of vegetation and the extent of building
construction can have an impact. Furthermore, the angle at which a cyclone hits the
coastline also determines the height of a
storm surge, with a higher storm surge associated with those cyclones that strike the
coastline at right angles. The storm surge is
most often highest when onshore winds are
strongest: in the right-front quadrant of the
hurricane mass (*see fig. 53*). For example, a
storm surge spawned by a hurricane moving northwestwardly and making landfall at
Corpus Christi will be most intense along
and up the coastline from (northeast of)
Corpus Christi. It is not uncommon to observe a storm surge of 10 feet or more associated with a strong hurricane as far as
100–200 miles from the point of landfall.
On beachfronts very distant from the point
of landfall, the storm surge usually is evidenced by an increase in seaweed, Portu-
gese men-of-war, and debris washed up on
the beaches. On the coastline nearer the
eye of the storm, considerable erosion of
some beaches and much deposition on others invariably occur.

Whereas the effects of a hurricane-
induced tide can be highly injurious to
open beaches, its impact on upper bays and
estuaries often is more catastrophic. This is
due to hurricane winds driving and channeling bay water to the extent that the
water level is raised at the downwind (or
landward) end of the bay while the level of
the bay water is depressed at the upwind
(or seaward) side of the bay. Water piles up
more so, by several additional feet, in the

upper reaches of bays and estuaries because
the area over which the water spreads is
small in relation to that along a smooth,
clean, flat coastline. The magnitude of a
storm surge entering from the Gulf increases if the estuary narrows down inland
from the mouth, while the height of the
surge lessens if the estuary is over fairly
flat land and expands out inland from the
mouth. A storm surge may behave in completely different ways in many of the bays
along the Texas coastline depending upon
the angle of approach of a tropical cyclone.
Moreover, the influx of saline water from
the Gulf has a detrimental impact on the
quality and productivity of fish and other
marine organisms in the coastal bays and
estuaries, although the extent of the harm
done to bay and estuarine ecosystems is
not yet fully known. On the other hand,
not all the effects of the storm surge are
always injurious. Bays are flushed of pollutants, and the surge may replenish sand on
beaches. In addition, heavy rains from the
tropical cyclone result in substantial freshwater inflow by way of rivers and streams
that contributes to the vitality and productivity of the estuarine food chains, such as
shrimp.

Torrential Rains

In addition to a swelling storm surge and
vicious straightline winds, a hurricane or
other tropical cyclone can disastrously affect Texas' coastal plain through a supera-
bundant rainfall and resulting flash floods.
The amount of rainfall from such cyclones
varies considerably and is dependent on the
size of the rain area, the rate of movement
of the storm mass, and the influence of
upper-atmospheric weather conditions beyond the periphery of the storm. As a general rule, total storm rainfall and, hence,
the worst flooding are greatest for broad
hurricanes that drift slowly, both prior
to and following landfall. Even tropical
weather disturbances so minor as not to be
assigned a name can drench the coastal
plain with phenomenally heavy rains. The
first tropical storm of the 1960 season provided the coastal bend section of Texas
with good rains prior to its landfall near
Corpus Christi around midnight on June

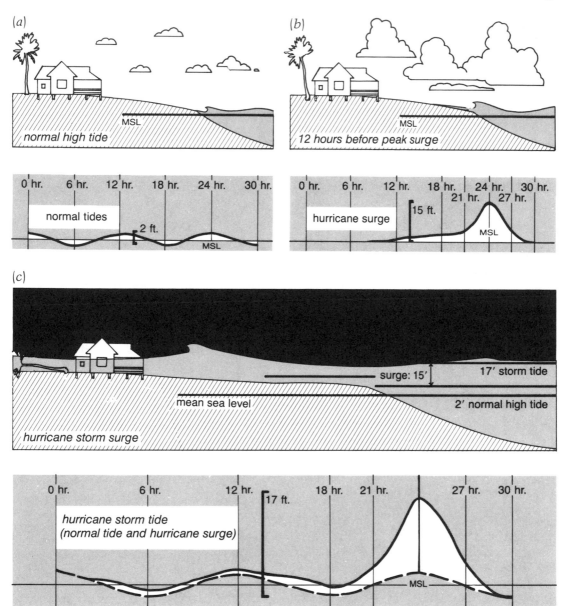

54. Effects of a typical hurricane storm surge on a Texas coastal area:
(a) On a normal day, the sea rises and falls predictably; (b) with the
approaching hurricane only 12 hours away, the swelling tide moves
farther up the beach, and breaking waves—some as high as 5–8
feet—crash ashore; (c) when the hurricane makes landfall, the "normal" 2-foot astronomical high tide is augmented by a 15-foot storm
surge, and this 17-foot "storm tide" assaults an area of coastline
50–100 miles wide. (U.S. Department of Commerce, When a Hurricane Threatens)

23, but then it dumped up to five times as much rainfall over three days as the remnants of the storm drifted through central Texas. Port Lavaca collected a bit more than 5 inches prior to storm landfall but was later inundated by 24 more inches, as serious local flooding from the torrential rains hit a five-county area extending from Aransas Pass to Freeport. That tropical cyclone, known otherwise only for its "modest" strength, caused flood damage in excess of $3.5 million and led to fifteen drownings.

Hurricane Cindy is long remembered for the disastrous flooding it instigated on September 16–20, 1963. Flood damage from Cindy's high tides was relatively light, but flooding from rains of 15–20 inches in a three-county region encompassing the Beaumont–Port Arthur–Orange metropolitan area sent water into 4,000 homes. At Deweyville more than 20 inches of a total storm rainfall of nearly 24 inches cascaded down on that Sabine River Valley city in only 24 hours. Damage to both property and crops in the extreme eastern section of Texas amounted to over $12 million.

Of the two types of flooding induced by tropical cyclones, the flash flood is most often the killer. Residents living in areas where drainage is incapable of carrying away excess water without overflow are in greatest jeopardy. Short-term, heavy rains can fill a dry stream bed in a matter of hours—if not minutes—and the resultant overflow inundates bridges, overpasses, and low-lying residential areas. On the other hand, river flooding usually is much slower to materialize. In fact, it may not begin until long after the hurricane or tropical storm has made landfall, and then it might persist for a week or longer. River flooding is frequently more costly in terms of property and crop losses than flash flooding because it covers a much more extensive area.

## Deadly Tornadoes

Besides the ever-present threat of river and flash floods, that part of Texas affected by a tropical cyclone must also contend with the likelihood of tornadoes, for these viciously destructive forces of nature are a common by-product of tropical disturbances. Most often they occur outside—from 50 up to 250 miles—the innermost portion of the cyclone where hurricane-force winds are present. Statistics gathered over the past several decades reveal that nine out of every ten tornadoes develop to the right of the direction of movement of the hurricane in an area bounded by radii with angles of 10°–120° from the direction of movement. Recent studies have concluded that tornadoes produced by hurricanes and tropical storms live for much shorter periods of time and are much smaller in size than those that dip out of murky skies on the Great Plains region. Usually these younger and leaner hurricane funnels cut a narrow swath or skip over the countryside, and damage done by them is mostly insignificant. However, if one of them should sweep through a heavily populated or industrialized area, the toll of casualties and the amount of property damage may skyrocket. Hurricane Beulah holds the record for most tornadoes generated by a tropical cyclone; that 1967 storm sent at least 115 tornadoes dancing across the southeastern half of Texas. Fortunately, they hit sparsely populated areas, so neither the death toll nor the total property damage was great. Hurricane Carla (1961) is the second-most prolific tornado producer (26) to plague Texas, though the great difference in tornado counts between the two hurricanes may be due largely to improvements made in observing and reporting tornadoes in more recent years.

Frequently, the misery invoked by a hurricane lingers for days, weeks, and even months. Vehicular movement is inhibited because of washed out, clogged, or flooded bridges and roadways. Consequently, the movement of emergency equipment and personnel, as well as food, water, and medical supplies, is impeded. Drinking water is scarce—if not altogether unavailable—and sewage and other wastes cannot be disposed of as usual. Disease may spread from drowned animals, and numerous other problems stem from living in emergency shelter conditions. Electric power lines downed by high wind or water present a real hazard, while direct communication is severed because of disabled telephone and telegraph lines. Many citizens discover an abundance of snakes, some lethal, who have been driven from their natural habitat by high water. These strong-swimming reptiles may be found almost anywhere, especially along roads, in trees, within remnants of buildings, and, worst of all, in homes.

### The Anticipation of an Onslaught

The northward shift of the sun, which culminates with the summer solstice on June 21, presages a certainty for Texas' coastal residents: the season of tropical cyclones. Initiated by the sun's poleward drift is the warming of the sea and lower atmosphere in the Caribbean and Gulf of Mexico, while winter's polar air recedes gradually toward the Arctic Circle. It is a time to look seaward, to pay closer attention to media weather statements, to make preparations for one or several of the behemoths of the sea that are a near certainty to appear on the distant horizon during summer's stretch run. If the current year has any semblance of normalcy, it will be marked by as many as ten tropical cyclones, six of which will develop into hurricanes. These will kill 50–100 persons along the nation's coastline between Texas and Maine and cause property damage of around $250 million.

Nominally, the "hurricane season" in the Gulf of Mexico extends from June 1 through November 30. Major tropical weather disturbances seldom materialize as early as June, but if one is born, the Caribbean and the Gulf of Mexico will likely be the birthplace. The principal area of origin usually shifts eastward during July and August such that, by September, the breeding ground for most Atlantic cyclones consists of a broad belt of tropical water extending from the Bahamas southeastward to the Lesser Antilles, then eastward across the mid-Atlantic to the Cape Verde Islands off the western coast of Africa. By mid-September, the focal point for the generation of most tropical cyclones reverts back to the Gulf of Mexico and the Caribbean Sea. Since tropical weather records for the

Texas coastline were begun in 1871, no cyclone has hit the coast of Texas any earlier than June 4 or any later than October 17. This is not to say that hurricanes and tropical storms cannot hit Texas at other times in the year, however, for cyclones have been observed in the Atlantic Ocean or in one of its appendages as early as February and as late as December; rather, it is to suggest that such untimely events are extremely rare.

### The Problem of Prediction

Forecasting more than forty-eight hours in advance the behavior of tropical cyclones—not to mention the timing and location of their initial generation—is an extremely arduous task, and the accuracy of those projections is little better than probabilities based on mere chance. To arrive at some fairly reliable basis for anticipating the likelihood of a tropical cyclone, the past is scrutinized to formulate climatological frequencies that may be construed as probabilities of cyclone occurrence. That is to say, based upon what we know has occurred over the past one hundred years or so, we reasonably assume that a similar trend is likely to apply in this or any other year. The following discussion of probabilities of a major tropical disturbance striking the Texas coast is provided in this context. The reader should remember that each tropical cyclone develops and behaves without regard to what its predecessors have done. Hurricanes obviously know nothing about historical averages.

Since the official Texas Gulf coast hurricane record was begun in 1871, thirty-nine hurricanes have made landfall in Texas. Seventeen more came close enough to the Texas coastline to cause damage on the mainland. During that same 112-year period from 1871 to 1982, twenty-six tropical storms crossed over the Texas coastline (*see Appendix D-2*), while thirteen others sideswiped some sector of it. From these records, it is seen that, on the average, the Texas coast experiences a hurricane every other year and a tropical storm every third year. We must use considerable caution, however, in applying these statistics because no regularity in hurricane or tropical

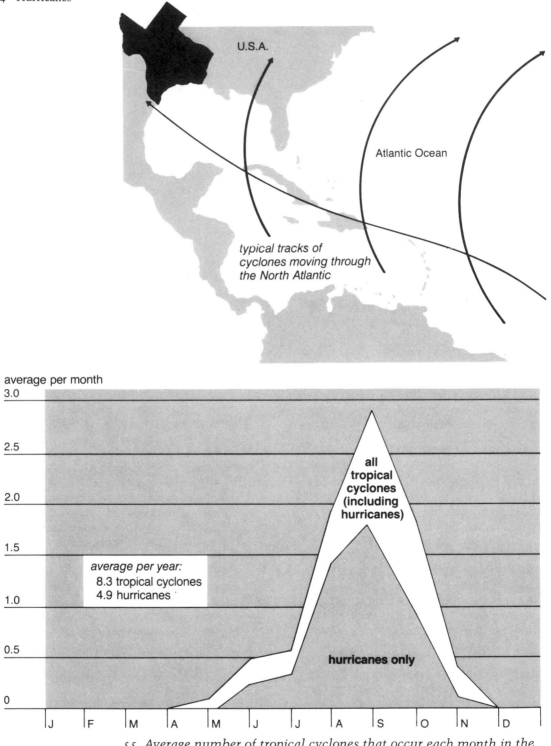

U.S.A.

Atlantic Ocean

*typical tracks of cyclones moving through the North Atlantic*

average per month

3.0

2.5

2.0

1.5

1.0

0.5

0

**all tropical cyclones (including hurricanes)**

*average per year:*
8.3 tropical cyclones
4.9 hurricanes

**hurricanes only**

J  F  M  A  M  J  J  A  S  O  N  D

*55. Average number of tropical cyclones that occur each month in the North Atlantic Ocean (including the Caribbean Sea and Gulf of Mexico); figures are based on observations made over the period 1886–1980. (National Oceanic and Atmospheric Administration)*

storm occurrence exists; some years have furnished two or more tropical cyclones, while numerous other years have been marked by relative tranquility in the Gulf of Mexico. Thus far the twentieth century has been characterized by many multiyear periods in which a hurricane and/or tropical storm occurred as well as numerous periods without any significant cyclones. The lengthiest period of nonactivity ended only recently when Hurricane Allen roared ashore near Port Mansfield on August 10, 1980; until then, no hurricanes had invaded Texas since Hurricane Fern on September 10, 1971 (Amelia, in 1978, was a tropical storm).

The tracks of most of the major tropical cyclones that have affected Texas in recent decades reveal that most of them impinge on the coastline at or very near right angles. In some unusual instances, cyclones approach the Texas coastline but then travel roughly parallel to the shoreline, causing damage along most, if not all, of the coast. In most instances, however, since the Texas coastline is lengthy, a hurricane or a tropical storm will not adversely affect all of it.

Based upon the 112-year historical record of tropical cyclone occurrence, there is about a one-in-eight chance that a particular sector of the Texas coast will be hit by the eye of either a hurricane or a tropical storm in any year (see fig. 56). The odds of being hit by a hurricane are about one-in-ten for most of the coastline.

Texas weather annals also suggest that the likelihood of two or more tropical cyclones hitting the same sector (with a length of about 50 miles) of the Texas coast in any one year is no more than one in twenty-five. On the average, 15 to 20 years elapse between extreme hurricanes striking the same sector of the Texas coast; for just any hurricane the average number of years between any two occurrences is 4–5 years, and for any tropical cyclone (hurricanes and tropical storms) the average is 3–4 years. The earliest major disturbance to hit the Texas coastline was the tropical storm that made landfall near Galveston on June 4, 1871. As illustrated in Appendix D-2, a surprising number of cyclones have entered

Texas in June. The likelihood of occurrence (about 1 of every 5) is about the same for July as for June and then peaks in August and September; the latter two months are marked by an equal number of cyclones to hit the Texas coast. The frequency of occurrence drops off dramatically in October. In fact, the latest cyclone to strike the Texas coast was the tropical storm that hit near Matagorda on October 17, 1938.

The Urgency of Alerting the Public

When a tropical cyclone is spotted (usually initially by weather satellite), the tempo of activity increases markedly at the National Hurricane Center (NHC) near Miami, Florida. Storm forecasters there receive and digest many bits of weather data processed and dispatched from the National Meteorological Center in Suitland, Maryland. They also scrutinize cloud photographs from orbiting weather satellites transmitted by way of the National Environmental Satellite Service in Washington, D.C. The cyclone is monitored continually, and statements are issued to the news media at 3–6 hour intervals. It is not until the disturbance grows into a tropical storm that the NHC assigns a name. Until very recently, feminine names were given to all tropical storms and hurricanes in the Atlantic. However, beginning in 1979, major tropical disturbances in the Atlantic–Gulf of Mexico were awarded alternating masculine and feminine names. Names to be assigned to future Atlantic and eastern North Pacific cyclones are given in Appendixes D-3 and D-4, respectively.

A hurricane emergency is initiated once the aerial reconnaisance supplied by hurricane hunter aircraft confirms that a tropical disturbance has progressed to tropical storm status. The NHC then activates the hurricane warning system and consecutively numbered advisories are issued systematically to the public through the news media. These special weather statements are issued at 6-hour intervals, or at approximately 5:00 A.M., 11:00 A.M., 5:00 P.M., and 11:00 P.M. (CDT), and they contain information on location, intensity, speed and direction of movement, and size of the storm. The advisories also invariably contain

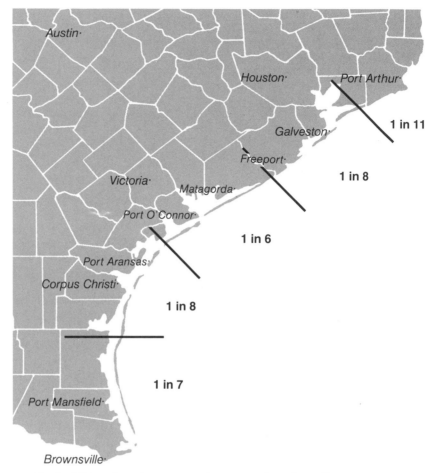

56. *Probability that the center of a tropical cyclone (hurricane or tropical storm) will make landfall on various segments of the Texas coastline in any year.*

projections on what changes (if any) in strength and movement are anticipated during the next 6–12 hours. Great strides have been made in recent years in forecasting the behavior of tropical cyclones. Evaluations of previous forecasts reveal that the error of 12-hour advance forecasts on landfall position averages less than 40 nautical miles. The normal forecast error for 24 hours is about 100 nautical miles. Much remains to be learned, however, before forecasters can provide longer-range forecasts (24 to 48 hours) with reasonable accuracy.

Previous experience in forecasting the time and location of landfall of tropical cyclones demands that meteorologists ex-

ercise caution in attempting to pinpoint when and where a storm will strike—unexpected late-course maneuvers are frequently made by some storms that threaten the Texas coastline. Such erratic behavior was demonstrated by Hurricane Beulah, which recurved sharply to the southwest after striking the southern tip of the Texas coast on September 20, 1967. Other cyclones undergo rapid intensification or deterioration just before they push inland. Hurricane Celia sustained rapid growth just prior to its invasion of Corpus Christi on August 3, 1970, whereas Hurricane Allen—the second-strongest hurricane in history for much of its lengthy trek across the

Caribbean and Gulf of Mexico—stalled and weakened just hours before it surged ashore near Port Mansfield on August 10, 1980. Not enough is known about these sea giants to allow weather experts to anticipate with much reliability these last-minute quirks in storm behavior and motion.

Hurricane/tropical storm watches and warnings for the Texas coast are issued by the NWS Hurricane Warning Office in New Orleans. A *hurricane watch* is defined as follows: "The first alert when a hurricane poses a possible, but as yet uncertain, threat to a certain coastal area, or when a tropical storm threatens the watch area and has a 50-50 chance of intensifying into a hurricane" (Henry, et al., *Hurricanes on the Texas Coast*). Small-craft advisories are usually issued as a part of a hurricane watch statement. Further intensification may warrant the issuance of a *hurricane warning*, defined as "a warning that, within 24 hours or less, a specified coastal area may be subject to (a) sustained winds of 74 m.p.h. (64 knots) or higher and/or (b) dangerously high water or a combination of dangerously high water and exceptionally high waves, even though winds expected may be less than hurricane force" (ibid.). The NWS weather bulletins also contain an estimate of storm tide and flood danger in coastal areas, some precautionary advice to small craft, a statement of gale warnings near the periphery of the storm, and any tornado forecast information.

The Challenge of Containment

Not all coastal residents can be persuaded to flee their homes when an approaching cyclone warrants a mass evacuation. Some choose to stay put no matter what advice is given. Also, no amount of boarding up ensures that a person's property and possessions will remain intact during some storms. Frankly, the estate of a Texas coastal resident is at the mercy of some of the more vicious hurricanes to strike the state. For this reason, it is essential to know what can be done, if anything, to alter the behavior of a cyclone so that its vigor is sapped or its course redirected. In the past three decades scientists have formulated and proposed numerous methods

for trying to reduce the hurricane's destructive forces. One idea involves the application of a chemical film one molecule thick over the surface of the ocean where the cyclone gains its strength. Supposedly, evaporation—the mechanism by which the storm obtains its vital source of energy—would be slowed significantly to inhibit growth. A second, and more plausible, hypothesis includes seeding the cyclone much later in its life cycle with silver iodide crystals to release prematurely its latent heat energy. This latter approach gained credence with many in the scientific community after World War II to the extent that a number of cloud-seeding experiments were conducted in the laboratory and later in the atmosphere to determine what effect the dispersion of silver iodide would have on towering cumuliform clouds that make up tropical cyclones.

Experiments initially begun in 1961 led to the organization of Project STORMFURY, a scientific endeavor of the National Oceanic and Atmospheric Administration to explore the structure and behavior of tropical cyclones in the western Atlantic Ocean and the potential for their beneficial modification. Over the following ten years, four Atlantic hurricanes were seeded with silver iodide crystals dispersed from aircraft of the U.S. Navy. Although STORMFURY involved the manipulation of only a few hurricanes, there were indications from the experiments that a reduction in wind speed can be produced from seeding. In none of STORMFURY's instances was there any intimation that seeding led to hurricane intensification.

What can be concluded from this limited attempt to tamper with hurricanes? In spite of our experiments, we still do not know just precisely how a hurricane will react to human interference. One severe limitation is our inability to formulate a sufficient number of assumptions. Scientists are still working to develop a credible model that demonstrates how a single cumulus cloud within a hurricane behaves, for example. To successfully modify a hurricane, researchers need to be able to predict what will happen with a hurricane thriving on its own and without any tampering from us. Until now, they have not been able to do that well enough. Unless, and until, a thorough understanding can be gained of the mechanics and dynamics of clouds that make up a hurricane, who is to say that scientists can recognize differences in the behavior of seeded and nontreated tropical cyclones? Put another way: if a hurricane reacts a certain way after it has been treated by cloud seeding, how can we know that it would have acted differently had we not intervened?

When one regards the incredible costliness of most hurricanes, and to a lesser extent tropical storms, there are few—if any—justifiable excuses for not pushing ahead to gain further clues as to how we can treat them to minimize their effect on our nation's coastlines. It is probable that, as meteorologists and other researchers learn more about these mammoth sea giants, more and better ways of controlling them will be conceived. The main hurdle in striving for successful hurricane-modification experiments is to differentiate natural fluctuations in hurricanes from those induced by our intervention. The difficulty would not be so great if hurricanes remained in a steady state for periods of, say, six or twelve hours; in reality the giant tempest is incessantly in a state of flux. It is a formidable obstacle but not an insurmountable one if wholehearted support is given to maximizing our opportunities for seeding hurricanes in the wide-open Atlantic.

# 5. Thunderstorms: Dynamos in the Skies

Early one summer evening in 1979 a family of seven labored hurriedly as they picked cucumbers in a patch within sight of Plainview, Texas. Their quickened pace was a reaction to the repeated claps of thunder that rolled across the flat landscape from a gigantic thunderstorm that towered almost directly overhead. The sight of sheets of slashing rain just a few hundred yards away and the distant smell of fresh rainwater carried by a stiff, refreshingly cool wind signaled the imminence of a deluge. At, seemingly, the last possible moment before the hard rains hit, the clan of seven hustled into a pickup that waited for them on a nearby roadside. The mother slid in behind the steering wheel and was joined in the cab by two of her five sons. At the same time the other three children leaped with their father into the exposed bed at the rear of the truck and huddled together next to the cab. The driver wheeled the truck onto the highway and sped toward Plainview, hoping to outrun the advancing storm. One too many minutes had passed, however, for the tumultuous black cloud soon caught up with the speeding truck. Suddenly, an unspeakably brilliant flash of light bolted to the earth from above, hitting the steel exterior of the vehicle not far from the four riders in the back. The mother and the children in the cab immediately perceived that something traumatic had happened. Only when the truck was stopped moments later did it become fully apparent to the mother that the lightning bolt had

struck and killed her husband and two of the sons riding with him in the exposed rear bed of the truck. The third son had been injured, but somehow he survived. The mother and the two children sitting with her up front had not been harmed at all.

It is no wonder that lightning has come to be symbolic of terror and mystery. Potent enough to split huge trees and discard large chunks of wood many yards away, lightning warrants utmost respect as a most impressive display of nature's awesome power. Many strange, even bizarre, happenings are associated with the freakish and unpredictable behavior of such a concentrated charge of atmospheric electricity. Lightning, however, is but one of several and often savage products of the tempestuous thunderstorm, that atmospheric behemoth responsible for causing extensive death and damage in Texas every year.

As a billowy cumulus cloud gone berserk, the thunderhead is a huge storm factory. Its foamy, puffy exterior belies the constant, violent turmoil sustained within its core. This giant of all clouds is infamously regarded as a generator of terrifying tornadoes and horrendous hailstones, destructive winds, and fearsome flash-flooding rains, as well as the killer lightning bolt. Yet for the hundreds of thousands of residents of the semiarid plains and plateaus of western Texas, the thunderstorm, while unwelcome as a hail producer or a begetter of funnel clouds, is a

*57. The towering thunderhead.*

cordially greeted guest in a water-deficient region. So dependent is western Texas on thunderstorms that, without the short-lived torrents of water that gush from the spring and summertime varieties, the vast region would be a desert wasteland.

## The Life Cycle of a Thunderstorm
Driven by an Updraft

All thunderstorms evolve through a cycle of stages. In its embryonic or infant stage, the thunderstorm is merely a cumulus cloud, having very little depth but typified by strong vertical currents of warm, moist air feeding into the growing cloud from near the surface. This characteristic updraft resembles hot air ascending in spiral fashion through a tall chimney (*see fig. 59a*). The column of warm, moisture-laden air, with a typical diameter of less than a mile and a rate of ascent of 15–30 mph, is in part the result of differential heating of Earth's surface. Because the landscape in any locale features variations in composition, color, shape, and texture, some parts of the terrain absorb more sun-

light and, hence, warm more rapidly and to a greater extent than others. A field of freshly plowed rich, loamy soil captures more heat from the sun than a nearby acre of ripe wheat; a pond or reservoir of water collects even less solar energy than the wheat field. Air above these good absorbers of sunlight is in turn heated by them and made to expand. The warmed air is less dense, and hence lighter, than surrounding air, so that it rises away from near the surface while cooler and more dense neighboring air rushes in to fill the void. Thus, a convective current is set in motion.

This vertical current of warmed air usually contains a fair amount of moisture that is invisible to the human eye until it condenses into droplets or ice crystals at an altitude of several thousand feet above the terrain. Condensation of water from the vapor to the liquid state takes place when the parcel of air, while rising and expanding, is cooled to its dew-point temperature. In other words, the temperature of that parcel of air falls to the point that the air can no longer hold all its moisture. Further cooling leads to some of the moisture being

58. *The forerunner of a thunderstorm, a cumulus cloud.*

"squeezed out" as droplets of water. When some of the moisture in this convective current is transformed from a vapor into a visible liquid (water) or solid (ice) state, a cloud forms. The process of condensation is dependent upon the presence of tiny particles, such as dust or salt. These miniscule bits of matter, called "condensation nuclei," serve as collection points for the formation of small cloud water droplets or ice crystals.

On a typical spring or summer day in Texas, thousands of small cumulus clouds develop as a result of differential heating of Earth's surface. Because their existence is dependent upon heating of Earth's skin, these "fair-weather" clouds do not appear until late in the morning or when the surface has been sufficiently heated by the rising sun to generate the convective currents. A setting sun, and hence a cooling surface, cuts off the cloud's source of energy, so most cumuliform clouds perish not long before dark. During much of the year addi-

*59. Three stages of the life cycle of a thun-
derstorm. (U.S. Department of Commerce,
Thunderstorms)*

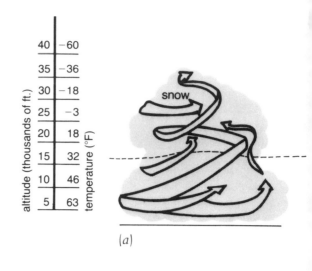

(a)

tional cumuliform clouds often form as a
result of lifting of warm, moisture-filled air
by an approaching cold front. The vast majority of them do not grow very large and
never attain the status of thunderstorm.
They perish only moments after they form
because either the updraft feeding the infant cloud diminishes or the cloud grows
into a layer of the atmosphere where air
that is very dry absorbs the incoming moisture to the extent that the cloud starves to
death.

For reasons not fully understood by scientists, some small cumulus clouds survive
during their ascent to a new environment
at higher levels of the atmosphere. The
taller these clouds grow, the speedier the
updraft becomes, and more and more water
is processed within the cloud; the cloud
prospers, growing to heights of 20,000–
25,000 feet above sea level. Additional
growth is made possible by the release of
heat when water in the vaporous state is
condensed into a cloud droplet; this added
energy fuels the updraft even more. Once
the top of the blossoming cloud pushes
through the freezing level, some of the incoming moisture is changed directly from
vapor to tiny ice crystals. A large portion of
the cloud moisture above the freezing level
remains in liquid form, however, as "supercooled water." If the cloud builds further,
say to an altitude of 30,000–40,000 feet,
more and more of the cloud water at this
higher level becomes ice. In thunderstorms
whose tops climb beyond 40,000 feet, all

the moisture in the top portion of the cloud
exists as ice.

When Rain Is Concocted

While the thundercloud grows taller, water droplets making up the bottom portion
of the cloud collide and join themselves together to form raindrops, which in turn are
enlarged by additional contact with other
cloud droplets. Ultimately, when a raindrop
enlarges to the point that its weight becomes too heavy to be supported by the
incoming air current from below, it falls
through the cloud and out the bottom as
precipitation. When rainfall in a thunderstorm is initiated, marked changes in
the cloud's internal circulation pattern occur suddenly. Up to this point, only an updraft was at work, feeding the cloud with
low-level moisture from which the cloud
derived its energy for additional growth.
Now, with the fall of the first batch of raindrops, a downdraft is set in motion (*see fig.
59b*). This cascading pool of cool, damp
air plunges downward to the ground, which
deflects it over an even greater area of
Earth's surface than that dampened by the
falling rain. The leading edge of this advancing wedge of cold downdraft, known as
a *gust front*, produces an abrupt windshift
and a short period of rapidly falling temperatures at the surface. The air being felt
by an observer on the ground is chilling
because it is air imported from the interior
of a cloud many thousands of feet above
the surface, where the temperature may be

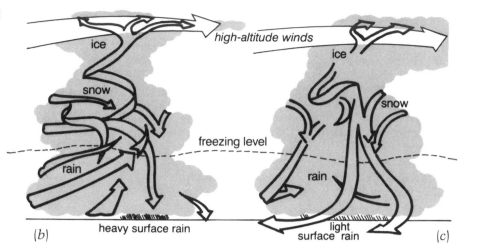

(b)   heavy surface rain

(c)   light surface rain

30°–50°F (17°–28°C) cooler than at ground level. In summer, the downrush of cool air may cause temperatures at the surface under or near the thundercloud to dip drastically in a matter of a few minutes. An extreme case is the 45°F (25°C) fall in temperature that occurred at Midland on July 9, 1979, when the cold downrush of a hail-producing thunderstorm sent the temperature plunging from 101°F (38°C) to 56°F (13°C) in only fifteen minutes.

Sometimes the advancing cold air from a thunderstorm is clearly visible to the ground-based spectator. Particularly in western Texas, where much of the terrain is covered by dry, exposed soil, the strong, gusty winds generated by the cold-air downdraft pick up and make airborne large quantities of dust particles. The dust cloud formed in that manner clearly outlines the advancing edge of the cold downrush. Quite often, when the air near the surface is moist, the onrushing downdraft acts as a cold front, wedging underneath this near-surface layer of moist air and lifting it until condensation of some of the moisture occurs. This leads to the formation of an *arcus*, or roll cloud, which then precedes the advancing thunderstorm. Though its appearance is particularly ominous, the roll cloud furnishes little, if any, rain.

## Maturity

The thunderstorm reaches its greatest height during the mature stage, when updrafts and downdrafts are both vigorous. In fact, this stage of the thunderstorm's life cycle is highlighted by a vigorous downdraft circulation that coincides with the region of heaviest rain—a strong downward rush of cold air that is due to the frictional drag force induced by the drops. The intricate circulation pattern within and near the thundercloud often gives rise to the formation of small flanking clouds, or turrets, that skirt the base of the parent cloud as appendages. Some of these offspring may grow to become sizable thunderstorms themselves, though the common fate of most of them is a short life without ever attaining the height of the parent cell. Close observation of some multicellular thunderstorms reveals that turrets form and dissipate in cyclic fashion, with each "pulsation" lasting 10–20 minutes. Still another distinguishing attribute of a mature thunderstorm is the spreading out of the top of the cloud as it approaches the tropopause. This feature, which resembles the top portion of an anvil, consists of ice crystals sheared off the top portion of the cumulonimbus by high-speed winds in the upper atmosphere (*see fig. 59c*).

Once the downdraft becomes the predominant and more extensive circulation mechanism within the thunderstorm, the massive cloud begins to decay. The falling rain and companion downdraft usually restrict, if not inhibit altogether, any additional growth of the thunderstorm cell by reducing the updraft that feeds the system. With much, if not all, of the updraft shut

60. *Dust is sometimes kicked up by the gust front. (National Severe Storms Laboratory)*

off, the amount of incoming moisture is lessened and condensation of water vapor decreases. With less rain falling out of the cloud, even the downdraft wanes. Gradually, the temperature of the thundercloud tends toward that of the ambient air. Sadly—particularly in the semiarid regions of western Texas—only a small fraction (say 20% or 30%) of the water vapor condensed in the updraft is left behind as rain on the ground; the remainder becomes cloud debris that ultimately vanishes in thin air.

The perishing thunderstorm may not be finished, however, for during its life it may have so disturbed the atmosphere around it that other thunderheads are born several miles away.

## Varieties of Thunderstorms

In the relatively warm, humid subtropical climate of Texas, there are two categories of cumulonimbus clouds that grow to become thunderstorms and assume impor-

61. *An arcus, or roll cloud, sometimes precedes an advancing thunderstorm. (National Oceanic and Atmospheric Administration)*

62. *The mature stage of a thunderstorm.*

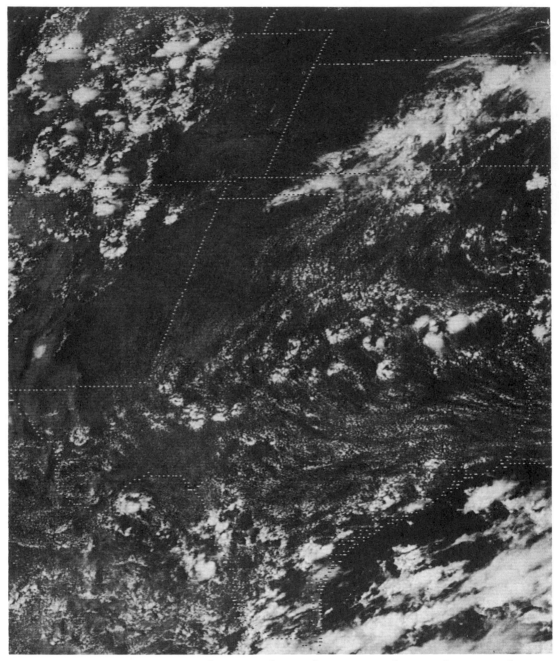

*63. A typical summer day with numerous clusters of air-mass thunderstorms and many hundreds of smaller "fair-weather" cumulus clouds. (National Oceanic and Atmospheric Administration)*

tance as major producers of rainfall: the air-mass thunderstorm and the localized severe thunderstorm.

## The Air-Mass Thunderstorm

The *air-mass thunderstorm*, a scattered-to-isolated cell of intense convection, is most common in Texas during summer. It is unlike the more organized and complex thunderstorms that occur in association with weather systems, such as cold fronts. The air-mass variety is also known as a "heat" thunderstorm, for it forms away from fronts and is due largely to differential heating of the land surface that sets an updraft in motion. If this convective cloud is nourished additionally with more moisture-laden air from near the surface, and if conditions higher in the atmosphere are not so dry as to inhibit vertical growth of the cloud, then the cumulus cloud may mature into a cumulonimbus, or thunderhead. Since most of them are due to intense heating by the sun of the moist layer of air next to Earth's surface, these air-mass thunderstorms are most numerous during daytime or prior to, during, and some time following the peak heating period of the day. Some may persist well into the evening hours, but most dissipate around if not before sunset. It is uncommon to see air-mass thunderstorms develop, independent of any frontal activity or upper-air disturbance, during the nighttime or prior to late morning. Because this variety is typically scattered or even widely scattered, predicting precisely where they will form is exceedingly difficult, if not impossible, with present-day prognostic technology.

On most days during summer, dozens of air-mass thunderstorms dot the landscape near the Texas coastline and sometimes extend for hundreds of miles inland. They first appear only a few hours before noon as small, puffy cumulus clouds. The vast majority of these myriads of cumuli live less than 20–30 minutes, growing to heights of only a few hundred feet above cloud-base level. Some, however, experience explosive growth, attaining heights of many thousands of feet above sea level and persisting for a half-hour or longer. A few of these grow even more and become full-fledged thunderstorms, replete with lightning, gusty winds, and a refreshing smattering of large raindrops. Probably at least half—and possibly as many as three out of every four—thunderstorms that develop and supply showers of rain on the Texas coastal plain during the three warmest months of the year are of the air-mass variety. Most of them provide insignificant amounts of rain and go undetected except by radar.

Another species of air-mass convection is the *mountain*, or *orographic, thunderstorm*. This variety of rainmaker is confined to the mountain ranges of the Trans Pecos region, where on most summer days one or several large thunderstorms will boom to lofty heights and send a torrent of rain cascading down onto a tree-decked hillside or a parched, dusty valley. Again, it is the sun that supplies the initial burst of energy in the form of heat that causes warm, moist Gulf air to flow upslope. Actually, when the moisture-laden Gulf air reaches the higher elevations, it is primarily the slopes of the mountains, which absorb more solar energy than the atmosphere, that initiate the thunderstorm updraft. The upward movement of the moist air begins when the mountain slope gives off some of its heat to the air immediately above it, thus warming that air and making it less dense than the surrounding environment. While thunderstorms that erupt in the Trans Pecos and High Plains in spring and early summer are the result of movement of the "Marfa front," or "dry line," that sharp gradient in low-level moisture is usually not present in mid and late summer. A very large percentage—about three out of four—of the thunderstorms in summer that give the Trans Pecos much of its annual rainfall are due to heating and orographic lifting of warm and highly moist Gulf air.

## The Localized Severe Thunderstorm

Whereas the air-mass thunderstorm usually is short-lived and hardly ever produces destructive wind or hail, a second type of thunderstorm, known as the *severe localized storm cell*, develops rapidly, persists for comparatively long periods of time, and often supplies flash-flooding rains and damaging hail. One element of a severe thunderstorm is the *squall line*, an arrange-

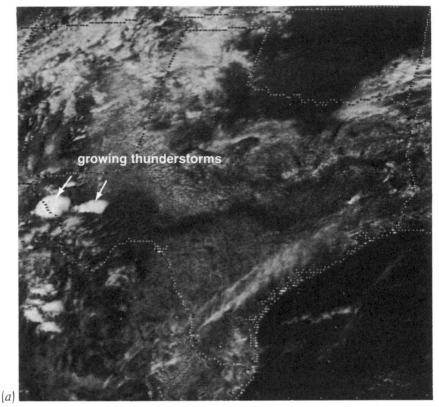

growing thunderstorms

*(a)*

64. *Large isolated thunderstorms, as seen in these satellite photographs, (a) start growing rapidly in the early afternoon heat in the
Trans Pecos in summer; (b) within only a few hours many of them*

ment of towering thunderheads in long
lines so close together that it is sometimes
regarded as a "line thunderstorm." Sometimes a squall line will develop well in advance of the leading edge of an advancing
cold air mass; hence, the phenomenon is
regarded as a "prefrontal" squall line. On
other occasions, the concentration of thunderstorms that compose the squall line will
mark the boundary of cold polar air as it
wedges underneath warm, moist air from
the Gulf of Mexico. The warm air is lifted
quite forcibly by the intruding cold air beyond its lifting condensation level, which
can be seen by a ground observer as corresponding to the dark, sharply defined cloud
base at the leading edge of the approaching
line of storms. This lifting of moist air
causes massive amounts of moisture to be
condensed out, giving the dark, foreboding
appearance that typifies squall lines. Squall

lines almost always bring strong, chilling
winds along with a slashing rain. Once
they pass, the atmosphere becomes tranquil again. In fact, squall lines sometimes
"use up" all the latent instability in the atmosphere to the extent that, if a cold front
follows a few hours later, it is unable to
generate a subsequent round of turbulence.

A second severe-thunderstorm element
is the *multicell storm*, which is made up of
several individual cells, each having its
own distinctive pattern of updrafts and
downdrafts. The majority of severe thunderstorms that produce hail, high winds,
and tornadoes in Texas each summer are
multicell systems. They frequently display
a peculiar kind of behavior in that they
move systematically toward the right of the
environmental air flow in the middle of
Earth's troposphere. During the lifetime of
most multicell thunderstorms, the contin-

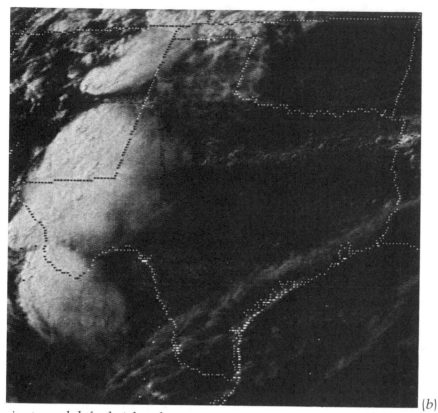

(b)

*rise to such lofty heights that strong upper-air winds shear off the tops of the thunderheads and transport them for hundreds of miles. (National Oceanic and Atmospheric Administration)*

uous generation of new cells on the right flank of the system, in concert with the dissipation of old cells on the opposite flank, is responsible for this drift toward the right of the prevailing wind.

Yet another very large and severe thunderstorm type is the *supercell*, whose organization is so pervasive that the entire storm behaves as a single entity. One distinguishing feature of a tornado-producing supercell is the hook-shaped radar echo that usually forms on the right flank. It is the recognition of this peculiar characteristic by NWS forecasters that often leads to the issuance of a tornado warning. Another unique trait of a supercell is the presence of an echo-weak region, or vault, that corresponds to the area of maximum updraft. In this so-called vault, updrafts with speeds of more than 50 mph propel water droplets far into the cloud in such haste that they do

not have time to grow large enough to reflect radar waves. Consequently, to the radar observer, the vault is that inner portion of a very intense thunderstorm marked by the conspicuous minimum level of radar echo. The image on the radar screen projects the appearance of an intense thunderstorm with a gaping hole in it. Recent investigations have revealed that the largest hail produced by a supercell occurs in a narrow band around the periphery of the vault. Better familiarity with the supercell should enable the weather forecaster to anticipate with greater confidence the likelihood of hail occurrence.

### A Majestic, Rampaging Marvel
Largely a Warm-Season Phenomenon

All sections of Texas witness the greatest number of thunderstorms in either spring or summer. While the majority of thunderstorm days are confined to the three months of summer in the western third of Texas, the number elsewhere is spread out a bit more evenly from spring to early autumn. The central third of Texas—including the Edwards Plateau and North and South Central Texas—sustains a relative "maximum" occurrence of thunderstorms during the latter half of spring, or from mid-April through the end of May. The number of thunderstorms drops from a peak of about six per month in April and May, levels off for summer, and then steadily lessens during autumn until the fewest (1–2 per month) are attained, on the average, in winter.

In the coastal plain, stretching in the south from the Lower Valley up through the Upper Coast region to southern East Texas, thunderstorms are almost always most numerous in summer and early autumn. Cool fronts seldom extend that far south at those times of the year, so the impetus for a high frequency (one thunderstorm day out of every 3–4 days) must be supplied by weather systems originating from the Gulf of Mexico. Indeed, influences ranging from minor tropical waves to full-fledged hurricanes contribute many of the thunderstorms that trek across the coastal plain from June through September. The influx of moist air cooled by the waters of the Gulf and then warmed by passage over the heated coastal plain in daytime is also a key factor in the development of mostly scattered and unorganized groups of thunderstorms that traditionally yield only modest amounts of rainfall. As in most other portions of Texas, thunderstorms in winter along the Texas coast are not common, though the frequency of occurrence, albeit small, is larger than that for any other region of the state.

Few areas elsewhere on Earth feature thunderstorms having the dimensions and ferocity of those that regularly roam the High Plains of Texas in spring and summer. Booming thunderheads that grow to heights of 50,000 feet or more and that unleash a menacing barrage of hail, gusty winds, and slashing rains are the prime ingredients in a typically tumultuous warm season in this sector of Texas. Locales in both the Panhandle and the far western Trans Pecos are buffeted by thunderstorms on one out of every three days in a normal summer. The frequency of so many thunderstorms in June, July, and August in the Panhandle is due to the tendency of numerous cool fronts to dip southward out of the central Great Plains into the northern reaches of Texas, then stall across the Panhandle for several days at a time. The leading edges of these weak polar air masses are almost invariably the focal point of widespread thunderstorm activity in summer. In fact, these slow-moving summer frontal intrusions account for more than half of all the thunderstorms that erupt in Texas' far northern sector in most any year. By contrast, autumn, winter, and the early half of spring are relatively sedate, with thunderstorms occurring an average of once each month. The supply of moisture usually is too scanty to support the development of but a few thunderstorms, though the frequency of frontal passage is high.

Forecasting Severe Thunderstorms

Severe, localized thunderstorms are some of the most difficult weather phenomena to forecast precisely. A thunderstorm is categorized by the NWS as "severe" when it is recognized as having either or both of the following characteristics: (a) winds with speeds of more than 58 mph or (b) large hail with a diameter of at least ¾ inch. One very important aspect of the forecasting performed by the NWS is the recognition of the probability that severe thunderstorms will occur and the issuance of special weather statements through the news media to forewarn the citizenry. To this end, the National Severe Storms Forecast Center (NSSFC), located in Kansas City, Missouri, continuously monitors weather conditions throughout the nation using a wide variety of weather data collected by radars, satellites, balloons, pilots, and ground stations. Meteorologists at the NSSFC then piece all this material together and from it ascertain when and where severe weather is apt to happen.

A *severe thunderstorm watch* states where and for how long the severe storm threat will exist. Watch areas usually are issued for rectangular areas whose sides may be as long as 200–300 miles. The watch is only an indication of where and when the probabilities for severe thunderstorms are highest. It should not be construed as meaning that severe thunderstorms will not occur outside the area or time frame specified in the watch statement. It is a good idea for persons living near or on the fringe of the watch area to be on the lookout for threatening weather conditions. When a watch is in effect, persons should be on guard for threatening weather and should be attuned to further information disseminated by NOAA Weather Radio or commercial radio or television.

A severe thunderstorm watch is not the same as a warning. Simply stated, a watch means that severe thunderstorms are possible or even likely. A *severe thunderstorm warning*, on the other hand, is issued by the NWS whenever a severe thunderstorm has been sighted or indicated by radar. The warning describes the area encompassing the severe thunderstorm, including, particularly, the downstream sector that stands to be affected by the thunderstorm, given its location, size, and direction and speed of movement. The reader should be advised of one precaution: a storm warning may not always be given by the NWS because a severe thunderstorm might not have been detected soon enough to warn those living in the area affected by it.

## Lightning

The sudden deaths of the three persons fleeing an electrical storm near Plainview in 1979 accentuate the fact that the lightning bolt is an exceedingly potent and capricious force of nature. Its erratic and deadly character explains why it ranks near the top of the list of those natural phenomena responsible for human fatalities. On the average, six people die from lightning every year in Texas, and nearly twice that many sustain serious injury. Nearly all lightning casualties occur during the months of April–September (*see fig. 65*), when thunderstorms are most numerous. Most of the lightning tragedies that have occurred in recent years could have been avoided had the victims been better informed of when and where lightning was apt to strike. Too, the number of fatalities could have been

reduced had the victims received more immediate and proper medical treatment.

### The Cause of the Flash

The lightning bolt is a gigantic electrical spark having an immense amount of power and lasting only a fraction of a second. Many bolts go unseen by humans. While Earth's surface may be struck, on the average, about one hundred times every second by lightning, not all lightning generated by the atmosphere ever reaches Earth's skin. Some lightning never leaves the confines of a thundercloud, whereas other strokes jump from one cloud to another. Though it is of extremely short duration, the lightning bolt is anything but simple. Though it is unpredictable and destructive, lightning is also necessary, for it plays a vital role in maintaining electrical harmony between Earth and its atmosphere.

Along with cosmic rays and storm processes within Earth's atmosphere, Earth is a source of the positive and negative particles that make the air highly conductive. In a sense, Earth's surface acts as a car battery by continually giving off these ions to the atmosphere. Were not some of this energy given back, Earth would exhaust itself of its electrical charge in a matter of minutes. It is believed that the myriads of thunderstorms that occur every day around the globe maintain this charge on Earth, and much of the work of replenishing Earth's supply of electricity is done by the lightning bolt.

In the same way that pressures vary at differing levels in the air and oceans, so too there is always a difference in "electrical potential" at various altitudes in Earth's atmosphere. For instance, there is a measurable difference, or potential gradient, in voltage between a person's head and feet. Long ago, Benjamin Franklin discovered that enough of a potential gradient exists between the ground and a kite flying several hundred feet up in the air to produce a spark. If this potential gradient is large enough, the energy yielded may perturb particles in the air so much as to make them luminous. This "glow" is a corona, or a point discharge also known as "St. Elmo's fire," and it occurs at the tips of grass, trees, and structures located in the vicin-

number of deaths and injuries

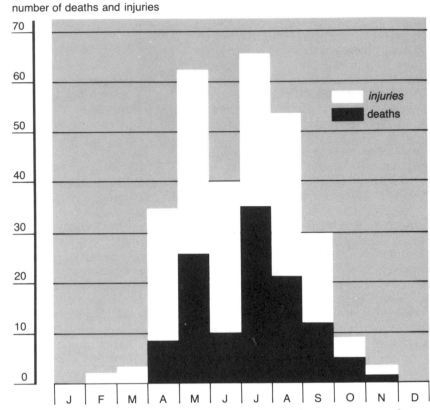

65. Number of deaths and injuries caused by lightning in Texas during 1959–1980. (U.S. Department of Commerce)

ity of storms where the electric field is substantial.

A thundercloud, simply stated, has electrical charges in two distinct areas within itself (see fig. 66a). Near the base of the cloud, there is a high concentration of negative charge. Opposite, or positive, charges predominate in the upper regions of the cloud. It is unclear how the thundercloud brings about this separation in electrical charges. Some physicists suggest that the distinction is due to the collision of ice and water particles within the cloud, while others believe the fracture of large raindrops is a primary contributor. Whatever the mechanisms might be, the effect on the motion of air within the storm cloud is important to its electrification. Because of this orientation of charges within the cloud, negative charge is fed back to Earth partially in the form of lightning strokes.

The lightning bolt comes into being when a charge travels toward Earth from the base of the thundercloud in a succession of steps. Just as this initial surge of current, which is established in a tiny fraction of a second and is commonly known as the initial "step leader," nears Earth's surface, other discharge streamers leap from the surface of Earth to intercept this downrushing streamer (fig. 66d). When the two invisible streams of current meet each other, a charged path, or conductive channel, is completed that allows the leader from the base of the cloud to reach Earth. Instantaneously, a large flow of electrical charge travels back up this freshly carved out channel, illuminating the branches of the leader track. This surge creates the brilliant flash that we see. Whereas the lightning bolt appears to emanate from the storm cloud, in truth it travels to the cloud

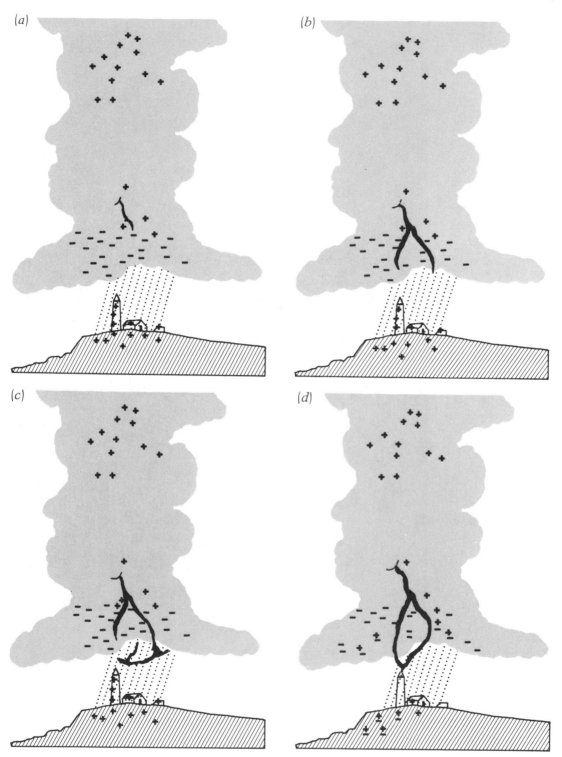

66. *Stages in the evolution of a lightning stroke.* (U.S. Department of Commerce, Thunderstorms)

*67. Streak lightning. (National Oceanic and Atmospheric Administration)*

from Earth's surface. This whole, elaborate series of processes consumes less than a second of time.

Although several lightning bolts may occur in the same vicinity in a matter of a few seconds, it is safe to assume that no two of them are ever identical. This is likely due to the fact that the charge in the sponsoring cloud is constantly changing, as is the temperature, pressure, and electrical makeup of the subcloud layer of air. That is not to say that lightning never strikes twice in the same place. Tall objects protruding from the ground, such as church steeples, transmission towers, and trees, become inviting candidates as the conduit for the flow of electrical charge between the cloud and Earth. These vulnerable ob-

68. *Combination of forked lightning and streak lightning. (National Oceanic and Atmospheric Administration)*

jects may be hit several times during a single thunderstorm event.

## Species of Lightning

The varieties of lightning are numerous. The most common kind in Texas is *streak lightning*, a cloud-to-ground discharge that appears to be concentrated in one channel (*see fig. 67*). Most strokes of streak lightning have branches, appearing as a photo rendition of a river with many tributaries appended to it. Often two or more of these branches hit the ground concurrently, causing what is popularly called *forked lightning* (*see fig. 68*). Whenever a strong wind blows at right angles to the observer's line of sight, a bolt of streak lightning may appear to be spread horizontally as a ribbon of parallel luminous streaks. Each successive stroke is displaced by small angular amounts and may appear as distinct paths to the human eye or the camera. This variety is referred to as *ribbon*, or *band*, *lightning*. *Bead lightning* (also known as *pearl* or *chain lightning*) is seen only occasionally when an observer is positioned relative to the lightning bolt so that his or her end-on view of a number of segments of the zigzag pattern gives an impression of unusually

bright pockets of light at various points along the lightning channel. Often in areas of the state where the far-distant horizon can be easily viewed, light is seen to be emanating from a great distance although the observer cannot actually see a lightning stroke or hear the clap of thunder. This luminosity is known as *heat lightning* and can be reflected or diffused by clouds, leaving a person with a sensation of a broad area of flashing light. Another fairly bright illumination, whose flash cannot be seen because it is obscured by clouds, is *sheet lightning* (*see fig. 69*). It is so labeled because it seems to light up a large portion of the cloud in which it occurs, giving the impression of a "sheet" of diffuse but bright light. It really is not a distinct type of lightning but rather is a manifestation of an ordinary variety of lightning that is hidden by clouds.

Unquestionably the most controversial variety, because of its rareness, is *ball lightning*. Some experts say that it does not exist, but a fair number of trained and reliable observers have supplied evidence to suggest that it is real. Those who have seen it say that it consists of a reddish, luminous ball that usually is about 1–2 feet in diameter

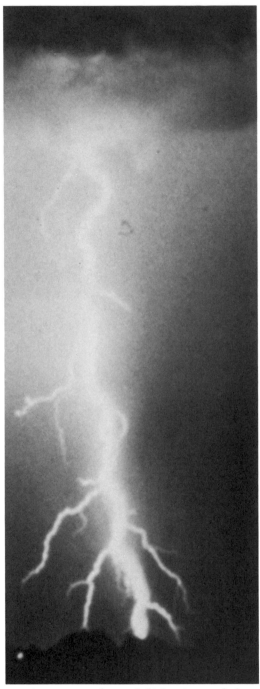

69. *A mixture of streak lightning and sheet lightning. (National Oceanic and Atmospheric Administration)*

that may zip across a floor, roll out of a tree, or float in midair. Hissing is sometimes heard, followed by a noisy explosion; but at other times, the phenomenon has been seen to disappear quietly. Those who contend that they have observed the phenomenon say that ball lightning (also called *globe lightning*) has a lifetime that may last up to several minutes and may be observed during or just moments after another type of lightning stroke has occurred in the vicinity.

Among the Greatest Dangers

Regardless of the type of lightning common to an area, people outdoors without adequate protection from buildings or vehicles are the most inviting targets. More than two-thirds of all confirmed lightning deaths involve campers, hikers, farmers, athletes, and beach enthusiasts. Persons who have their minds set on finishing a round of golf when the rumble of thunder can be heard are extremely vulnerable. So are hikers who seek the cover of a lone tree on a hilltop or swimmers who delude themselves by thinking that water provides protection.

It is true that one of the safest places to be during an electrical storm is inside an automobile, not because protection is afforded by the rubber tires but because the occupants are surrounded by the car's metal body. Lightning may strike an auto, but, if it does, the electricity travels easily from one area to another on the car's body. It is less likely that the current would take a shortcut through the interior and injure the passengers. Probably the greatest danger to a motorist during a thunderstorm is not electrocution but being blinded by the brilliant flash of a lightning bolt that hits nearby. Some drivers have been involved in serious accidents after being temporarily blinded by a lightning flash, losing control of the automobile (especially at night), and striking an oncoming auto or a fixed object off the roadway. On the other hand, it is possible to be hit by lightning while getting into or out of your automobile as happened to a minister and his son near Murchison in Henderson County one afternoon in May 1979: although the boy was only slightly injured, the father was killed while getting into their car when a lightning bolt hit a nearby tree and jumped to the automobile.

The nature of many of the deaths resulting from lightning over the years in Texas points to the great danger of outdoor sports when thunderstorms are in progress. A tally of athletes who died from lightning in Texas in just the past few years reads thusly: a young man in San Angelo killed by lightning while holding an aluminum bat during a baseball game at a family reunion on an afternoon in late May; a Terrell high school lad felled by a lightning stroke from a leaden sky in the middle of an afternoon football practice in early September; a jogger in north Houston who, when suddenly stunned by a bolt of lightning, crumpled to the pavement and died on a hot, sticky day in July; and an aspiring golfer felled by lightning on a municipal golf course in Harlingen on a sultry evening in August. Spectators of outdoor athletic events are vulnerable too. Swimming while a thunderstorm is going on in the vicinity can be exceptionally hazardous, for lightning can—and often does—hit water. The current of electricity sent traveling in all directions through the water will pass through the human body with the potential to at least render the victim unconscious.

People at work outdoors and those who seek what is apparently adequate shelter may be at the mercy of the lightning bolt. More than a few people have been struck and killed by lightning while seeking shelter under a tree. An 11-year-old Keller girl died in a Dallas hospital ten days after being struck by lightning while sitting in a swing under a tree during an electrical storm. On a late summer afternoon in 1977 a man seeking refuge under a tree in a pecan orchard was hit by a lightning flash; the bolt first hit the tree, then traveled to the back of the neck of the victim and down his body, burning holes in his clothing and knocking the soles of his shoes off. Others have died while exposing themselves in wide-open areas. Two men—and the horses they were riding—were slain by lightning while herding cattle across an open meadow in Montgomery County near Dobbin during a late-morning April thunderstorm. In another similar instance, a youth died from being hit by lightning near

Dubina in Fayette County while hauling hay one August evening in 1978.

Even people secluded inside their homes are not altogether safe from lightning. Some homeowners have a false sense of security in believing that their television antenna affords them an ample amount of protection from lightning. To the contrary, most residential TV antennas provide no protection at all and many even invite lightning into the home because their ground connections are inadequate. When lightning strikes an outside antenna whose ground is not sufficient, the surge of electricity meets too much resistance at ground level. As a result, the current piles up and then may jump to a better-grounded conductor, such as a water pipe. A house fire may then ensue when the current passes through a wall in the house.

Although lightning is less likely to enter your home along telephone lines than through power circuitry, you can purchase lightning arresters for use where the telephone line enters the house. Since the current and voltage in a typical telephone circuit are relatively low, the protective device can function at a lower voltage than that which operates on the power lines. That is not to say that it is altogether safe to use a telephone when lightning is around. While there are some known isolated incidences of people being burned by lightning while on the telephone, a more widespread hazard is acoustic shock. A person's hearing may be affected by the loud click made in the telephone receiver when lightning strikes the circuit. To play it safe, delay making a phone call until the storm has passed.

It seems logical that aircraft, while flying into and near electrically charged clouds, would face the greatest lightning-induced hazard of any vehicle. It is true that airplanes, while airborne, are frequently struck by lightning. Evidence of these hits consists of small burns, pits, or holes on such inviting targets as wing tips, rudders, and the nose of the fuselage. Aerials that protrude into the air are especially vulnerable. However, pilots worry much more about turbulence than lightning. The stresses exerted on the structure of an aircraft by violent updrafts and downdrafts have been known to tear off wings. The battering absorbed by hail can also be a serious problem. On the other hand, lightning most often merely disrupts communications with the ground, although it can cause some structural damage and it can be an irritant to the pilots. Airplane crashes directly attributable to lightning are extremely rare.

## The Lucky Survivors

There have been some rather bizarre instances in which people were struck by lightning but somehow survived the experience. In fact, available statistics suggest that more people struck by lightning live to tell about it than die from being hit. Most likely, however, survivors are never "hit" directly by lightning, for a direct strike leaves severe burns. Most persons injured—and even many who die—by lightning have few, if any, marks on their bodies; the current that passed through them was relatively small. The ability of the human body to accommodate an electrical current is very small; merely a small fraction of an ampere of current lasting only a fractional second can so affect one's central nervous system as to kill. Most lightning injuries result from the victim's being stunned or knocked unconscious. Those persons hit while standing under a tree, for instance, do not receive the brunt of the lightning bolt. The amount of electricity that passes through their bodies is but a small portion of the total amount of current that emanates outward along the ground from the tree. They are more likely to be harmed by the concussive effect of an exploding tree or by flying chunks of tree trunk the instant a lightning bolt registers a hit.

## Lightning's Effect on the Environment

The amount of damage wrought by lightning to timber resources and structures is enormous. Every year hundreds—perhaps even thousands—of lightning-caused fires bring destruction of a hardly estimable dimension to timber, other vegetation, and wildlife. Without fail every year, homes, farm buildings, power facilities, fuel storage installations, and even aircraft become frequent targets of the lightning bolt. Prop-

erty damage to homes and business establishments alone perennially runs into millions of dollars. Few structures are immune to the stroke of lightning. Objects susceptible to this type of natural violence range from roving livestock to underground circuitry and ships at sea. Grass fires triggered by lightning—like those that razed thousands of grazing acres one evening in August 1980 in Clay and Jack counties—are common throughout much of the year.

Explosions induced by lightning are not infrequent either. One of the most notorious incidents of an oil storage facility being struck by lightning occurred at a dock in Nederland on April 19, 1979. A flash of lightning struck a Liberian oil tanker, igniting fumes from a recently emptied oil tank onboard the ship. One man died in the ensuing fire, another drowned when he was blown into the water, and sixteen others aboard the ship at the time of the explosion were injured; the ship subsequently burned and sank. A thunderstorm on a cold January night in 1974 produced a lightning bolt that struck a large oil storage tank at Port Neches. The resulting fire, fed by strong winter winds, persisted for 24 hours and consumed 150,000 barrels of crude oil. Even subterranean objects are susceptible to lightning: about 5,000 feet of telephone cable were burned near Mount Pleasant one afternoon in March 1976, disrupting telephone service for two days and costing $275,000 worth of repairs. An untold number of livestock are also taken every year by lightning. The repercussions of a flash of lightning can be preposterous; for example, a bolt struck and shattered a tree in Kilgore before dawn one day in November 1977, and fragments of that tree blew across the street, smashing windows and damaging the roof of a church building.

Of course, lightning is not all bad. Although the extent of the contribution is not known, lightning is a help to farming. Crops need nitrogen to grow, and lightning supplies some of that natural fertilizer. Though nitrogen makes up about 79% of all the gases in Earth's atmosphere, it is of no use to plants unless and until it is joined with other chemicals. The lightning bolt is the catalyst for bringing some of the nitrogen in the air together with oxygen to form a nitric oxide gas, which is then dissolved and transported as nitrate to Earth's surface by precipitation. The soil absorbs the nitrate and then furnishes it to plants.

Thunder: A Product of Lightning

Thunder results when the electrical current that has begun to flow in a new channel suddenly heats up the air. This very intense and rapid heating explodes the air out from the lightning channel, thereby compressing adjacent air, which in turn instigates sound waves. The loudness of thunder depends upon the magnitude of the electrical current as well as the rate at which the current builds up in the lightning channel. Louder thunder occurs when the current builds up more rapidly to a larger magnitude.

Thunder usually can be heard for distances up to 10 miles away, although it is possible to hear thunder as far away as twice that distance. On the other hand, at times it cannot be heard even at fairly close distances. The distance between the lightning and the observer determines the amount of time that elapses between seeing the lightning and hearing the thunder. This leads to a rule of thumb by which you can gauge your distance from lightning: since sound travels at a speed of about 1,090 feet per second, or about 1 mile every 5 seconds, count the number of seconds between these two events, then divide by five to get the distance in miles. For example, if the sound of thunder follows the flash of lightning by 10 seconds, it may be assumed that the lightning flash is about 2 (10 ÷ 5) miles away. Similarly, the duration of thunder produced by a particular lightning bolt depends upon the distance from the observer to the closest and the farthest parts of the lightning flash.

Variations in the sound of thunder are due to several factors. Thunder that is generated by a single bolt of lightning is usually perceived as a mixture of claps and rumbles because the various elements of the lightning channel are positioned differently with respect to the observer. That means that thunder's sound waves would be perceived differently by observers at contrasting locations. In addition, the sound of thunder will be altered by the medium in which the thunder travels. The thunder signal can be scattered, attenuated, and refracted by characteristics of the atmosphere, such as its viscosity and temperature. Furthermore, the wind may have a pronounced effect on the sound of thunder sensed by the observer. Since sound waves propagate downwind faster than they do upwind, the observer's position relative to the source of thunder and the direction from which the wind is blowing could determine whether or not he or she even hears the sound of thunder.

## Hail

Like many other broiling afternoons in late summer, billowy cauliflower-type clouds bespeckled the sky from one end of the distant High Plains horizon to the other. Scores of these cumuliform clouds served as mute testimony to the presence of ample moisture in a lower atmosphere that stewed from the searing heat of the overbearing, relentless sun. Nearly all of them lived but for a few fleeting moments, their only contribution being numerous brief spells of shade for the sun-drenched landscape. A few of the more ascendant domes unleashed a terse shower of rain that did little more than whet Earth's thirst for more. Then, an hour or so past noon within sight of the New Mexico border, a lone thunderhead went berserk. The massive cloud churned uncontrollably and evolved quickly into a rare supercell. It began to spit out small hail profusely near Hereford while razing the terrain at a speed of 20 mph. Within two hours the gigantic thunderhead had given birth to numerous corollary cells that fed on the same source of energy as the parent cloud: torrents of moist air near the ground flowing northwestward upslope from a source region rich in moisture from the Gulf of Mexico. This intense thunderstorm complex pushed onward across the Texas High Plains, destroying hundreds of thousands of acres of crops with an incessant barrage of hail. Hailstones as large as softballs tore holes in the roofs of homes in Lamb County; some of the stones, propelled by winds estimated at 80 mph, crashed through storm windows and screens and dented floors. Hail persisted for forty minutes in parts of Hockley County, covering the terrain to the extent that it looked "like a blanket of snow" to one observer at the scene of the catastrophe. The hail covered U.S. Highway 385 near Dimmitt with a glaze of ice that sent automobiles skidding out of control.

Unquestionably one of the worst—if not the most cataclysmic—hailstorms ever to devastate Texan territory left a swath of damage and destruction more than 200 miles long and up to 40 miles wide, stretching from Deaf Smith County in the northern High Plains to Glasscock County in the extreme southern sector of the region. The most extensive hailstorm in many years left some 150,000 acres of cotton, corn, and other High Plains crops "looking like fields in the dead of winter." Another 550,000 acres of crops were damaged by the hail, most of which was no larger than marbles. In all, farmers lost on that "Black Friday" in August 1979 crops valued at an astounding $200 million. To make matters even worse, most of the cotton crop that was destroyed or heavily damaged was judged by area farmers to be "the most promising cotton crop in many years."

That nature selected the far-ranging, fertile cropland of the Texas High Plains to be the target of such a devastating onslaught underscores the fact that the sector of the Texas economy that stands the most to lose from hail is agriculture. Hail damage to crops in some parts of the Panhandle and plains of western Texas is a virtual certainty every year, with losses running into the tens—and sometimes hundreds—of millions of dollars annually. Hail not only restricts the quantity of crops produced every growing season but also affects the quality of crop yields. While food growers suffer the most from crushing hailstorms, not to be ignored is the great amount of loss and damage that occur each year to livestock and personal property in all regions of Texas.

## One of Nature's Eccentricities

Of all the forms of precipitation engendered by the thunderstorm, hail is by far the most peculiar. Stated simply, hail forms when either (or both) large raindrops or snow pellets (also known as *graupel*) are

repeatedly propelled up and down in a cloud, thereby allowing them to grow at the expense of millions of tiny cloud droplets. It is believed that graupel serves more often as a hail embryo than do frozen raindrops. As they are transported about within the thundercloud by its forceful updrafts and downdrafts, these embryos grow larger with each successive layering of opaque ice (rime), a result of the impact of supercooled cloud droplets onto the embryos.

Each time additional water freezes onto the hailstone, a bit of heat energy is released. Consequently, the temperature of a growing hailstone may be several degrees warmer than its cloud environment. This is an important characteristic of the hailstone, for the temperature of the hailstone's surface in turn has a lot to do with its rate of growth. If the hailstone's surface has a temperature below freezing, the collected cloud droplets readily freeze and the surface of the stone remains relatively dry. However, if the temperature is right at the freezing level (32°F, 0°C), the collected water will not freeze immediately. The surface of the stone stays wet, and some of the water may be shed as the stone moves around inside the cloud, thereby limiting its growth. During its lifetime, a hailstone may sustain several "wet" and "dry" growths as it passes around within a large thunderstorm having a varying temperature and liquid water content. In this way, a hailstone develops the layered structure that is often observed when it is sliced into pieces (*see fig. 70*).

Hailstones assume a variety of shapes and sizes. Spheroidal stones are the most common, and it is this type that often exhibits the layered interior structure resembling an onion. It is not uncommon, however, for hail to be either conical or otherwise generally irregular in shape. Some of the irregularly shaped hail consists of chunks or clusters of smaller hail elements frozen together. By the standard established by the World Meteorological Organization, for an ice particle to be classified as hail, it must have a diameter of at least 0.2 inch (5 mm)—anything smaller than that is classified as an ice pellet. Small hail or ice pellets may be found in almost any thunderstorm and, for that mat-

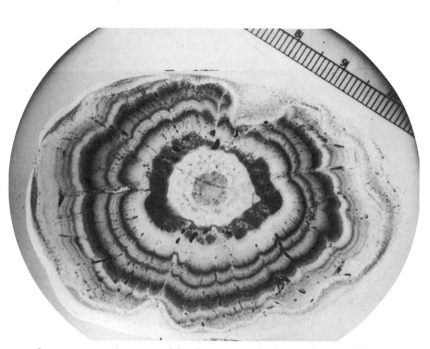

70. *Thin section of a typical hailstone, with the scale in millimeters. This is a negative, which means those areas that look darker are, in nature, really whiter. Air bubbles in the ice make parts of the ice look opaque. (National Center for Atmospheric Research)*

ter, in many towering cumulus clouds. Large hail, by definition, has a diameter of more than 0.8 inch (2 cm). It is usually found within, alongside, and underneath large convective clouds (cumulonimbus) whose tops may extend to 50,000 feet or more above sea level. In fact, some of the very prolific "hailers" that blossom in the High and Low Rolling Plains of Texas soar to levels of 60,000 feet or even higher. These hail-producing thunderheads typically have diameters ranging from 5 to 10 miles, while the diameter of the hailing area at any one instant is usually on the order of 1 to 3 miles.

Most of the hail observed in Texas is no larger than marbles, ranging in size from a small fraction of an inch to about one inch. However, hail the size of golfballs or even tennis balls is not uncommon in much of the state, particularly in spring and early summer. Indeed, seldom a year passes without at least a few occurrences of hail as large as baseballs. Some of the largest hailstones observed in recent Texas weather history fell in San Antonio in May 1946;

these stones, larger even than Texas grapefruit, struck the ground with such momentum that they bounced upward and broke second-story windows in some buildings.

Hail intensity varies considerably among regions of Texas on most any "hail day." One disastrous hailer can wipe out a dozen farms while leaving hundreds more between them unharmed. One measure of a hailstorm's intensity, other than the appearance of crops pummeled by the hail, is the depth of hail accumulation on the ground. Most Texas hailstorms fail to deliver enough hailstones to cover completely the ground. Some hail is plentiful enough to do that, however, and in a few instances, the hail may accumulate to astounding depths. One such notable incident occurred in the far northern portion of the state on May 28, 1970, when hail up to the size of baseballs severely damaged wheat and other crops in a 450-square-mile area of eastern Carson County. That aspect of the storm is not so uncommon for the Texas High Plains, however; what is striking is that the hail piled up to 18 inches on Texas

FM Road 294 so that only one lane could be used for several hours. Phenomenal hail-fall accumulations are not confined to the "hail alley" of western Texas, however. Motorists on Texas Highway 323 near Henderson had to contend with small hail that was bumper deep after a severe thunderstorm pummeled Rusk County in May 1976. Yet, hail depths of more than an inch or two are rare in central and eastern sections of Texas on most hail days during spring.

## Hail's Painful Consequences

Much of the annual loss due to hail is almost always concentrated in a few storm days. Indeed, a single massive hailstorm, like that which ravaged the southern High Plains in August 1979, may account for a very large percentage of any year's total statewide. Obviously, most—if not all—of the loss sustained when a hailstorm strikes a swath of the flat west Texas countryside will consist of damage to crops. On the other hand, a hailstorm hitting in the more densely settled areas of eastern Texas likely will result in much greater cost to personal and business property and a lesser amount to crops. One such hailstorm whose legacy consisted almost solely of losses to residential and business property pounded the city of Texarkana on April 22, 1978. In a

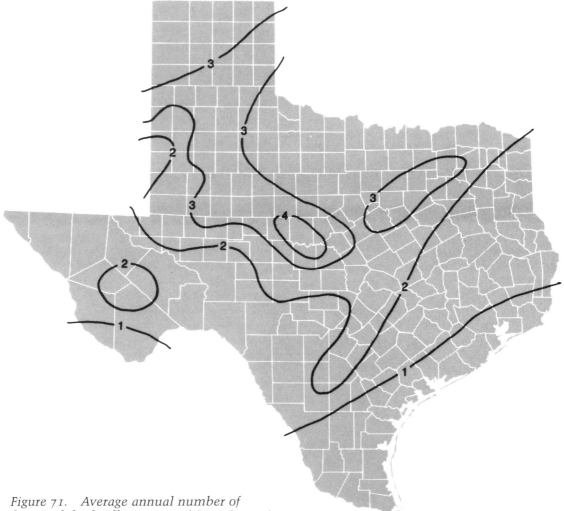

Figure 71. Average annual number of days with hail. (Illinois State Water Survey)

mere twenty minutes the ferocious storm caused damages totaling $10 million to residences and businesses and another $5.4 million to automobiles. Hailstones as large as 2¼ inches were measured in the northern sector of the border city. At the same time on this memorable hail day, numerous other hailstorms were ravaging scattered communities from the Low Rolling Plains to East Texas; one of the hail-producing thunderstorms, whose top was gauged by weather radar to have extended to 70,000 feet, caused many millions of dollars in damage to property in Mansfield.

Very few crops can survive a hailstorm without sustaining damage, but the crops most easily hurt by hail are fruits, for they lose their value from even slight bruising by hail of any size. Luckily for the Texas food economy, much of the fruit crop is grown in the state's southern quarter, where hail seldom occurs. Tobacco is also highly susceptible, but very little is grown in Texas. Soybeans, barley, rye, sugar beets, sorghum, potatoes, and other vegetables are also vulnerable to hail. By and large, however, most of the crop loss sustained by Texas growers each year that is attributable to hail is made up of wheat, cotton, or corn. In fact, Texas leads the nation in sustaining the greatest average annual loss to such crops as wheat and cotton because of hail.

Loss of or damage to property on account of hail consists primarily of disfiguration to fixed structures (such as homes, business establishments, and automobiles) and injury (or even death) to livestock and trees. Much of the loss incurred in residential areas during hailstorms is made up of damage to roofs of homes. Of course, windows may be broken, paint may be chipped off, and siding may be damaged, but these types of losses usually are small in comparison to the damage inflicted on roofs. Even human beings sometimes become casualties. Injuries to the head are common, especially when hailstones reach golfball size or larger. Obviously, a safe place to be during a hailstorm is within a stout, fixed structure like a home; still, seclusion in one's home may not guarantee safety from the ravages of hail. Although she was not seriously injured, a woman living in Com-

fort was knocked unconscious one day in April 1970 by a hailstone as big as a baseball that came through the window of her home and struck her on the head. The advice given to residents during a lightning storm also applies for hailstorms: confine yourself to the interior portions of your home away from doors and windows.

The threat of hail is an excellent reason for staying away from large thunderstorms while in an aircraft. Most pilots can relate one or more experiences of having encountered large hail not only within a thundercloud but near one as well. It is not uncommon to discover while in the air that a thunderstorm can and often does eject hailstones of varying sizes into clear air as much as five miles away from the main storm cell. The amount of damage that hail can do to an aircraft—whether a light version or a larger and heavier model—is stupendous. A classic case of hail causing an extensive amount of damage occurred near Carswell Air Force Base in Fort Worth in 1959. A B-52 jet bomber, part of the Strategic Air Command fleet, penetrated a severe thunderstorm while cruising along at 8,000 feet when suddenly it encountered hailstones the size of baseballs. The ten-man crew lost control of the huge aircraft momentarily, but before they could bail out, the jet passed just as abruptly out of the storm cell. It was not until the crew took the plane to 23,000 feet for an assessment of the damage that the full impact of the hail outburst was realized. The huge hailstones had shattered the windshield and had ripped off the radome, which sheltered the radar's antenna assembly. The leading edges of the wings had been hammered almost flat, while numerous holes were discovered in the wings and the engine nacelles (coverings). The plane's electrical system had momentarily failed during the encounter with the hail because pieces of metal from the body of the plane had been swallowed by the plane's engines. Although the 47-second confrontation with the hail failed to disable the giant bomber, it did batter the $8 million jet extensively, and it took several months to refurbish the B-52 for duty.

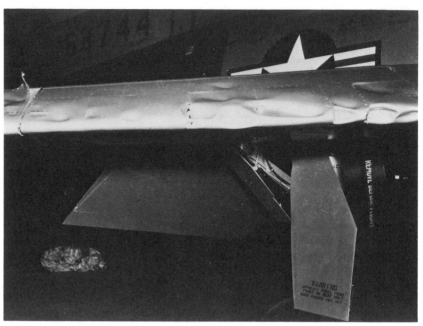

*72. Damaged B-52 jet bomber. (National Severe Storms Laboratory)*

When and Where to Expect Hail

No portion of Texas is safe from hail. In fact, most places in the northwestern quarter of the state are struck by hail three or four times, on the average, every year. Hail is most common in the southern Panhandle, the southern High Plains, the northern Edwards Plateau, and parts of the Low Rolling Plains and North Central Texas. On the other hand, hail is comparatively scarce in the coastal plain and in the Big Bend area of far western Texas. A few, or even several, years may pass between the occurrence of hail in these latter areas.

Since much of the hail is generated by very large and intense thunderstorms induced by the rapid movement of cold fronts, hailstorms usually are short-lived. Most outbreaks last no more than a half hour and oftentimes no more than ten or fifteen minutes. Particularly in spring and early summer, hail may accompany prefrontal squall lines that precede by one or several hours the arrival of cooler, drier air from the west or north. At other times, in the absence of prefrontal squall lines, hail-producing thunderstorms may not form until the cold air arrives. Though it is fairly uncommon, hail may occur more than once on the same day in the same locale—first with one or more prefrontal squall lines, then later from frontally induced thunderstorms accompanying the invasion of cold air.

Some hailstorms, especially those that form in the far western portion of Texas, appear to have no association with cold fronts. Most of the locales in the mountain ranges of the Trans Pecos observe one or two incidences of hail in most years during late spring and summer as a result of orographically induced thunderstorms. These massive, towering thunderheads form when warm and quite moist air transported into the region from the Gulf of Mexico is forced several thousand feet up the slopes of the Davis, Guadalupe, Chinati, and Chisos mountain ranges. The development of these potent but often isolated thunderstorms may be abetted by the intrusion of desertlike air from the interior of Mexico. This very dry continental air mass acts in much the same way as a cold front by wedging underneath and lifting up the dense and more moist air from the Gulf.

Since thunderstorms are scarce during winter, hailfall also is rare. The lowest monthly hail frequency is in January, when hail is likely to be seen at any one locale only once every 15–20 years. For winter as a whole, hail is most rare in the Panhandle and the Trans Pecos. Even though the frequency of cold fronts is usually at a peak during this season, the thermodynamic state of the west Texas atmosphere is not conducive to the type of intense convection needed for the production of hail.

With the advent of spring and its concomitant increase in atmospheric instability, the incidence of hail increases markedly in every section of Texas. By the month of May, the hail season normally has peaked. During this month, nearly every community from the High and Low Rolling Plains to North Central and East Texas will be pelted at least once by hail. An ample amount of near-surface moisture and much warmer temperatures set the stage for numerous eruptions of booming thunderstorms on those frequent occasions when strong, fast-moving cold fronts sweep through the state. Farther south between the Balcones Escarpment and the Gulf of Mexico, hailstorms are also most frequent at this time of the year, although the frequency is not as high as in the Panhandle and the Red River Valley. Residents of such cities as Amarillo, Lubbock, Midland, Abilene, and Wichita Falls can count on two or three hailstorms during the three-month period of March, April, and May. The frequency of hail occurrence is about half that much in places like San Angelo, Austin, Waco, and San Antonio.

Once the hot and mostly dry weather of a typical Texas summer takes hold in June, the number of hail-bearing thunderstorms wanes markedly in all but the extreme northern sector of the state. Vigorous cold fronts that push quickly through most of Texas are highly uncommon during June, July, and August, so the "forcing mechanism" essential for the eruption of massive thunderheads with the capability of yielding hail is missing. Besides, the broad subtropical, high-pressure ridge, in assuming a quasi-stationary position over Texas in summer, keeps the upper atmosphere nearly devoid of moisture, thereby ensuring that any convective clouds that form will not grow tall enough to attain a hail poten-

tial. Weak cool fronts often enter the northern extremity of Texas in summer and usually ease southward through the Panhandle, creating enough turmoil in the lower half of the atmosphere that thunderstorms—even a few with the capacity for hail—are very common. Most inhabitants of the High and Low Rolling Plains will see hail about once, on the average, each summer. In fact, for those living in the Panhandle, two or three hail outbreaks are the rule in the typical summer. Elsewhere in Texas, chances for hail during the three hottest months of the year are very remote. One or two summers in every ten years may provide a few incidences of hail in the Trans Pecos and North Central Texas.

When summer's heat gives way to the refreshingly cool breezes of autumn, the possibility of hail increases for most Texas residents. With an increasing number of cold fronts entering and pushing completely through the state early in autumn, the atmosphere grows much more unstable. Thunderstorms are more rampant and widespread than during summer in the eastern two-thirds of Texas, and quite a few of them grow severe enough to unleash a barrage of hail. The likelihood of hail occurrence is not nearly as great as during the spring, but it is significant nonetheless. Only in the High Plains, where thunderstorms abruptly become rather scanty with the onset of autumn, is there a lessening of the potential for hail. Hail can occur as late as November, but when it does fall that late in the year, it is almost always extremely spotty.

The individual desirous of avoiding hail altogether should seek out one of the locales in the state's southern tip, for it is in that section of Texas that hail is exceedingly uncommon. In such communities as Corpus Christi and Brownsville, hailfall is fairly rare anytime of the year. Those very seldom occasions when hail does occur almost invariably come during spring. The primary reason for hail being so scarce in this sector of the state is the relatively high freezing level in the atmosphere above the coastal plain. Whereas the typical freezing level in the hail-prone High Plains during the warmer half of the year may be as low as 11,000 or 12,000 feet (above mean sea level), in the coastal bend and Lower Valley regions the freezing level is usually higher than 15,000 feet. This means that, in southern Texas, a lower percentage of the total amount of cloud mass lies above the freezing level; hence, there is less cloud volume in which the hail embryo can circulate and grow to an extent that it can reach the ground as a hailstone. A secondary consideration is that temperatures in the subcloud layer (between the base of the cloud and the ground) are invariably warmer in spring in the extreme southern portion of the state than in areas farther north. A hailstone falling out of a thunderstorm would be subjected to more melting in this warmer layer and, consequently, would have less of a chance of reaching the ground intact as a hailstone. Even though hail is exceedingly rare in southern Texas, it can happen. One most notable incident took place on May 11, 1971, when hailstones 4 inches in diameter and chunks of ice 5 × 7 inches across and weighing nearly a pound each smashed automobile windshields, greenhouses, and roofs and windows of homes in Brownsville. Damage to property in a three-county area including the city of Brownsville amounted to $2.8 million, including the total destruction of 80 acres of honeydew melons in Starr County.

Hail Suppression

Can anything be done to prevent or inhibit the growth of hail in a storm cloud? In the past two decades, a sound, modern technology involving the seeding of hail-bearing clouds has been devised to deal with the problem. This technology was applied in parts of western Texas in the 1950s and again during most of the 1970s. Although proponents of the cloud-seeding activities were convinced that they were successful in reducing the incidence of hail, opponents feared that the cloud seeding was responsible for reducing rainfall in the area. No conclusive scientific evidence has yet been supplied supporting either of the two stances.

Previous efforts to suppress hail in Texas have involved the deliberate seeding of large convective clouds to trigger or inten-

sify the ice phase in those clouds. This intervention in the cloud's natural processes is meant either to stop outright the cloud's production of hailstones or to make many more small hailstones (instead of a relatively "few" large stones) that will eventually melt into large raindrops as they fall through the relatively warm layer of air between the cloud and Earth's surface. Silver iodide has been the most commonly used nucleating agent because it has a crystalline structure that closely resembles that of natural ice. The silver iodide, when injected into a cloud believed to have the potential for producing hail, serves as surrogate ice crystals by attracting a large portion of the multitudes of droplets within the cloud. These droplets freeze into ice when they make contact with the silver iodide crystals. Theoretically, if the number of these hailstone embryos having silver iodide as their nuclei can be increased to a point that they compete for the available supply of supercooled water within the cloud, there is likely to be many more, but much smaller, hailstones. These would melt during their descent and ultimately reach the ground either as much smaller, and hence less damaging, hailstones or as nothing more harmful than large raindrops.

Answers to questions relating to the utility and effectiveness of hail-suppression efforts remain as elusive as ever. Studies made by proponents and opponents of the High Plains projects, as well as nonpartisan, government-financed investigations, have been inconclusive. To verify or refute the utility of hail-suppression programs necessitates that these efforts be designed and carried out as careful, systematically controlled experiments. The controversy over cloud seeding's effectiveness—or lack of it—has raged long enough, and the potential benefits to be realized if seeding is proven effective as a suppressant warrant finding an answer.

# 6. The Tornado: Our Atmosphere's Most Savage Creature

The wailing of the siren came not as a complete surprise to thousands who were about to enjoy their evening meal. After all, for more than an hour the news media had filled the air with severe storm statements that warned all of Wichita County of impending weather violence. Word had spread throughout the Red River Valley that a line of approaching thunderstorms had unleashed a barrage of tornadoes farther west at Vernon and Lockett. Furthermore, the sky overhead had grown ominous in a matter of minutes. Yet, in spite of these sundry intimations that a city embedded in the middle of "tornado alley" would soon be lashed by another in a customary series of spring squalls, the thousands of inhabitants of Wichita Falls, hardened by previous weather calamities, were shocked at what they saw: a gigantic black cloud hanging beneath an inky overcast and plunging straight into the city from the wide-open plain to the southwest. To those who peered out windows and doors on that evening of April 10, 1979, and beheld the massive funnel expanding and throwing huge pieces of debris high into the air, the horror of the moment defied expression. As the mammoth tornado churned and bore down on the southern sector of Wichita Falls, hundreds fled to the nearest refuge and took shelter under tables and counters, in closets, bathrooms, and hallways. Others, trapped in their vehicles at busy intersections, either leaped from them and fled or vainly tried to extricate themselves from the jams and drive away. Many thousands who took appropriate life-saving action survived. Hundreds more were spared too, although they sustained serious injuries. But for more than two score, some of whom abandoned homes that would not be touched, one of the most massive tornadoes in recorded weather history meant sudden death.

The monumental twister took less than thirty minutes to desecrate a broad swath of the Red River metropolis. With a breadth of 1½ miles, it whipped across 8 miles of residential area in the southern sector of Wichita Falls, leaving forty-two dead among the litter and debris. More than half of those killed were in their automobiles when the twister lifted and spun the vehicles in the air as if they were bottle caps. The tragedy was all the more terrible because some of the dead had vacated homes never touched by the twister and had tried to flee from the storm in their autos. Nearly two thousand other persons were left hurting, while thousands more agonized inside at the sight of three thousand homes demolished and another six hundred heavily damaged. In all, the damage and destruction wrought by the merciless tornado cost the citizenry of Wichita Falls an estimated $400 million. Moreover, the tornado lived long enough to invade Clay County east of Wichita Falls, where it caused forty injuries and property damage of $40 million. The catastrophe culminated an unforgettable day in much of the northern

Low Rolling Plains of Texas, for two other huge tornadoes that afternoon had wrought havoc in several other communities west of Wichita Falls. One of them crashed through the southern edge of Vernon around three o'clock, leaving in ten minutes' time eleven persons dead and sixty more injured, with several hundred homes destroyed or damaged extensively. A second major tornado with a continuous ground track of almost 60 miles that extended far into Oklahoma hit the town of Harrold, causing extensive damage and one death.

## An Incomparable Spectacle

No other atmospheric convulsion is as terrifying as the tornado. Spinning in corkscrew fashion from beneath a murky sky, the phenomenon has to be the most alarming spectacle that nature can create. The awesomeness of a black column of rapidly gyrating air slinging wreckage in all directions inspires far more than reverential fear; rather, the eerie, sinuous funnel evokes nothing less than trepidation—and often panic—in those few who have beheld one. The word *tornado* originates from the Latin word *tornare*, which understatedly means to twist or to turn. The Italian *tornare*, the Spanish *tornear*, and the French *tornade* all mean the same thing. What these descriptors identify is a small yet violent body of winds packing a concentrated wallop more powerful than that of any windstorm on Earth. Though tornadoes usually affect relatively small areas, they strike faster and with more savagery than any other storm.

Meteorologists make a distinction between a funnel cloud and a tornado. Popularly referred to as a "cyclone" or a "twister," the violently rotating column of air pendant from a large thundercloud is regarded throughout its life history as a funnel. A tornado is, simply stated, a small-scale circulation that makes contact with the ground. Tornadoes occur in every month, though the preponderance of them occurs in spring and again in late summer or early autumn. Most reported funnel clouds are spotted visually by trained weather watchers, and many of them are "sighted" by NWS personnel manning weather radar fa-

cilities. Still, it is probable that the NWS weather-observing network fails to detect quite a few of the tornadoes that form, especially those that are short-lived and that occur in sparsely settled regions of the state.

### Evolution of the Twister

The tornado is one of several atmospheric phenomena not fully understood. It is known that, as with the formation and growth of hurricanes, conservation of angular momentum is highly instrumental in bringing twisters into existence. Tornadoes are the product of large, intense thunderstorms that grow in an excessively unstable atmosphere. They seem to evolve from very strong updrafts, or highly energetic flows of moist air from near the surface that feed into the base of a growing thunderhead. Apparently, buoyant forces that drive the thunderstorm enhance the convergence of this low-level moist air in toward the center, or core, of the updraft. In the same way that ice skaters quicken the rate at which they spin by drawing in their arms, air entering a column from different directions is made to accelerate rapidly. This means that, as air near and beneath the thunderstorm is directed into the updraft, the speed of the air may easily be tripled or quadrupled. Recent research suggests that many tornado-producing thunderstorms have updrafts with a pronounced rotation within a deep layer of the atmosphere extending upward from cloud base for several thousand feet, a fact that undoubtedly contributes to the development of a twisting funnel that suspends itself beneath the base of a cloud near where the updraft enters the cloud.

An examination of satellite photographs supplies evidence of the unusually vigorous updrafts that nourish tornado-producing thunderclouds. Fig. 73a is a photograph in the visible wavelength—the same view we would see if we were aboard the weather satellite taking the picture—of a large mass of intense thunderstorms in central Texas being generated by an upper-atmospheric low-pressure area. From this photograph we might deduce that rainfall is occurring within much of the area of North Central Texas covered by the clouds, which appear from above as solid white masses. Even a trained eye would have difficulty pinpointing the precise locations where intense thunderstorms were producing hard rains and possibly other forms of turbulence. A different kind of photograph, that of infrared wavelengths (not visible to the human eye), shown in fig. 73b, provides considerably more detail about this broad band of intense convection. The visible photograph shows us both low and high clouds having essentially the same appearance (either white clumps or white strands), mainly because both types of clouds reflect sunlight, thus exhibiting a white complexion. But since these two kinds of clouds are at much different levels in the atmosphere, and hence they have greatly varying temperatures, the infrared makes a clear distinction between the two. The infrared photograph reveals to us several very distinct thunderheads that help make up the broad band of cloudiness. The infrared view allows the weather analyst to locate very intense thunderstorms, which may have the potential for generating tornadoes and other kinds of severe weather. Indeed, the mass of thunderstorms labeled "T" produced a tornado at Waco at about the time the photograph was taken.

Texas: A Prime Spawning Ground

Texas, by virtue of its location between the Gulf of Mexico on its southeastern flank and the Rocky Mountains on its western periphery, is a prime area for the formation of tornadoes, particularly in the spring, early summer, and autumn. Often during these periods, three main flows of air converge on Texas to establish the type of atmosphere conducive to the formation of tornadoes (see fig. 74). As a cold front pushes into the state from the higher elevations to the northwest, the flow of near-surface, warm, and moisture-laden air from the Gulf of Mexico is accelerated. At the same time, high in the atmosphere (between 10,000 and 15,000 feet), very warm and much drier air pours northeastward out of Mexico. It is where the very dry air intrudes into or over the lower, more moist air from the Gulf that a very sharp gradient of moisture is established, and it is in this mixing zone that very intense thunder-

storms erupt with the potential of spawning tornadoes. The presence of a very strong jet-stream wind even higher in the atmosphere exacerbates this unsettled situation. Towering thunderheads that grow almost explosively upward are exposed to an intensifying and shifting wind as they reach levels of 30,000 feet, and even higher. It is this shearing wind that enhances the likelihood that a thunderhead will produce one or more funnels.

## Texas' Most Unforgettable Twisters

Two calamities head up the list of tornadoes that have wrought unforgettable havoc on some segments of the citizenry of Texas (see Appendix E-1). Of the hundreds of tornadoes that have ripped their way through rural hamlets, small towns, and bustling cities, these two are the most prominent because they were the most virulent of them all. Few people living today witnessed the black twister that smashed through the town of Goliad on May 18, 1902, extinguishing the lives of 114 of its citizens and injuring more than double that number. With damage amounting to $50,000—a vast sum of money to those living around the turn of the century—a large part of the town was decimated.

Equally lethal was the savage tornado that tore through the heart of Waco on the afternoon of May 11, 1953. Having carved out a path of destruction 23 miles long that knifed through the downtown section of a city that, at the time, contained about 85,000 residents, the twister finally ascended near the community of Axtell, leaving grim rescue workers to discover a death toll that matched precisely that of Goliad 51 years earlier. The impersonal tally of grief and woe revealed the tornado's legacy to be, secondarily and in addition to the 114 who died, almost 1,100 injured, 196 buildings demolished or in need of removal for safety's sake, 850 homes wrecked or partially destroyed, 376 other buildings declared to be unsafe, and about 2,000 automobiles ruined or severely damaged. The total property loss to the McClennan County seat amounted to more than $51 million. Ninety-four of the dead were found within a two-block area including

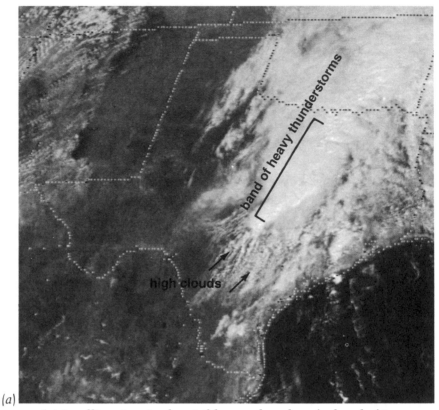

73. (a) Satellite view, in the visible wavelengths, of a band of intense thunderstorms in North Central Texas early in the afternoon of October 13, 1981. (b) Satellite view of the same band of thunderstorms in

the square around City Hall. Scores of people were buried under tons of rubble. Ambulance sirens wailed almost continuously far into the evening, ferrying to hospitals those who had been found alive and pried loose from the piles of debris that filled the streets of the roped-off downtown section. Broken power and telephone lines and ruptured gas mains made rescue attempts even more treacherous. The rain that fell throughout the evening would have been scorned as yet another handicap had it not been an aid in putting out numerous fires. For hours, rescuers brought out injured survivors, some of whom had endured only because workers piped oxygen down through cracks leading to would-be graves far below the tops of the heaps of debris. After many hours in which the air had been pierced repeatedly by screams and cries for help, those agonizing sounds ceased, and every

body retrieved thereafter was lifeless. Area hospitals were filled to overflowing, and many of the injured had to sit or lie on floors in corridors. As if nature was not content to confine its fury to Waco on that day in May 1953, another intense thunderstorm spun out a second deadly tornado that rent asunder the Lakeview section of San Angelo, killing 11, injuring 159, and wiping out 320 homes.

The tornado that gave Lubbock the darkest day in its history was like the one that hit Waco in that it caused immense destruction by plowing through the very hub of the sprawling urban center on the Texas south plains. In sculpting out a path about 8 miles long and as much as 1½ miles wide that began near the Texas Tech University campus and ended minutes later at the Lubbock Municipal Airport, the broad tornado caused damage and destruc-

band of thunderstorms

T

(b)

*the infrared, which locates the most intense thunderstorms; the thunderstorm labeled "T" formed a tornado at about the time this photo was taken. (National Oceanic and Atmospheric Administration)*

tion over 15 square miles—or about one-quarter of the whole city. The loss of property totaled an incredible $135 million. Twenty-six persons died when the tornado snaked through the downtown section of Lubbock not long after dark on the eleventh day of May, 1970. As with most tornadoes, nearly all of the deaths stemmed from either flying debris or the collapse of structures. The tornado exerted such force that numerous large and tall office buildings were damaged extensively. About 80% of all plate-glass windows were smashed in the downtown area. Some cars were flattened to within two or three feet of the ground. At least five hundred persons were injured; likely the injury toll was much higher, for crowded hospitals were forced to turn away scores of people with cuts and bruises seeking admittance.

The Lubbock tornado of May 1970 was one of two twisters regarded as the largest ever witnessed in Texas. The other, with a path of devastation measuring 1½ miles wide, slashed through the town of Higgins in Lipscomb County in the far northeastern corner of the Texas Panhandle on April 9, 1947. The same tornado also struck and totally wiped out the little community of Glazier in Hemphill County. In fact, the tornado ripped out a path having a length of 221 miles, shredding a part of Kansas and Oklahoma as well as the Texas Panhandle. The twister's death toll in Texas amounted to 68, with another 201 seriously harmed. Damage totaled $1.55 million. Few tornadoes ever live as long—more than six hours—as the one that demolished Glazier, Texas.

As the most densely populated area on the banks of the Red River and situated smack in the middle of the battle zone where cold polar air often clashes violently with warm, volatile Gulf air, Wichita Falls has borne an inordinately high number of weather calamities over the past several decades. Renowned as one of the locales delineating "tornado alley," the city has been scourged several times by blockbuster tornadoes that roared through the gently rolling hills of an area that has had to adapt to sweltering summers, bone-chilling winter winds, and vicious spring thunderstorms. None of the twisters, of course, had an impact like the tornado that devastated the southern sector of the city in April 1979, but another very large tornado that razed neighborhoods on the opposite side of

town in April 1964 left death, destruction, and misery on a scale almost as lamentable as that which superseded it fifteen years later. It ripped up a 6-mile-long swath of northwestern Wichita Falls at midafternoon on April 3, 1964, killing 7 residents, hurting 111 others, and costing the community $15 million worth in property damage.

Whereas the Waco, Lubbock, and Wichita Falls tornadoes wrought incredible destruction in portions of those cities, other twisters have been responsible for nearly, if not totally, eradicating all of several smaller communities. When the path of a twister happened to trek through the little town of

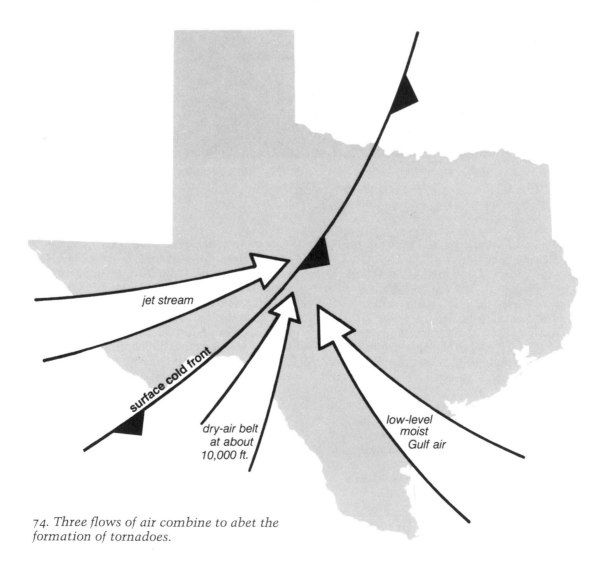

74. *Three flows of air combine to abet the formation of tornadoes.*

*75. The tornado that razed parts of Wichita Falls, including Sheppard Air Force Base, in April 1964. (Texas Department of Public Safety)*

Bellevue, in Clay County, on April 26, 1906, practically everything was demolished, 17 people were killed, and 20 were injured, while damage amounted to $300,000. Much the same thing happened to Melissa, Texas, on April 13, 1921; the little town in northern Collin County was practically destroyed when a tornado tore through it one evening, killing about a dozen people and hurting several scores of other residents. Add Rocksprings to the list of locales virtually wiped out by tornadoes. The county seat of Edwards County was leveled by a twister that hit suddenly on April 12, 1927. Seventy-two persons died there, about 200 were harmed, and property damage exceeded a million dollars.

The greatest single outbreak of tornadoes in Texas weather history occurred in conjunction with Hurricane Beulah, an enormous tropical cyclone rated as the third-largest hurricane the nation has ever seen. For as much as twelve hours before and subsequent to Beulah's landfall near Brownsville on September 20, 1967, a multitude of at least 115 tornadoes roamed across most of the southern half of Texas. One of them hit near Palacios, killing four persons and injuring six others. Other Beulah-inspired twisters caused extensive damage in Burnet, Louise (in Wharton County), and New Braunfels. The more than 100 tornadoes engendered by Beulah easily erased the previous record total of 26 hurricane-related tornadoes established by Carla.

Very often a tornado that inflicts death and desolation in a locale is accompanied by numerous other twisters that also cause harm nearby. When nature succeeds in assembling the right atmospheric conditions for fomenting a tornado, the end result is sometimes not a single tornado but rather a whole horde of them. If several of the twisters hit populated areas, the cumulative effect can be enormous. On May 6, 1930, a twister tore through Ennis and several other small towns in Hill and Navarro counties, killing 41 people, while a second tornado hit Kenedy and two other towns in Karnes and DeWitt counties, taking another 36 lives. The sum total of deaths for the day—77—is second only to the 114 dead exacted by the tornadoes that struck Waco and Goliad. As shown in Appendix E-1, 76 people lost their lives when a wave of tornadoes roared through Sherman and two smaller communities on May 15, 1896, while other one-day fatality totals of at least two score included 74 at Rocksprings and surrounding locales in April 1927, 68 in several Panhandle communities in April 1947, and 42 in a five-county area of northeastern Texas on April 9, 1919.

## Always an Untimely Phenomenon

Tornadoes often are difficult to discern from other low-hanging cloud forms. "Scud" clouds suspended beneath the bottom of a thunderhead usually exhibit a wind-torn motion, but they should not be mistaken for funnel clouds. You can be an effective tornado spotter if you are aware of the things for which you should look and listen. Look for any protuberance having a rotary motion near the bottom of a dark cloud. Organized rotary motion is the key to distinguishing a funnel cloud from harmless other clouds. Discernment of a tornado may sometimes be made more difficult when it descends from the base of a thundercloud without developing a visible funnel of its own. In such instances, you can ascertain its existence by noticing debris or dirt being picked up in a spinning motion from the ground. Do not confuse

these rotating whirlwinds with "dust devils," a phenomenon hardly ever associated with clouds. As a rule, if the tornado persists for any length of time, the material from the ground will be lifted higher and higher toward the bottom of the thunderhead, thereby "filling in" the spinning funnel to the extent that it becomes readily recognizable.

Of course, at night, it is exceedingly difficult, if not impossible, to see an approaching tornado. Often, brilliant flashes of lightning will afford a glimpse of one. Even in daytime, a tornado may not be visible because of heavy rain or nearby tall buildings that obscure one's vision. Nonetheless, a tornado can be detected, for its violently spinning column produces a distinctive roar that often can be heard for several miles. Some witnesses describe the sound of a twister as that of a loud freight train or a large jet aircraft. While the precise sound generated by a funnel cloud or tornado may be in the ear of the listener, it is known that the noise of a funnel cloud increases as the funnel approaches the ground and that it is loudest when the twister is moving along the ground.

## Texas: First in Tornado Occurrences

Official NWS statistics reveal that Texas, on an average basis, experiences more tornadoes each year than any other state. The tally shows that nearly 3,600 tornadoes were reported in the Lone Star State during the period 1953–1980, a number that translates into an average number of tornado occurrences of about 128 per year, or almost 5 tornadoes per 10,000 square miles. Actually, in terms of tornado density, Oklahoma leads the nation with an average of about 8 tornadoes each year per 10,000 square miles. The Panhandle and south plains of Texas, however, are not far behind with a density of about 6.5 tornadoes per 10,000 square miles.

The frequency of tornado occurrence and the death and injury toll exacted by tornadoes during the 30-year period ending in 1982 are given in fig. 77 and fig. 78. "Average" numbers of tornadoes and tornado days are based on nothing but arithmetic sums of the monthly and yearly totals of

76. *This twister ripped through west and south Dallas on April 2, 1957, killing 10 persons, injuring 200, and causing $4 million in property damage. (Texas Department of Public Safety)*

observed tornadoes and tornado days. Some months in some years may have more than a score of tornadoes, while the same months in other years may have less than half a dozen.

Due largely to the toll of 114 lives lost in Waco in May 1953, Texas sustained more tornado-related deaths than any other state in the nation except Mississippi during the period 1953–1980. With spring furnishing more than half of the year's tornadoes, it is reasonable to expect that season of the year to contain, on the average, the greatest number of deaths and injuries from tornadoes. Indeed, of the 374 deaths attributa-

ble to tornadoes during the 30-year period ending in 1982, exactly 300 of them—or slightly more than 80%—took place in April and May. In only one out of every seven years have tornadoes failed to cause fatalities. Whereas more deaths from tornadoes have occurred in May, injuries have been far more numerous in April. This oddity in the statistics is due to the "bias" provided by the inordinately high number of deaths that resulted from the Waco tornado of May 1953. Exclude the 114 fatalities assigned to that single twister and, on the average, May is marked by fewer tornado-related deaths and injuries than

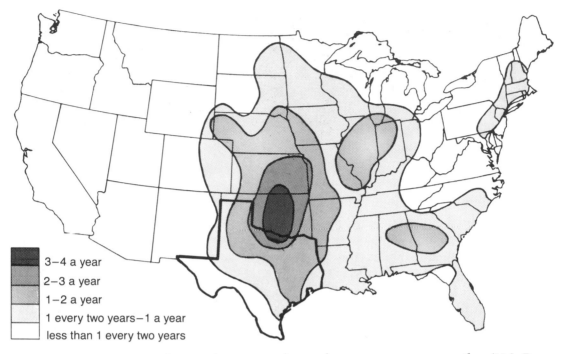

3–4 a year
2–3 a year
1–2 a year
1 every two years–1 a year
less than 1 every two years

*77. Average frequency of tornadoes per 2,500 square miles. (U.S. Department of Commerce)*

April. Nearly 100 people are hurt each April by twisters, and 58 more are injured in May. For the year as a whole, on an average basis, about 200 people are hurt by tornadoes in Texas, while about a dozen other people lose their lives.

### A Mostly Springtime Event

On an average basis, more than half of all tornadoes that strike in Texas every year occur during late spring or early summer. Six out of every 10 are observed in either April, May, or June. With an average of 36 tornadoes, May is by far the most unruly month of the year: on the average, one out of every three days in the month will be marked by the occurrence of at least one tornado. April and June virtually tie for second with an average of about 20 tornadoes apiece. August and September feature an average of 9 and 10 tornadoes, respectively. This slight rise in frequency after a noticeable decline in July is likely due to the influence of tropical cyclones that often send tornadoes skipping over the landscape in late summer and early autumn. During

winter in Texas, tornadoes are uncommon but not rare.

Tornadoes may, and often do, occur anywhere in Texas during spring. At other times of the year, however, certain regions of the state are more vulnerable than others. In the southern third of Texas—south of a line connecting Del Rio, San Antonio, and Houston—nearly all tornadoes appear during spring or during the active hurricane season that runs usually from mid-July through September. As spring blends into summer, the scene of the most unsettled weather shifts northward to include most of the northern two-thirds of Texas, excluding the Panhandle and Trans Pecos. While the frequency of tornadoes is highest for the Trans Pecos during summer, tornadoes so seldom occur in that western, semiarid region that its total number for the season pales in comparison to those sustained farther north in the High Plains. In July and August, and in the absence of any significant tropical disturbance affecting the Texas coastal plain, tornadoes are almost always confined to the Panhandle. This is

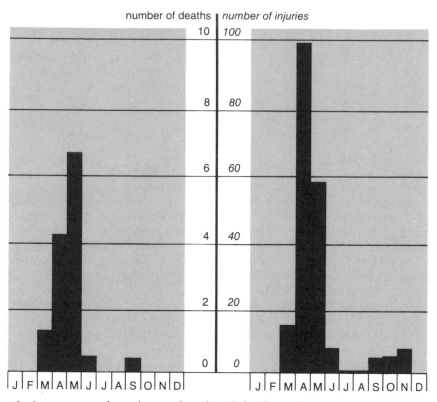

number of deaths | number of injuries

78. Average number of tornado-related deaths and injuries in Texas (based on data for the period 1953–1980). (U.S. Department of Commerce)

because the triggering mechanisms necessary to instigate twisters—namely, the intrusion of cool air and a vigorous enough wind flow from the southwest in the middle atmosphere—usually can be found only in the northern periphery of the state during what is usually the state's most dormant season. When invasions of cold air become commonplace in autumn, the frequency of occurrence of tornadoes increases notably, with most tornadoes from nontropical weather systems in August and September located in the northern half of the state.

The majority of tornadoes that strike in Texas occur during the afternoon and early evening hours, with the greatest frequency of occurrence confined to the period from about 3:00 P.M. to 7:00 P.M. Some tornadoes strike later in the evening, while relatively few occur after midnight. Tornadoes are infrequent, too, in the late-morning hours. Nocturnal tornadoes are most apt to occur during spring. In springtime, however, do not rule out the likelihood that a twister will happen at any hour, day or night. When the coastal plain is being affected by a tropical weather disturbance—such as a hurricane or a lesser tropical storm—twisters are prone to occur at any hour of the day or night. Factors that influence the timing of twisters in such circumstances include the immensity of the tropical system, the time it makes landfall, and how rapidly it dissipates once it moves into the interior of the state.

### Traits of a Texas Tornado
Consistency in Movement

Most tornadoes travel along the landscape with much haste, at speeds ranging between 15 and 35 mph. Unless a community is sprawled out over many square miles, a tornado will have wrought its injury to the locale in a matter of minutes. As a general rule, tornadoes in spring and autumn move on the ground faster than those that occur in summer. This is because the steering currents at the 10,000–15,000-foot level, which dictate the speed and direction of movement of tornado-bearing thunderstorms, are usually more vigorous in the months preceding and following winter than during summer. At least three out of every four tornadoes that afflict Texas move in a southwest-to-northeast direction. Paths of most tornadoes extend for only a few miles and most often are merely a small fraction of a mile wide. Quite a few of the tornadoes that occur in Texas each year carve out paths less than one mile long. Because winds at some distance from the funnel can be potent, the width of the area of Earth's surface affected by the tornado may be two to four times as broad as the width of the funnel itself.

The flow of air around the vortex of a tornado is most generally in a counter-clockwise (cyclonic) direction. Since no wind-measuring device has ever survived a direct hit by the core of a twister, the speed of tornadic winds cannot be known precisely. Estimates, using Doppler radar and photogrammetry and based upon special engineering studies of tornado damage, place wind speeds in the neighborhood of 200–300 mph. In some of the more extreme cases, tornadic winds may exert a force as much as 800 pounds per square foot. A 250-mph wind exerts a force 25 times that created by a 50-mph thunderstorm wind. A tornadic wind turns what are normally harmless, inert objects into lethal missiles: sand and gravel are propelled like shotgun pellets, and straw becomes nails that lodge an inch or more into tree trunks.

The destructive capability of a tornado stems not only from its strong rotating winds but also from the huge pressure differential that exists from the rim to the center of the funnel. No one yet knows the precise magnitude of this pressure variation, for an instrument package usually does not survive in the midst of a funnel.

Nearby pressure measurements reflect an abrupt drop in barometric pressure of 0.50–1.00 inch in the vicinity of the twister, though undoubtedly at the vortex of the funnel the pressure reduction is far greater than that. Although the sudden reduction in pressure brought about as the core of the funnel passes nearby does cause considerable stress on buildings, recent research suggests that the force exerted by high-velocity winds is primarily responsible for the devastating impact that tornadoes have on structures. For this reason, it avails little to rush around opening doors and windows in a house during a threatening situation. It is not only unnecessary but also dangerous, for you could be hurt by flying debris. Tornadic winds flowing over the roof of a building cause a strong updraft, which may be potent enough to lift off the roof; then, the walls soon crumble. Garages and carports are the most vulnerable parts of a single-level house when a tornado is nearby.

## Freak Incidents

Freakish things often happen during the passage of a tornado. There are documented cases of chickens being plucked of their feathers by the strong, vacuum-cleaning effect of tornadic winds. Blades of grass and straw have been driven through telephone and fence posts. Motorists have been lifted off highways and then dashed down in the middle of wheat fields. Heavy railroad cars have been hoisted from tracks and inverted hundreds of yards away in valleys and ravines. One of the most bizarre happenings in Texas' colorful weather history, an event verified by the U.S. Weather Bureau, took place in the northeastern corner of the Texas Panhandle near the town of Higgins in April 1947. The owner of a home in rural Lipscomb County, upon hearing the loud trainlike noise of an approaching twister, opened his front door and was lifted hundreds of feet into the air over the tops of nearby trees. A visitor in the man's home then went to the same front entrance to check on his friend and was also carried high into the air, but on a slightly different course. After a few very anxious moments,

79. *Damage wrought by a tornado in Plainview, 1973. (Texas Department of Public Safety)*

both men were lowered to the ground several hundred feet away from where the house had originally stood. Unharmed but understandably shaken, the pair proceeded to try to walk back to the house but the persistent strong wind forced them to crawl. When they ultimately got back to the site of the house, they found nothing but the foundation. Sitting on the floor was a lamp and a couch containing the owner's terrified but unharmed wife and two children.

Far more common than these illogical events are the less freakish sights that always attest to the vengeance of a tornado. The scene left behind in the wake of a twister resembles in many ways the devastation wrought by a heavy demolition squad. Fallen trees with their root systems displayed and virtually defrocked of their leaves lie across streets. Other trees nearby, still decked out in full foliage, stand erect as they were in the prestorm period. Houses, furniture, and automobiles are left strewn along the storm's path, sometimes piled up as much as a dozen feet. Within sight nearby are other homes and vehicles standing as they were, not disturbed at all by the tornadic blitz. Vehicles of all types are rolled over and crushed as if they were nothing but small toys. Where office buildings once stood are piles of broken boards, shattered glass, chunks of concrete, and gnarled pipes. Telltale signs of a twister's strike in the country often consist of uprooted trees, some stripped of their foliage and limbs, and fields of crops looking as if they had been scathed by a bulldozer.

## Varieties of Shapes and Colors

The majority of tornadoes are funnel shaped—broad at the top where they are attached to the base of a thunderstorm but tapering to a small diameter at the end touching the ground. The funnel sometimes undergoes strange twists and stretches that make it appear as a swaying trunk of an elephant or a writhing snake. Some of the time, winds at cloud level are swifter than those at the surface of the ground and the upper portion of the tornado moves along faster than the bottom; the pendant cloud is stretched to the extent that it resembles a slender string or rope. On other occasions, the tornado may hang straight down, and, as the diameter of the base of the funnel expands and grows to become as broad as the top portion, the tornado takes on the appearance of a huge, thick pillar or column.

Tornadoes manifest different shades of black or gray in part because of the diffraction of light by the water droplets that make up the dangling, twisting cloud. A tornado will appear light or dark depending upon the amount of light available to the observer and the type of material picked up by the tornado as it cuts across the landscape. If you spot a tornado between yourself and the sun, for instance, it is apt to be darker than if sunlight were directly hitting the side of the tornado visible to you. If the

*80. A typical tornado. (National Oceanic and Atmospheric Administration)*

soil or debris lifted by the tornado is dark colored, the funnel will most probably have a swarthy appearance, possibly even black.

### Waterspouts

Closely akin to the tornado is the waterspout, a phenomenon that occurs over water. This variety of the severe local storm is so called because, in the same way that a tornado stirs up sand and soil as it rolls across the landscape, the waterspout draws up spray from the body of water over which it moves. The most common kind of waterspout observed in Texas' coastal waters builds downward from a towering cumuliform cloud. Initially, the base of the cloud lowers toward the surface of the water, which at the same time is disturbed by vigorous winds blowing across the water's surface. About the time the whirling cloud material forms a distinct funnel, spray is thrust upward out of the sea. When the dipping funnel reaches the sea surface, water droplets from the cloud are merged with the saltwater spray to form the waterspout. Ordinarily, waterspouts are not nearly as intense as their landlubbing cousins, though some have produced wind speeds in excess of 100 mph. With an average size of 100–150 feet in diameter, the waterspout is considerably smaller than the tornado. Boats have been known to pass through weak waterspouts and sustain little or no damage, although some of the larger, more intense waterspouts are capable of destroying small craft. One such notable incident occurred in San Antonio Bay around noon on May 8, 1980. A waterspout ripped off the cabin from a shrimp boat, then overturned and sank the craft. Two of the three crewmen aboard were injured but rescued, while the third person was never found.

From almost any vantage point along the Texas coastline, at least a score of waterspouts can be seen in some summer months. The waterspout is much like the dust devil in that its formation is enhanced by unstable conditions—often typified by high near-surface temperatures and humidities—in the first few thousand feet of the atmosphere above the sea surface. For this reason, waterspouts are most apt to occur near the Texas coastline in summer, though sometimes a few form as early as mid-spring or as late as mid-autumn. Most of them live less than half an hour.

Occasionally, a waterspout occurs in interior sections of Texas when a tornado moves over a body of water, such as a large lake. In such instances, waterspouts can be just as devastating as their counterparts on land.

## The Urgency of Tornado Recognition

Weather violence and the response of citizens to it each year bear out the fact that a quick alert is essential if lives are to be saved. The NWS has become commendably adept at recognizing those conditions in the atmosphere that are prone to generate tornadoes. However, technology has not yet developed to the point that we can always be certain that a timely alert will be issued by the NWS and that the whole citizenry will receive proper notice. Until a totally dependable detection system has evolved, the NWS must rely upon the involvement of hundreds of trained storm reporters, or SKYWARN "spotters," for quick, accurate detection of tornadoes from their infancy to their expiration. These volunteers possess a thorough knowledge of local severe storm characteristics, and their ability to provide notice of developing weather hazards has saved countless lives.

### The Limitations of Tornado Forecasts

While the services of spotters are invaluable, the need for an improved capability of predicting the formation of tornadoes is not diminished. The limited amount of weather data, combined with limitations in our knowledge of what causes tornadoes, prevents weather experts from predicting the precise time that tornadoes will form, but these inadequacies do not stop the NWS from trying to project the likely time and place of tornado occurrence. By continuously analyzing conditions at all levels of the atmosphere within and adjacent to Texas, experienced meteorologists at the National Severe Storms Forecast Center in Kansas City, Missouri, gain considerable insight as to the probability that tornadoes will develop. The technology for predicting tornadoes has evolved to the degree that forecasters can now specify areas (usually a rectangle about 120 miles wide and 100 miles long or more) where most tornadoes can be anticipated to occur. Most tornado watches are issued with a "lead" time (i.e., the interval between time of issuance and the beginning of the valid watch period) of about one hour, and watches ordinarily cover time periods of five–six hours.

Remember that a watch is not the same as a warning. A *tornado watch* suggests that people go about their business as usual, but, since weather conditions are expected to become unstable enough to possess the potential for producing tornadoes, persons should check the skies periodically and stay within reach of special weather statements disseminated by the NWS through the news media and NOAA Weather Radio. On the other hand, a *tornado warning* informs the populace that a tornado has actually been sighted, either by a public spotter or by weather radar. The tornado warning contains information on the location and movement of the tornado, the area through which it is expected to move, and the interval of time during which it will pass through the area that is being forewarned. It also urges people in the vicinity of the detected tornado to take immediate cover, and it advises other persons near the reported site of the tornado to be prepared to seek safety.

No Cause for Complacency

There is a tendency for people to grow complacent about the threat of tornadoes, if their area has not been affected by one for several years. Most communities are included in at least several tornado watches and warnings every year, but relatively few of them are ever appreciably affected by a twister. Consequently, some people deduce that, because tornado advisories covering their area of concern seem to be followed repeatedly by an absence of tornadic activity, those alerts needlessly alarm the public and should be ignored. In reality, no weather alert is given without sufficient cause. The tornado advisory is not meant to guarantee that you will be affected but rather to signal the potential danger that actually exists. No preparedness program can be effective if the citizenry does not have a due regard for severe storm advisories and if it is not capable of taking immediate action to ensure its safety.

Unfortunately, little attention has been devoted to the issue of how a tornado might be influenced. Whether or not the capability exists for modifying thunderclouds to alter, if not suppress, the development of tornadoes is not known at this time. Some hypotheses have been offered by elements of the scientific community as to how the behavior of a tornado might be controlled. However, modifying all large thunderstorms thought to have the potential for generating tornadoes would seem to be an impossible goal when one considers that most thunderstorms sizable enough to engender tornadoes undergo essentially explosive growth in a short period of time and are scattered all over the countryside. Unless research is vigorously pursued, however, we will not even know of any potential influence that we may have on tornadoes or the parent thunderstorms that beget them. Such a concentration of our investigative resources is justified by the continuing, serious threat to life and property posed by this most dangerous weather phenomenon. Until that capability is identified and cultivated, we must be resolved to know and to follow tornado safety rules, realizing at the same time that we still are very much at the mercy of nature's whims.

# 7. Heat Waves and Drought: Legacy of a Texas Summer

The event recurs with striking regularity—so much so that the precise moment can be predicted years in advance. June 21 of each year marks the summer solstice, that instant when the direct rays of the sun reach their northernmost point on Earth's globe: 23½°N latitude, also known as the Tropic of Cancer. The occasion is regarded popularly as the time at which the sun is farthest north. By this juncture the atmosphere over Texas has already been made to simmer enough to make it sultry. The buildup of heat will persist for several more weeks, causing daytime temperatures to peak sometime in July or early August. Not until September will the sun's torrid grip on Texas be relaxed perceptibly.

For some Texans the cessation of what is invariably a sweltering summer comes a bit too late, for, during an "average" summer, thousands of people suffer from too much heat and sun. The severity ranges from acute sunburn suffered by untold thousands of outdoor activists to heat stroke that may, in some especially torrid years, result in scores of fatalities. In some especially scorching summers, more Texans may die from heat exhaustion and stroke than from any—or all—other elements of the family of natural hazards. Heat waves exact an uncertain toll on humanity by putting great stress on the body's capacity to perform its normal, life-sustaining functions. The most likely victims of a siege of intensely hot weather are the aged and the infirm—those with weak or diseased hearts that succumb to the demands of an accelerated biological pace necessitated by the heat.

## Perennially Hot

With Texas' geographical location not far from the Tropic of Cancer, summers here are destined to be hot. As the searing rays of the sun become more vertical with the approach of the summer solstice, the land slowly but steadily warms under their influence. The polar jet stream is forced to retreat far northward, and the infiltration of refreshingly cool and dry Canadian air practically ceases to occur. The parade of vigorous midlatitude cyclones that contributed to an uproarious spring shifts poleward also, leaving Texas without any large-scale means of ventilating its atmosphere by ushering in fresh doses of cooler air. In the absence of cool fronts, warm, muggy air that swarmed into the state from the Gulf of Mexico at semiperiodic intervals in winter and spring becomes a climatic fixture in summer. Occasional intrusions of very arid air from the Mexican and U.S. southwestern deserts inflate temperatures even more, particularly in the western half of the state. In short, the traditional weather pattern of a Texas summer is one of monotony: mild to warm muggy nights and repressively hot days.

The Heat Wave of 1980

There is no more vivid example of the tragic and costly penalty paid for living in the Texas heat than the epic killer heat wave that engulfed the state in the summer of 1980. The very severe and prolonged siege of heat caused human suffering of a magnitude rarely experienced; at least sixty people within Texas expired from heat stroke, while scores of others died from complications of other medical problems aggravated by the excessive heat. Thousands of Texans suffered from heat cramps, heat shock, and heat exhaustion as daytime temperatures shot above 100°F (38°C) in parts of the state on dozens of successive days. A superabundance of sunshine resulting from a virtual absence of clouds spanning a six-week period from June to mid-July brought about extremely large evaporation rates that led to parched and cracked soils, plunging river and reservoir levels, and depleted groundwater supplies. Crops not nourished often with irrigated water suffered greatly, and some even withered away to become total losses. Cotton yields in many areas were reduced by one-fourth. A 25% reduction in grain sorghum yields was the largest in a quarter of a century in the southern half of Texas. Estimates of livestock and crop losses there

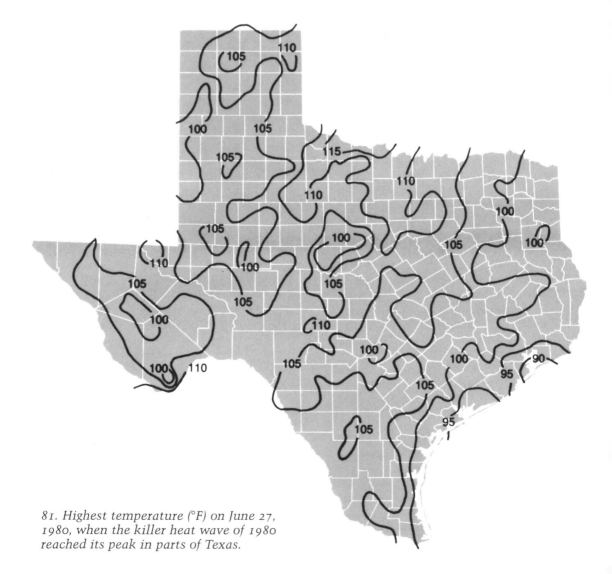

81. Highest temperature (°F) on June 27, 1980, when the killer heat wave of 1980 reached its peak in parts of Texas.

ranged from $250 to $500 million. In heavily populated Harris County, streets cracked and buckled, about 100,000 acres of corn and soybeans were either lost or badly damaged, and power consumption reached record high levels as the citizenry struggled to stay comfortable.

The heat wave of 1980 seldom slackened during the months of June and July, leaving many sections of Texas with some of the hottest temperatures ever witnessed. Thermometers repeatedly measured afternoon temperatures well above 110°F (43°C) in the Low Rolling Plains and North Central Texas. In fact, the last week of June 1980 brought the hottest weather ever inflicted upon Dallas–Fort Worth (113°F [45°C] on June 26 and 27) and Wichita Falls (117°F [47°C] on June 28). The heat wave contrived to scorch all of Texas throughout the month of July, leaving such cities as Abilene, San Antonio, Houston, El Paso, Austin, and Dallas–Fort Worth with the highest average monthly temperature for July in recorded weather history. In much of the northern half of Texas, daytime temperatures shot above 100°F (38°C) every day in July. One good measure of the intensity and staying power of the 1980 heat wave is the record number of 100° days registered during that summer (see Appendix B-10); triple-digit readings were measured at Dallas–Fort Worth, for example, on 65 of a possible 92 days during the three summer months of June, July, and August 1980.

Characteristics of a Texas Heat Wave

To some who spend the season in Texas, summer is merely one massive and lengthy "heat wave." Yet, in the true sense of the term, "heat wave" connotes a weather anomaly brought about by a marked change in the hemispheric wind pattern. The huge dome of hot, dry air that perennially drifts northward out of the tropics during spring and envelopes Texas in summer is, in some years, especially strong. This "subtropical ridge" of high pressure in the upper atmosphere, whose surface element is popularly known as the "Bermuda High," may take hold with such a ferocity that the oft-welcome blessing of fair weather (particularly after a wet spring) turns into the dreaded curse of a searing heat wave with

its concomitant severe drought. When this abnormally potent high-pressure cell becomes entrenched, its descending currents of air destroy any opportunity for a significant cloud cover. Day after day may pass with nary a cloud visible in the sky. On other days, numerous "fair-weather" cumulus clouds may dot the sky, but the subsiding air within the high-pressure dome squelches any appreciable vertical growth of the clouds. Another regrettable aspect of the behavior of this large dome of hot air is its tractability. Once it settles into position, the subtropical ridge often budges only a little for periods lasting as long as several weeks at a time. Any potential relief, in the form of weak maritime low-pressure systems migrating out of the Pacific Ocean into the western portion of the nation, is shunted northward into Canada.

The time-series plot, shown as fig. 83, of weather conditions at Dallas–Fort Worth on a day during the heat wave of 1980 reveals the combination of weather ingredients that typically produce phenomenally hot conditions in summertime in Texas. With high pressure prominent at all levels of the atmosphere over north Texas on June 26, 1980, a sky devoid of clouds—thereby allowing a maximum influx of solar radiation—teamed up with an arid desertlike southwesterly wind to force the temperature at midafternoon to a level never before seen in the Dallas–Fort Worth area. Characteristically in summer, when daytime temperatures climb to the "normal" level of near 100°F (38°C) in North Central Texas, winds emanate from the south or south-southeast, transporting just enough moisture to maintain a dew-point temperature in the low 70s and thereby helping to prevent the air temperature from climbing very far above the 100° level. However, when the wind shifts into the southwest, drier and hotter desertlike air is channeled into the region and the air temperature soars far beyond typical daytime levels. Note in fig. 83 that the dew-point plunged shortly after noon from the upper 60s to the mid 50s in a few hours, during which time the air temperature shot up from the vicinity of 100°F (38°C) to 113°F (45°C). This is stark evidence of the fact that air that is less moist heats up much more

*82. A common scene in summer as viewed from an orbiting satellite: not a cloud in the sky over Texas. (National Oceanic and Atmospheric Administration)*

rapidly and to a greater extent than air high in water content. This cycle of shifting wind and fluctuations in dew-point and air temperature was repeated often in the summer of 1980, and it is a phenomenon that can be identified almost anytime that summer heat becomes very intense.

When and Where Temperatures Are Hottest

Heat waves intense enough to kill are most common in Texas from mid-June through late August. Often, telltale indications of a budding heat wave are manifested first in the western sector of the state, where very hot temperatures well above 100°F (38°C) may occur frequently in May. A full-fledged siege of abnormally intense heat likely will be in full swing in the Trans Pecos sometime in June or early July. In fact, in nearly every year, whether marked by a pronounced heat wave or not, the temperature in this westernmost region peaks sometime in late June or early July. Fig. 84

reveals that in a typical summer the hottest daytime temperatures, averaged on a weekly basis, occur from about June 16 to July 7 at El Paso. Daytime highs then slowly tail off for the duration of summer, the end of which is characterized by daytime highs around 92°F (33°C) in Texas' westernmost metropolitan area. Elsewhere the hottest weather in summer takes place a few weeks later in the season. On an average basis, daytime high temperatures usually peak in the low 90s in the High Plains either in mid or late July. Average daytime high readings for Amarillo drop off in late summer as rapidly as they rise in late spring and early summer. From the Low Rolling Plains and Edwards Plateau to East and Southern Texas, hottest summer weather, on the average, occurs during the first ten or fifteen days of August.

In many years, the hottest temperature of all is felt in one or several points in the semiarid basins and valleys of the Trans Pecos. To many, the name "Presidio" is

*83. Hourly weather conditions in Dallas–Fort Worth on June 26, 1980, when the area's hottest temperature (113°F) in its 84-year weather history was registered.*

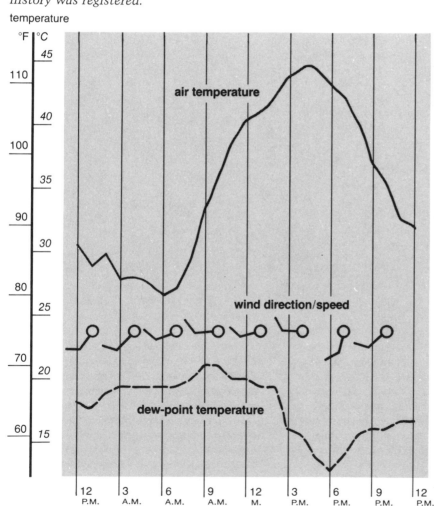

temperature

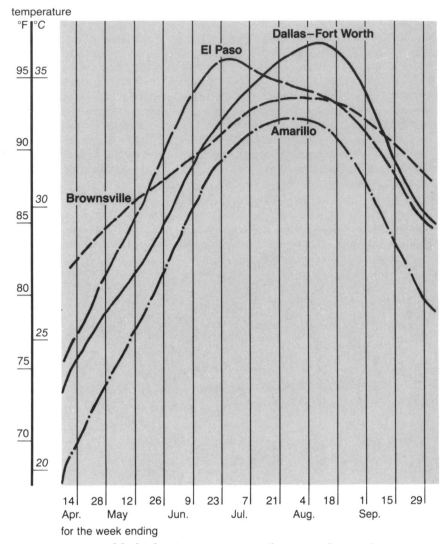

*84. Average weekly high temperature, April–September, at four cities in Texas.*

synonomous with hot weather, for the tiny community nestled in the Presidio Valley on the northern banks of the Rio Grande experiences 100° heat on scores of occasions every year. Its daytime high temperature is broadcast consistently in summer by television and radio as the hottest reading anywhere in Texas. Presidio does collect, on the average, over eighty 100° days every year (*see Appendix B-9*), beginning in April and sometimes lasting well into October. Often, however, midafternoon temperatures in parts of the Low Rolling Plains

or North Central Texas exceed those of the Trans Pecos, especially in the latter half of summer. In two of every three years, some location in the Trans Pecos—with a temperature in excess of 110°F (43°C)—garners the distinction of being the hot spot in Texas (*see Appendix B-8*). These very torrid extremes in the Trans Pecos most often occur in late June; by contrast, when the hottest point is elsewhere in Texas, the temperature—usually between 110°F (43°C) and 115°F (45°C)—normally takes place in July or early August.

If summer as a whole is considered for its heat content, the hottest climate may be found in Southern Texas. Parts of the Trans Pecos, Low Rolling Plains, or North Central Texas may register many of the hottest, single-day maximum temperatures in most years, but week in and week out the weather is apt to be persistently hottest in the state's southern extremity. In most years Southern Texas will endure the highest average monthly temperature of any region within the state. Almost invariably, July or August will be the warmest month, with mean temperatures often in the mid or upper 80s. Daytime temperatures almost always reach well into the 90s or low 100s

throughout the whole summer, while nighttime lows rarely dip below the mid or upper 70s.

### The Scourge of Drought

Drought is unlike any other type of weather phenomenon in that it is distinctly insidious. It creeps upon us slowly and gradually, leaving us to discover almost abruptly its terrible reality. It leaves its victims debilitated and frustrated, for as much as any other aspect of nature's behavior, it rudely demonstrates our utter helplessness in its presence. Day upon day, with the last dash of rain a distant memory, we peer into

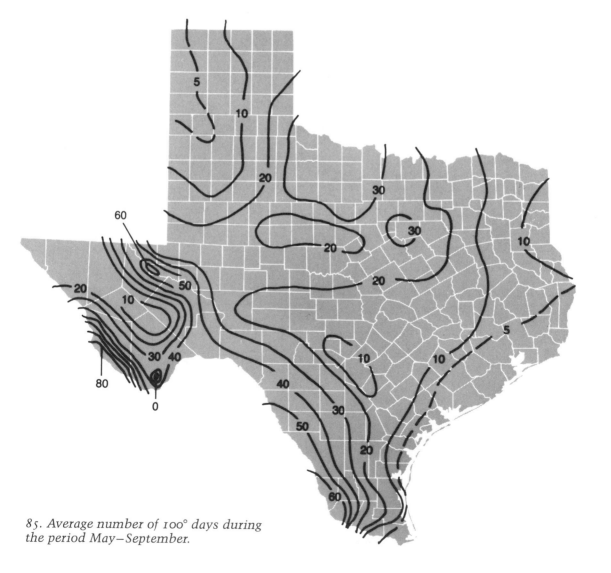

85. *Average number of 100° days during the period May–September.*

a sun-splashed sky, knowing full well that it holds tons of invisible water vapor that are more than ample for our needs. We gaze upon pastureland so badly scorched as to be no longer edible for livestock, stock tanks and reservoirs whose levels sag lower with each passing day, and acres and acres of crops that wither and wilt into oblivion.

## Defining Drought

Drought is an enigma in part because, more so than any other weather phenomena, it is difficult to define. We can measure its effects: crops fail, the price of most foodstuffs skyrockets, water rationing is invoked. However, the aspect of drought that is difficult to recognize is its incipiency. As a rule, it is much easier to identify its effects long after it has become established than to discern its presence while it gradually grows more and more severe. The fact that it is commonly used to describe a wide range of different dry-weather conditions accounts for the difficulty in categorizing it. Depending upon the context in which the word is sometimes used, drought can assume some entirely different meanings. To wit, permanent drought has reference to a condition in which precipitation is never sufficient to meet needs, whereas invisible drought suggests a borderline inadequacy of rainfall—not quite enough to satisfy the

86. *Drought, occurring on the average in one of every four months, is not uncommon in the Guadalupe Mountains and adjacent areas of the Trans Pecos.*

needs of crops each month, with the result being reduced yields at the end of the growing season.

The concept of drought varies with the individual, depending upon geographical location, vital interests, and the impact of dry weather on those interests. Farmers discern the existence of drought when they notice the effects on their crops of a lack of moisture during critical periods of the growing season. To them, agricultural drought, or that condition when rainfall and soil moisture are insufficient to support the healthy growth of crops and to prevent extreme crop stress, is the most relevant classification of drought. On the other hand, water engineers get concerned about drought when lake levels or stream flows fall to threateningly low levels. Their primary interest is in hydrologic drought, which is a long-term condition of abnormally dry weather that ultimately leads to the depletion of surface and groundwater supplies, the drying up of rivers and streams, and the cessation of spring flows. Suburbanites might not notice the prevalence of drought until water shortages in their community necessitate a cutting back of lawn watering.

Aside from the derivation of an all-fulfilling definition of drought, the choice of an appropriate measure of drought has been the source of much controversy for many years. If drought is discerned to exist, how may it be quantified? Numerous climatic classification schemes and indexes have been devised to categorize the intensity of drought. One criterion having much usefulness for comparing historical droughts is the Palmer Index. With this indicator, the duration and severity of Texas droughts widely separated in both time and space can be evaluated. By taking into account the amount of moisture required to have normal weather for a specific area, the Palmer Index describes departures from this normal condition in terms of a numerical index. Positive values indicate wetter than normal conditions, whereas negative values represent varying intensities of drought. The Palmer Index will serve as the basis for categorizing in this chapter those droughts that have had a major impact on the Texas economy in the past fifty years.

## What Brings About A Drought

Most, if not all, of Texas is destined to sustain periodic droughts. One foremost reason for this inevitability is Texas' proximity to the Great American Desert of the southwestern United States. Indeed, the western extremity of Texas—where rainfall averages less than 10 inches annually—can be regarded as being on the periphery of this vast desert. Desert regions expand and contract intermittently, so that areas on the fringe are engulfed from time to time by the spread of droughty weather that so often typifies them. The more distant an area is from the edge of a drought region, the less accustomed it is to the long and hurtful dry spell that results when the desert expands to include it. In these unprepared areas—one of which includes most of the western half of Texas—a severe or extreme drought can be nothing short of calamitous. The amount of water extracted from Earth through evaporation and transpiration exceeds by far the meager amounts of rainfall that may accompany a lengthy period of severe or extreme drought. On the other hand, the great desert rarely, if ever, spreads far enough east to encompass all of Texas. Even in those years when drought is detected in the eastern sector of the state, rainfall usually is ample enough to negate the amount of moisture yielded through evapotranspiration.

More than just being due to a deficit in precipitation, the cause of a drought can be attributed to terrestrial and extraterrestrial influences. Since the sun must ultimately be recognized as the sole source of energy that drives our atmospheric machine, it follows that quirks or anomalies in the sun's behavior doubtlessly affect our planet's climate in some fashion. One theory that has gained increasing attention—if not altogether acceptance—is that one solar phenomenon, known as the sunspot, can be linked directly to those prolonged periods of little or no rain on Earth. Sunspots are enormous magnetic storms that swirl within the torrid, gaseous atmosphere of the sun, and they appear to the protected eye on Earth as relatively small, dark areas on the sun's surface. Sunspot activity runs through an irregular cycle, usually lasting around eleven years but sometimes varying in duration between ten and fifteen years; during a cycle, the number of sunspots changes from a minimum to a maximum and then back to a minimum. On an average yearly basis, that number may range from near zero to as many as two hundred. Some weather analysts, by matching the sunspot cycles with occurrences of drought on Earth, allege that a definite positive correlation exists between the two phenomena. However, evidence suggesting a definite cause-and-effect relationship between the two is not overwhelming.

A key terrestrial factor in causing drought in Texas is the behavior of the vast subtropical high-pressure cell that drifts latitudinally with the passing of the seasons. As explained earlier, it is this large-scale weather system that, when it becomes entrenched in summer over the southern U.S., may produce many days of rainless and searingly hot weather. Though the precise cause and effect remain a mystery, apparently changes in the positioning and strength of this high-pressure cell are brought about at least in part by variations in the amount of insolation received from the sun. This huge ridge or mound of air sometimes shifts out of the Pacific or Atlantic oceans and assumes an almost stationary position over North America. Drought from mid-spring to early autumn in Texas is often due to a lack of rain days caused by the dominance of this high-pressure cell. Storm systems with the potential to bring rain to Texas and migrating out of the extreme northern Pacific or from Canada are shunted eastward away from Texas, and the storms' potential for manufacturing inclement weather is realized somewhere other than Texas.

The behavior of the vast high-pressure system that allows the atmosphere to heat up to excessive levels in summer may be related not only to the sunspot cycle but also to the eruption of volcanoes. Nearly invisible clouds of sulfuric dioxide gas are injected high into Earth's atmosphere by these volcanic eruptions and become entrapped within the stratosphere, where they encircle the globe and limit the amount of solar insolation reaching Earth's surface. This scientific notion has been given added

credence in recent months as a result of the eruption of the El Chichón volcano in the southern Mexican state of Chiapas in March and April 1982. Within four months of the initial series of volcanic blasts, satellite measurements indicated the massive cloud of sulfuric acid particles had spread over a broad area of at least 20° latitude stretching from Panama into the central Atlantic. While this was happening high in the atmosphere, Texas was enduring one of the driest and hottest summers of the century. Numerous localities in the Texas coastal plain suffered through the driest summer since 1917, and drought rapidly worsened from moderate to severe intensity. At Brownsville—or only about seven hundred miles from the El Chichón volcano—temperatures climbed to 90° or above on 118 straight days, the lengthiest string of such torrid weather in the 105-year history of that far southern city. We have yet to find indisputable evidence that the volcanic eruption caused or exacerbated the uncommon warmth that typified the Texas summer of 1982, but the idea that the two phenomena may be related seems too plausible to be dismissed.

Particularly during the cooler half of the year, drought may become established as the result of too frequent intrusions of polar air. Where rain occurs in Texas in autumn, winter, and spring is often governed by where invading cold air from the north or northwest encounters warm, moist air from the Gulf of Mexico. If the air enveloping Texas is not moist enough when a cold front arrives, the front passes without setting off precipitation. If the air covering Texas is only marginally moist, an incursion of cold air will trigger only scattered and mostly light precipitation—not the variety to arrest the formation or spread of a drought. The circulation pattern in the upper atmosphere may be structured such that it does not elicit much of an inflow of low-level air from the Gulf. It matters not that the number of triggering mechanisms (cold fronts) is ample; rather it is the absence of enough moisture that inhibits the development of rainstorms. Most often in winter, the infusion of cold polar or Arctic air is so frequent that little opportunity is afforded for southerly winds to transport

*87. Declining reservoir levels are stark evidence of a severe drought.*

sufficient amounts of moisture back into Texas before the arrival of the next front.

Great Droughts in Texas' History

Every decade thus far in the twentieth century has been marred by one or more serious droughts in Texas. Unmistakably, the most calamitous drought to strike Texas in recorded weather history was the severe-to-extreme drought that afflicted every region of the state in the early and mid-1950s. That "super drought" was the most notorious of all not only for its intensity and vast coverage but for its persistence as well. The incipient stage of the drought first developed in late spring of 1949 in the Lower Valley, while in western portions of Texas the drought materialized a few months later in autumn. Several months of deficit rainfall in most of the state promoted the gradual development of a severe drought that was fully manifested in nearly all of Texas by midyear 1951. By the end of the following year, water shortages in many areas had become alarming: Lake Dallas, a critical source of water for many in the Dallas–Fort Worth metroplex, stood at only 11% of capacity. Generous

spring rains in 1953 brought a brief respite from the drought to the northeastern quarter of Texas, but elsewhere the drought relaxed very little. The drought deteriorated to become "extreme" in the Trans Pecos that year, where rainfall for the whole year amounted to less than 8 inches in many areas. In fact, rainfall was so scanty that Imperial, Texas, measured a paltry 1.95 inches for the whole twelve-month period (*see Appendix C-4*).

This most memorable of all Texas droughts, after slacking off in 1953, grew worse again in 1954 and ultimately was climaxed in 1956 as the most intense ever. Cattle and sheep raisers battled desperately to survive. Ranges went bare and stock water became critically short or nonexistent on many farms. The flow of many rivers and streams dipped to near or even below record levels. The flow of the Guadalupe River near New Braunfels was deficient in 35 out of 36 months during 1954–1956. Some streams slowed to a trickle, while others ceased to flow or dried up completely. Many wells declined to record levels. Farmers sharply increased the amount of acreage of crops grown under irrigation

**wetter than normal**

| | |
|---|---|
| 4 *very much* | |
| 3 *much* | |
| 2 *moderate* | |
| 1 *slight* | |
| −1 *mild* | |
| −2 *moderate* | |
| −3 *severe* | |
| −4 *extreme* | |
| **drought** | '51  '52  '53  '54  '55  '56  '57  '58  '59  '60  '61  '62  '63  '64  '65 |

*88. Drought and wet-weather conditions by season of the year (January–March, April–June, July–September, October–December) for the period of 1951–1982 in the High Plains. (Palmer Index data, U.S. Department of Commerce)*

in an effort to avoid being wiped out altogether. The drought reached its utmost worst in the late summer of 1956, when it was gauged on the Palmer Index scale at −6.53 in North Central Texas. Finally, in late winter and spring of 1957, the most devastating drought ever ended abruptly throughout Texas. Rains ranging from "slow soakers" to "gully washers" began falling in February, and within two months all regions in Texas had seen the last vestige of drought erased.

The drought of the Dust Bowl era of the 1930s, while the greatest weather disaster in American history, was not as intense or prolonged in Texas as the extreme drought of the 1950s, but that is not to say that it proved to be much less catastrophic. Farmers who moved to the prairies and plains of Texas in the last few decades of the nineteenth century had plowed under natural, deep-rooted grasses, which they had found there upon their arrival. The grasses, which were capable of surviving

prolonged droughts, were replaced by wheat, corn, and cotton, thus baring the topsoil to the mercy of the relentless winds that buffeted Texas and stirred up tons of dust. The first fingers of the Dust Bowl drought crept eastward out of the nation's southwestern desert into the Texas High Plains early in 1933. With the region collecting only about half of the normal rainfall, the drought steadily worsened until it reached "extreme" magnitude in July 1934. The Dust Bowl drought in the High Plains and the Trans Pecos, which was the region of Texas most adversely affected by the drought, did not ease up perceptibly until appreciable rains came during the summer of 1935. Before the drought finally ended in the High Plains in late summer 1936, thousands of head of livestock had been lost to starvation and suffocation (from the dust). Crops that somehow survived the onslaught were withered and stunted. The remainder of the western half of Texas suffered a great deal also from the drought of the 1930s, while

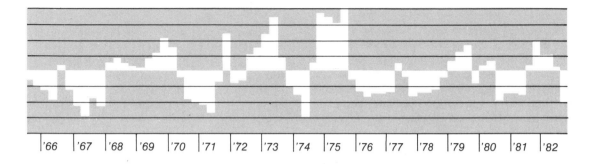

'66 | '67 | '68 | '69 | '70 | '71 | '72 | '73 | '74 | '75 | '76 | '77 | '78 | '79 | '80 | '81 | '82

in the eastern and southern sectors of the state the drought hit in spurts and was short-lived and merely moderately intense.

Numerous other very serious but relatively short-term droughts have plagued Texas since the granddaddy of them all was erased in 1957. Many of these droughts ended when very substantial rains were generated by tropical weather systems at the end of hot, dry summers. For instance, before fledgling Tropical Storm Amelia drifted ashore and later produced killer floods in the Texas Hill Country in mid-summer 1978, the central third of Texas had suffered from a moderate-to-extreme drought for nearly a year. In only a matter of two or three days, however, Amelia spawned enough flash-flooding rains to totally quash the drought and leave Texas' midsection with far more water than it needed. Unfortunately, Amelia failed to eliminate a severe drought in North Central and East Texas, which had been in progress since the start of 1978. It remained Tropical Storm Debra's task to lessen the drought in that sector about a month later.

Rainmaking to Alleviate Drought

For hundreds of years people have been trying to make rain during periods of drought and intense heat. In fact, it is when drought has reduced the availability of water to a dangerously low level that our interest in rainmaking efforts seems to be heightened the most. That is unfortunate, for it is during drought spells that most techniques for artificially stimulating the precipitation processes in clouds are least—if at all—effective. This is because most modern rainmaking procedures depend upon the presence in the atmosphere of clouds that

are conducive to the formation of precipitation. Cloud seeding with such materials as dry ice or silver iodide, the most popular rainmaking method in use, is totally ineffective when skies are cloudless—a condition that often exists during spells of drought and intense heat. Even if clouds are present, their depth or body temperature may not satisfy the demands of a rainmaking methodology.

It is conceivable, however, that even during prolonged periods of little or no rainfall enough of the right kinds of clouds may exist to warrant human intervention to boost the clouds' naturally occurring process of manufacturing rain. This rationale was the basis for "Project T-Drop," an emergency cloud-seeding program performed in June 1971 by government scientists to provide relief to much of the southern half of Texas in the throes of an extreme drought. Project T-Drop was designed to increase rainfall over large portions of Texas by injecting silver iodide, salt, and ammonium nitrate into clouds deemed to be suitable candidates for seeding.

In spite of the intensive cloud-seeding effort, June rainfall in North Central and East Texas was little more than scanty and considerably less than normal for early summer. Farther south, however, rainfall was bountiful in Southern Texas and in parts of the Edwards Plateau. Torrential rains near the end of June amounted to 3 to 9 inches near the Rio Grande between Laredo and Del Rio, while other less substantial but meaningful rains sent streams and rivers in much of the southwestern half of the state on a rampage. Were these mixed results related to the seeding campaign? Possibly, although no hard evidence was generated afterward suggesting that the heavy rains in the Winter Garden area were the direct result of seeding performed in the vicinity. Whether the intervention had a positive impact or not, the facts at the end of the endeavor were that extreme drought continued to grip much of the central and eastern thirds of Texas while the drought in the southern sector of the state was eased markedly.

Predicting Drought

If we are far from being able to make it rain during a drought, can we foresee one well enough in advance to plan for its eventual arrival? Unfortunately, weather experts are no more in harmony about the predictability of drought than they are about its definition. Therein lies a monumental obstacle in forecasting the long-range likelihood that a drought will strike. Weather observers are reluctant to predict the onset, continuation, or cessation of a drought partly because drought is so difficult to define in the first place. What might constitute a drought in one area may not even be a climatological anomaly in another. Disagreement on drought definitions, however, is not the primary reason for our inability to predict accurately drought occurrences. Rather, our lack of confidence in long-range weather predictions is a reflection of two critical deficiencies. One is the lack of information about the behavior of the atmosphere over the oceans and of the upper atmosphere over continents. This inadequacy is due to a sparcity of upper-air, continental stations that monitor conditions in the higher levels of the atmosphere and to a paucity of sea-surface weather installations.

A second reason for the failure of forecasters to foresee drought is the complexity of Earth's atmosphere and oceans and the scientists' inability to understand them satisfactorily. Much is yet to be learned about the impact of the sun's rays on the development and movement of the subtropical high-pressure system that dominates Texas' weather for much of every summer and sometimes produces virtually cloudless skies for days, if not weeks, on end. What is understood even less clearly is the impetus for the strengthening and positioning of this "blocking" ridge of high pressure. Why is the ridge stronger and more assertive in some summers than in others? Once the ridge takes hold, how long will it maintain its grip? What will be required to weaken the system or to get it to shift its position to allow a rain-producing, inclement spell of weather to enter Texas? Full, viable answers to these queries are as yet elusive. Understanding more fully the behavior of the sun and its impact on the weather of our planet will lead to increased ability to foresee and predict accurately the onset and cessation of droughts. While we still grope for such an understanding, the effects of continental heat waves remain detestable and destructive.

At the present time, long-range forecasts of temperature made by the NWS have a success rate that is only modestly higher than pure chance. Evaluation of these forecasts in recent years suggests that monthly and seasonal temperature predictions have an accuracy of about 60% (as compared with 50% for pure chance). Thirty-day and three-month forecasts of precipitation are less accurate than those for temperature and only slightly better than chance. While obviously there is ample room for improvement in forecasting drought conditions, the slight skill now demonstrated by weather experts intimates that they do have some understanding of the long-range behavior of Earth's atmosphere. However, this skill is yet so limited as not to justify the farmer or engineer's confidence when critical management decisions must be made. As we wait for the day when reliable drought predictions become a reality, we must in the interim focus on providing the best available information (on the likelihood of drought) to those who supply our water and food.

# 8. Snow, Cold, and Ice: Winter's Beauty and Treachery

The landscape could have been mistaken for a stretch of Canadian tundra. To any onlooker the scene consisted solely of one solid blanket of white extending in all directions to the distant horizon. Virtually every nook and cranny of the Texas High Plains lay under a snow cover varying in depth from as little as 2 inches in the extreme south to 2½ feet in places farther north. The phenomenally deep layer of snow was the aftermath of a blinding snowstorm that tormented west Texans for four days, snarling all forms of transportation by blocking roadways, railways, and runways with huge drifts of snow. The blizzard struck so abruptly that travelers by the hundreds were stranded in cars and buses and had to be rescued by the National Guard and train crews. The deaths of twenty persons were attributed to the record-setting wintry siege, while livestock by the thousands perished from exposure. So extensive and intense was the bitter blast that snow-removal operations were still in progress in parts of the Panhandle nine days after the last snowflake had been deposited.

No blizzard has paralyzed the northwestern half of Texas like the storm that blitzed the region on February 1–6, 1956. That cold wave was not only the deadliest winter storm to afflict Texas in modern times but also the most prolific snowstorm. The week-long siege deposited a total of 61 inches of snow at Vega in the western Panhandle. Most of that mammoth sum fell in less than four days, thereby set-

ting an all-time record for the most snowfall produced by a single snowstorm in Texas (*see Appendix A-1*). A blanket of snow 33 inches deep covered the ground at Vega on the morning of February 6. Meanwhile, the same snowstorm was producing a record amount of snowfall for a single 24-hour period—24.0 inches—at nearby Plainview. Yet, Plainview's storm total of 31.8 inches was surpassed not only by Vega's astonishing sum of 61 inches but also by an aggregate of 36 inches at neighboring Hale Center. Other cumulative snowfall totals for the first week in February 1956 that were phenomenally high included 26.5 inches at Hereford, 18.0 inches at Canyon, and 16.8 inches at Lubbock. Surprisingly, farther north in the Panhandle, where snowfall is almost always the most substantial, the snow cover measured only a few inches; only sixty miles north of the point where 61 inches of snow were observed, a scanty storm total of 3 inches was registered at Dalhart.

## Nature's Pristine Offspring
### The Snowflake

The singular nature of snow has mystified people for centuries. Not all snow crystals have six sides, a trait inherent in the atomic structure of ice; rather, some have five sides, others as few as three. The shape and rate of growth of snow crystals are determined solely by conditions in the environment. In much of Texas where

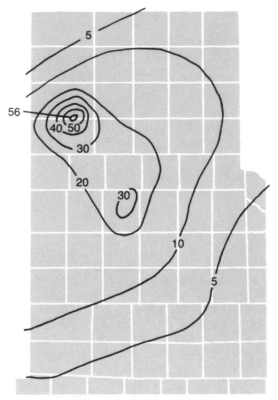

89. *Snowfall totals (inches) for February 1– 6, 1956.*

snow falls with temperatures barely below freezing, snow consists largely of hexagonal plates. Ice needles, a product of an environment with a temperature range of 27°–23°F (−3°−−5°C), are most common in the High and Low Rolling Plains and the Trans Pecos. As a general rule, smaller, more elementary crystals are more common in the Panhandle, where cloud temperatures and moisture content are lower, while farther south and east, where comparatively warm cloud temperatures and more plentiful moisture are present, the crystals are larger and more complex.

The shape of a snow crystal may undergo numerous changes during its fall from a cloud to Earth's surface. As it drops through air warmer than itself, a snow crystal will very often sustain rapid growth through sublimation, which is the transition of a snow crystal from the solid phase directly to the vapor phase without first passing through an intermediate liquid phase. Often the flake will pass through a layer of supercooled droplets and will be coated with rime. These droplets remain at subfreezing temperatures as long as they do not contact anything, but they freeze the instant they touch something solid like a falling snow crystal. Sometimes the snow is of the soft variety because an abundance of these supercooled droplets coat the crystal (a process known as "accretion") to the extent that it becomes graupel, or pellets that are white, opaque, approximately round, and easily crushed.

A snow crystal gets its start if a certain nucleus of matter having a suitable molecular structure is available. These nuclei may consist of a tiny speck of dust from a cornfield, a salt particle from the sea, particulate matter from the exhaust of an automobile, or a small segment of another snow crystal. If the cloud environment is right—that is, if the temperature is ideal and if enough vapor is present—then a snow crystal will likely grow from the sublimation of tiny cloud water droplets onto the nucleus. Regardless of the form a snow crystal initially assumes, its ultimate shape will be determined by such factors as changes in the pressure, temperature, and humidity of the air in which it moves.

Understanding the processes by which a snow crystal forms and later becomes a segment of a larger snowflake has led to efforts in other states to alter these processes to benefit humanity. This form of weather modification is intended to furnish additional snow and, hence, augment the amount of water reaching the surface of water-deficient areas of the country. No cloud seeding for snowfall augmentation has yet been attempted in Texas, but experimentation goes on elsewhere, especially in some Rocky Mountain states. While results vary, the potential for using such substances as silver iodide to effectively transform a great deal of cloud vapor into snow is unmistakable.

Evolution of a Texas Snowstorm

Many factors enter into the prolongation or cessation of a snowstorm. Major determinants are the strength, position, and rate

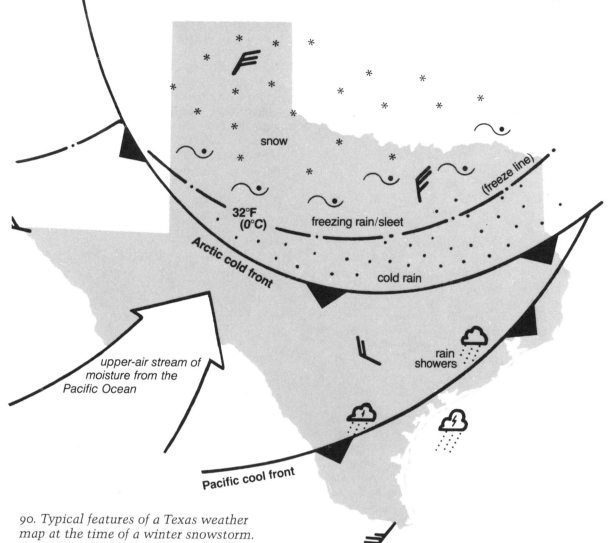

snow

32°F
(0°C)

freezing rain/sleet

Arctic cold front

cold rain

(freeze line)

upper-air stream of
moisture from the
Pacific Ocean

rain
showers

Pacific cool front

90. *Typical features of a Texas weather
map at the time of a winter snowstorm.*

of movement of the low-pressure cell often
found in the middle levels of the atmosphere.
This upper-air storm, embedded in a strong
circulation pattern of high-level westerly
winds, propels cold polar or Arctic air into
Texas and helps import mid- and high-level
moisture into the state from the Pacific
Ocean and Gulf of Mexico (*see fig. 90*). As a
general rule, if the upper-air low maintains
its vigor or intensifies and is slow to move
across the southwestern portion of the
United States, the snow that begins falling
likely will last for some time, and chances
are good that accumulations will be substantial. On the other hand, if the system is

moving along rapidly, the spell of snow will
probably be short-lived, although snowfall
amounts might still be appreciable. Of
course, more than a few upper-air storms
lose their vitality as they migrate eastward
out of Arizona and New Mexico every
winter and never develop into a major
snowstorm.

The usual sequence of events coinciding
with the invasion of an Arctic snowstorm
into Texas is as follows: a thick cloud
cover develops prior to the passage of the
front, and drizzle or light rain may fall for
several hours; with the wind gradually
shifting from southerly to westerly, the

temperature begins to plummet; once the wind veers more northerly and the temperature falls more rapidly, light rain or drizzle changes into either sleet or freezing rain (this spell of frozen precipitation may last from less than an hour to several hours, depending upon the nature and rate of movement of the invading air mass); thoroughfares become glazed with a thin sheet of ice, and travel becomes increasingly hazardous; finally, the intruding cold air has chilled the temperature aloft enough to allow freezing of the cloud water, at which time snow forms. When snow first begins to fall, it may be mixed with sleet, freezing rain, or just a cold "liquid" rain. Almost always, the other forms of precipitation do not persist if snow continues to fall for a period lasting one or a few hours.

Snowfall Intensities

The occurrence of falling snow is best categorized by the intensity of the snowfall. The NWS reports a *very light* snow when scattered flakes do not completely cover or at least wet an exposed surface. Snow is regarded as *light* when the visibility caused by the falling snow is equal to or greater than 5/8 of a mile; *moderate* when the visibility is between 5/16 and 10/16 of a mile; and *heavy* when the visibility is less than 5/16 of a mile. When snowfall sticks to the surface, the amount of accumulation is measured by many of the more than six hundred cooperative observing stations maintained in Texas by the NWS. These snowfall measurements customarily are made once daily, at around dawn or sunset, and the data are published in the U.S. Department of Commerce's *Climatological Data: Texas*. The NWS defines a heavy snow as either (*a*) an accumulation of 4 inches or more within twelve hours or (*b*) an accumulation of 6 inches or more in twenty-four hours.

When an intense upper-air winter storm approaches Texas, it may send blustery winds, plunging temperatures, and a driving snow in the form of a *blizzard*. Blizzards affect the northwestern quarter of Texas at least once in most years, while their occurrence elsewhere in the state is rare. Technically, a blizzard is a severe winter weather condition in which strong

winds (with speeds of at least 32 mph) accompany a great amount of falling or blowing snow for at least three hours; characteristically, in a blizzard the "blinding" snowstorm limits visibilities to less than a small fraction of a mile. Moreover, a *severe blizzard*, as the term suggests, is even more abrasive. For a blizzard to qualify as such, winds must be or exceed 45 mph, the air temperature equal to or less than a bone-chilling 10°F (−12°C) and visibilities lowered to near zero by a heavy snow. Severe blizzards are not uncommon in the Panhandle, the number averaging no more than one or two in some winters and none in others. It is very rare for severe blizzard conditions to extend deep into Texas.

Through its elaborate and farflung network of weather observations, the NWS monitors closely the development of winter storms and issues through the news media various statements and advisories to forewarn the populace of impending and potentially hazardous weather. A *winter storm watch* covers the possible occurrence of the following weather elements, either separately or in combination: blizzard conditions, heavy snow (or just light snow in areas where snow is rare), the accumulation of freezing rain or freezing drizzle, and/or heavy sleet. A watch gives a longer advance notice of the potential for the occurrence of a winter storm event than does a warning; for this reason, a watch has somewhat less chance of verification. A *winter storm warning* notifies the public of a "high probability" that severe winter weather will occur. This severe winter weather entails the same elements mentioned in conjunction with a winter storm watch. Whereas a watch is intended to alert the public of the possibility of bad, wintry weather, a warning is notice that snowy or icy weather has actually materialized or is in the process of evolving and that the public should take necessary precautions. The NWS may even issue specialized advisories intended to forewarn particular segments of the economy. One such statement is a "stockmen's advisory," which usually describes inclement weather that does not constitute a serious enough threat to warrant the issuance of a winter storm warning. The stock raisers' advisory typically contains information for ranchers or live-stock owners that is helpful to them in the protection of their stock.

## When Snow Fails to Materialize

More often than not, snow-loving residents of central and southern Texas are tantalized by the threat of a meaningful snowfall but end up getting nothing more substantive than a cold rain. At the surface, conditions seem ripe for at least a little snow: a thick, water-laden deck of low clouds overcasts the sky and the temperature of the air near ground level hovers only a few degrees above freezing. This set of circumstances is very common in the eastern half of Texas during winter, yet snowfall seldom materializes. Why does nature seem to incur great difficulty in supplying snow to locales like Austin and Houston? Seldom is the "problem" due to a lack of moisture in the air. It may not even be blamed on the temperature of the air at or near ground level, although, if snow does fall, it will not remain as an accumulation on the ground if the temperature at ground level is not at or below the freeze mark.

Obviously, one essential ingredient for a munificent snowstorm is a plentiful supply of moisture. Abundant low-level moisture from the Gulf of Mexico is seldom lacking when a strong winter storm pushes into Texas because the intruding upper-level storm system from the west accelerates the rate at which moisture rushes in from the Gulf in advance of the approaching cold front. Southerly winds intensify over Texas as a major winter storm comes rolling out of the Rockies. The nature of the circulation about the winter storm also ensures that ample moisture is propelled into Texas at higher levels in the atmosphere from the Pacific Ocean. A manifestation of this bounteous reservoir of moisture in the sky is the thick, dark overcast that frequently accompanies a strong cold front. Light rain and drizzle may fall for hours on end—if not for several days—prior to, at the time of, and long after the cold air arrives.

The key reason for a prolongation of rain or drizzle and little, if any, sleet or snow is the lack of depth in the dome of cold air pushing into the state. Cold air plunging southward into Texas behaves much like a large mound of sand when dumped on a flat surface. The sand spreads out in all directions, but, the farther the spread of the sand, the more shallow is the depth of it on the periphery of the pile. When mounds of very cold air from Canada spill across the Great Plains into Texas, points farther north in the state usually are covered by a much greater "thickness" of cold air than are locales in the south. The thickness of cold air over a given locality determines whether the precipitation that falls to the surface is snow, sleet, freezing rain, or just cold rain. To get snowflakes to the surface, the source of the precipitation—the overshadowing deck of clouds—must lie within the newly arrived mass of cold air (assuming the temperature of the air at that level is at or below freezing). Otherwise, if the dome of cold air is too shallow, the clouds bearing the moisture lie above it, where temperatures are above freezing (*see fig. 91*). The water droplets in the clouds obviously then do not freeze into snowflakes. If the cold air is thick enough, precipitation falling as rain from the cloud layer will freeze on the way down, and the result is sleet at the surface. Or, if the precipitation does not freeze during its descent, perhaps the cold air has been around long enough to freeze the surface of the earth so that the rain, upon impact, turns into ice. In that instance, a potentially hazardous condition of freezing rain (or drizzle) is set in motion. Even if initially the layer of cold air is not thick enough to support the formation of snow, snow may materialize later on if a reinforcement of Arctic air enters the state from the north.

## Snow's Impact on the Environment

Aside from the delight that it brings to children and other snow-loving folks, the accumulation of snow provides benefits for crops and other plants. Notwithstanding the fact that substantial snowfalls seldom occur anywhere in Texas but in the Panhandle, to the prairie or High Plains farmer a layer of snow means the addition of valuable nitrates to the soil. Other nutritious elements—like potassium and sulphates—whose sources are industrial pollution, atmospheric gases, dust, and ocean spray, are delivered to the earth by the falling snow.

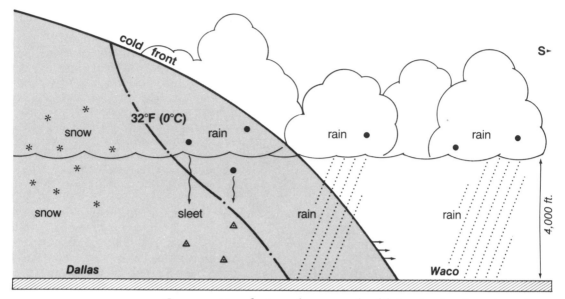

91. *Cross-sectional view of a mass of cold Arctic air triggering rain, sleet, and snow as it moves southward through Texas.*

Significantly, a blanket of snow also provides a degree of warmth to the soil, holding in the heat emanating from Earth's crust and warding off the chill of a frosty air mass. During much of winter in the High Plains of Texas, when the region is without a snow cover, the soil may be frozen to a depth of several inches. This bare soil sustains large fluctuations in temperature during the course of a day and a night, for it may be heated by direct sunlight for a short while in daytime, then sheltered by an intermittent cloud cover, and ultimately allowed to cool steadily with the setting of the sun. A layer of snow, however, brings moderation to these changes in temperature and also keeps the minimum soil temperature higher than it otherwise would be. At least one study has revealed that a snow cover of 6 inches ensures that frost penetrates less than an inch into the soil, while a field without the protection afforded by a snow blanket may be frozen to a depth of 6–12 inches. Snow can be advantageous to the grower in that, depending upon the hardiness of the plants or crops, it may limit winterkill or stunted growth.

Though the snow cover even in the northern extremity of Texas usually amounts to no more than a few inches on any single occasion, the quantity of water supplied to the soil often can be appreciable. Snow is especially welcome in this sector of the state, for it is in winter that the High Plains sustains its seasonal precipitation minimum. Much of the water equivalent gauged during winter at various weather stations in the region stems from snowfall. For instance, it is common for snow to constitute at least half of the total water equivalent of about 0.5 inch of precipitation that is normal in January for cities like Dumas, Miami, and Spearman. The adage that 10 inches of snow are needed to supply the equivalent of 1 inch of liquid water usually—but not always—applies to the semiarid High Plains. Since some of the snow that occurs in this northernmost sector is of the "dry" variety, nearly 6 inches of the crystalline substance are often required to give a half-inch of water to the soil.

Naturally, the rate at which snow melts will determine how much of it infiltrates the soil. If the snow falls on soil and a plant cover having relatively warm temperatures (i.e., readings at least several degrees above freezing), melting of the snow begins immediately. However, if the ground and the layer of air next to it have temperatures at or below freezing, the delay in snow melting may last several hours, or even a number of days in places like the Texas Panhandle. How much of the snow cover contributes to runoff depends upon the terrain and soil characteristics as well as the rate of snow melt. Furthermore, a critical factor is the rate of sublimation. In the semiarid Texas High Plains, where the relative humidity of the lower atmosphere is often quite low in the wake of a snowstorm, a substantial amount of the snow cover is lost to the air through this process.

A warm wind, not direct sunshine, is the most effective means for removing snow. If the wind happens to be quite dry, then much of the water content of the snow cover is carried away into the atmosphere and very little enters the soil. Sunlight is not a major factor in melting snow because of the high albedo, or reflective power, of a snow cover; in other words, snow tends to reflect nearly all the incoming solar rays, so that sunshine contributes very little to melting the snow. Yet another factor in snowmelt runoff is the temperature of the soil underneath the snow cover. If the soil is frozen, there is very little melting and runoff of water under the layer of snow. In Texas, subsurface soil temperatures usually do not remain at or below freezing but for a few days at a time. Consequently, the melting process is active most of the time. Since the process is a slow one, most of the snowmelt enters the ground to become groundwater rather than runoff.

Snow on the ground can also be a nemesis to farmers. Snow mold is a plague dreaded by cereal growers. More than occasionally, when a snow comes early to soil that is unfrozen, various types of fungus become active and threaten young crops. These fungi thrive from the ample moisture supplied by the snow and from temperatures that are maintained near—but above—the freeze level. Most of the time, plants are affected only modestly; even still, they may later respond only feebly to the warmth of spring because their leaves have been blighted by a deficiency of nitrogen. As a result, harvesting is delayed and, further-

92. Meltwater from snow is sometimes appreciable enough to replenish soil moisture in the Panhandle, but it is rarely substantial enough to pose flooding problems. (Texas Department of Highways and Public Transportation)

more, the containment of weeds becomes more of a formidable challenge.

Great Snowstorms of the Past

The history of Texas winters is marked by more than a score of severe snowstorms and blizzards that roared across portions of the state with a vengeance nearly equal to that experienced anywhere in the country. While the severe blizzard of February 1956 cost more lives and supplied more snow than any other winter storm in modern Texas weather history, another blizzard that followed a year later—in March 1957— left an equally indelible impression on thousands of Panhandle residents. Ironically, both memorable blizzards came near the end of the most devastating drought to plague Texas in modern times. Indeed, moisture supplied by the March 1957 blizzard was part of a lengthy spell of inclemency that gradually erased extreme drought in much of Texas during spring of that year. To some inhabitants of the Panhandle, repercussions from the March 1957 blizzard were more severe than those of the previous year's blizzard, in part due to winds that gusted to 80 mph. As with many other major snowstorms that plague the Panhandle in winter, transportation was halted and many communities were isolated without power or means of communication. Ten people died during the ordeal, and more than 4,000 others were marooned. Unlike its predecessor of thirteen months, this snowstorm was far less copious; few localities south of Lubbock received any snow at all, and snow depths of one-half to one foot were confined to the two northernmost tiers of Panhandle counties.

Another intense blizzard that cost livestock owners dearly blasted the Panhandle in February 1971. Fierce 60-mph winds whipped up snowdrifts as high as 12 feet. Two young men died from asphyxiation while huddled in their automobile a mere 400 feet from shelter at a campground, and the body of an older man was found buried in a snowbank in Amarillo. A six-engine, ninety-car freight train balled up in a massive snowdrift and derailed near Borger the day after the great blizzard ended. Thirteen thousand cattle died during the ordeal, with most of them either smothering or being trampled to death as a result of their bunching together to stay warm. Three thousand hogs also perished, and overall livestock losses totaled $3 million.

Few winter storms in this century have had as much impact on the farming economy of central and eastern Texas as the farflung, severe snow-and-ice storm of January 8–11, 1973. A frigid blast of bitterly cold Arctic air knifed through Texas and far into the Gulf of Mexico, leaving a blanket of snow as deep as nearly a half-foot in the piney woods of East Texas and 3 to 4 inches on Galveston Island. The numbing cold cost cattle owners $25 million from the loss of 150,000 head of cattle and another $25 million due to the surviving cattle losing weight. In the Texas Hill Country, the combination of an icy wind, sleet, and snow destroyed 25,000 turkeys, many of which were fully grown. Fruit and ornamental trees in the Upper Coast were damaged heavily, and cotton losses were substantial. Oddly, even though temperatures in the Lower Valley plunged far into the 20s, damage to the citrus crop was minimal, due in no small way to the hardiness acquired by the trees from previous cold spells that winter and from an abundance of moisture in the soil.

Whereas the February 1956 snowstorm remains unparalleled in Texas weather history as the most prolific wintry siege ever, the whole winter of 1982–1983 distinguished itself as one of the most extraordinary snow seasons of all time. A trio of cold waves in December 1982 produced a record amount of 18.2 inches of snow (and an all-time high melted precipitation total of 2.61 inches) at moisture-starved El Paso, while the following month bestowed a mammoth 25.3 inches of snow at Lubbock—16 inches of which fell in one 24-hour period. Then a near-record sum of 13–18 inches blanketed the Panhandle in February, thereby closing out a winter that, surprisingly, was not in the least an uncommonly chilly season in most of Texas.

Of note, not because of inordinately large snowfall accumulations but rather due to their untimeliness, are several autumn and spring snowstorms that smote the Panhandle of Texas as early as October and as late as May. An extraordinarily early and intense winterlike storm surged out of the Rockies on October 30, 1979, and supplied the northern extremity of Texas with 50-mph winds and snow depths of 3–6 inches. More than 9,000 cattle perished from exposure, costing livestock owners at least $3.5 million, while about $1 million worth of grain sorghum was destroyed. Weather annals reveal that a snowstorm struck the Panhandle as early as October 8, though stock and crop losses from that modest blizzard in 1970 were not of the magnitude of the mid-autumn storm of 1979. By contrast, the most memorable late-spring snowstorm left snowflakes as big as half-dollars and in depths of as much as a half-foot in

the northern periphery of the Panhandle on May 2–3, 1978. Twelve inches of snow at Stratford nearly doubled the amount ever observed previously in May in that city. The heavy snowfall that struck Stratford was just one of a bizarre series of events illustrating the vagaries of Texas Panhandle weather. Just one day after the 12-inch snow coated the city, with temperatures only in the 40s, a massive thunderstorm swept through, generating a tornado and pummeling the town with 3-inch hail and a slashing rain.

93. Snowfall totals (inches) for January 8–11, 1973.

## The Likelihood of Snow

Since the prospect of snow depends largely upon the depth of the cold polar or Arctic air mass invading Texas, it is obvious that, the closer a locale is to the source regions of these air masses (viz., the Rocky Mountains and the central Great Plains states), the better the chance of a meaningful snow accumulation. Longtime snowfall statistics bear this out by indicating that inhabitants of the Panhandle and adjacent Red River Valley can expect at least one appreciable snow accumulation every winter. In fact, the issue is not "whether," but "how often" and "how much." Farther south, in the state's mid-section, the likelihood of snow is much reduced.

Average amounts of snowfall for each month of the colder half of the year are given in Appendix F-1 for various locations in Texas. A word of caution, however, is offered to the user of snowfall statistics. Snowfall averages are strictly arithmetic sums of observed snow accumulations for a given period of time divided by the length of that period—they are nothing more than mathematical means. To say the average amount of snow in January in Dallas is 1 inch is not to suggest that, in any year in January, there is a strong likelihood that Dallas will collect "about" an inch of snow. That average value of 1 inch is the mathematical mean of snowfall totals registered over a thirty-year period; in many months of January, no snowfall was observed, while in some, totals of 3–4 inches or more were measured. What is true of Dallas' snowfall history is also valid for many other areas of Texas: some years afford one or several snowfall accumulations that are highly appreciable, while numerous other years elapse without even a trace of snow. Only in the northwestern quarter of Texas, where snowfall accumulations occur every year, can credence be placed in the "average" snowfall value for a particular locale. Even in those instances, the user of average snowfall data should understand that the variance about the mean is sizable.

Winters in which some sector of Texas other than the Panhandle receives the most substantial amount of snow are exceedingly few. By virtue of its poleward projection away from the heart of Texas and northward into the nation's breadbasket, the Panhandle lies in the path of more than a few of the major winter storms that intensify over the Rocky Mountains and subsequently roar into the Great Plains. A person can count on one hand the number of winter seasons in this century when snow fell in Texas fewer than a half-dozen times and amounted to only a few inches. The winter of 1950 was the leanest bearer of snow of the twentieth century, for it yielded a total of 1 inch or less in most of the Panhandle. Typically, nearly all portions of the Panhandle collect over a foot of snow during the course of a winter. As one would expect, that sector of the Panhandle which, on the average, receives the highest cumulative total of snowfall over the winter is the northern tier of counties from Dallam in the west to Lipscomb in the east. Ordinarily, snowfall in this northern extremity of the state is not confined to the winter season. More often than not, at least one storm will deposit one or several inches of snow in November, and seldom does March pass without at least one snowy spell. Snow can also be seen on the ground as late as April, particularly in the northern reaches of the Panhandle, in one out of every three years. About an equal amount of snow can be expected on the average in each of the three months of winter (December, January, and February). In this region where, characteristically, at least a half-dozen significant snowfalls occur every winter, accumulations of snow are more common than ice.

Cumulative snowfall totals are not nearly as substantial in the southern half of the High Plains and the Low Rolling Plains as in the Panhandle in an average winter. However, rare is the winter that does not yield at least one snowfall of several inches. Snow accumulations average from 5 to 9 inches from November through March in the Low Rolling Plains, while in the southern High Plains, in the vicinities of Plainview and Lubbock, average seasonal amounts are a little more than that. Snow is likely to be the heaviest in late January

94. *A stretch of High Plains highway near Palo Duro Canyon is coated with snow and ice, a major cause of interruption in transportation in winter. (Texas Department of Highways and Public Transportation)*

and early February, and the number of days with snow is approximately the same for both months.

The grandeur of a landscape smothered by a thick blanket of snow is nowhere as breathtaking as in the mountains and valleys of the Trans Pecos. The view of the jutting peaks of the Guadalupes and the tree-dotted Davis range bedecked with snow is one of indescribable splendor. Unfortunately, that beautiful scene is far from common. Snowfall is sporadic, due in part to insufficient moisture in the atmosphere and to the high elevation of the terrain. Much of the "fuel" for a meaningful snow in this nearly arid region comes from the Pacific Ocean. If a large, upper-air storm moving toward Texas from the Rocky Mountains is not intense enough to haul in vast quantities of cloud water from the Pa-

cific across mainland Mexico, there is too little moisture for perceptible snow cover. Often, however, moisture in the air is sufficiently ample for a noteworthy snowfall, but what is lacking is a thick enough layer of cold air. The planetary circulation pattern combines with the force of gravity to propel most really cold masses of air through Texas from the north to the south, or from higher (the High Plains) to lower elevations (the Texas coastal plain). Dense Arctic air—like water—flows with much greater ease downslope than upslope. Consequently, the bulk of an invading mass of very cold Arctic air flows south and eastward into the central and eastern sectors of the state, while the higher elevations in the west receive only a relatively thin layer. On the few occasions when conditions are right for a widespread, bountiful snow cover, accu-

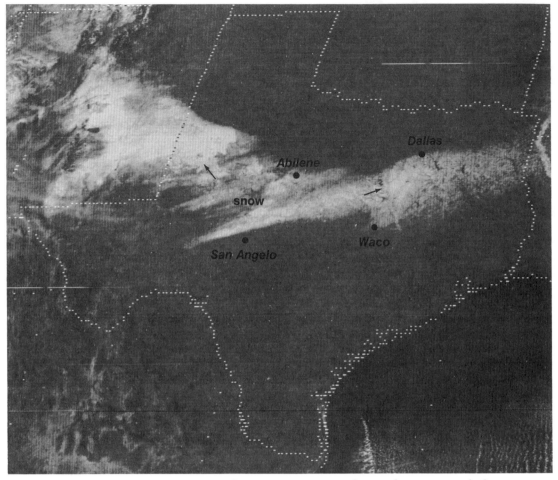

(a)

95. A strong midwinter snowstorm dumped uncommonly heavy amounts of snow in a swath across central Texas on January 12–13, 1982. Satellite views of the snow cover reveal that (a) the blanket of snow extended all the way from the Texas–New Mexico border to

mulations typically amount to 2–5 inches in the northern and central sections of the Trans Pecos. Occasionally—perhaps once every decade—the snow blanket in the Guadalupe and Davis mountains will be as much as 6–12 inches. One portion of the Trans Pecos almost always untouched by snowstorms is the Presidio Valley, where only one measurable snow cover was observed during the 28-year period ending in 1982.

Most of the Edwards Plateau and North Central and East Texas receives one or a few spells of snowfall every winter. Periods of snow normally last only a day or so, and

amounts ordinarily are a few inches at most. As a rule, the farther south in these three regions, the less is the probability of a measurable snow cover at any time. When snow does occur, amounts are apt to be most substantial in January and February. These regions usually serve as the boundary between snow to the north and ice or cold rain to the south. Spells of sleet or freezing rain are as common as snow, slickening roadways for one or two days in every two or three weeks in the winter.

Ice is more common than snow for residents of South Central Texas and the Upper Coast. With the layer of cold Arctic air not

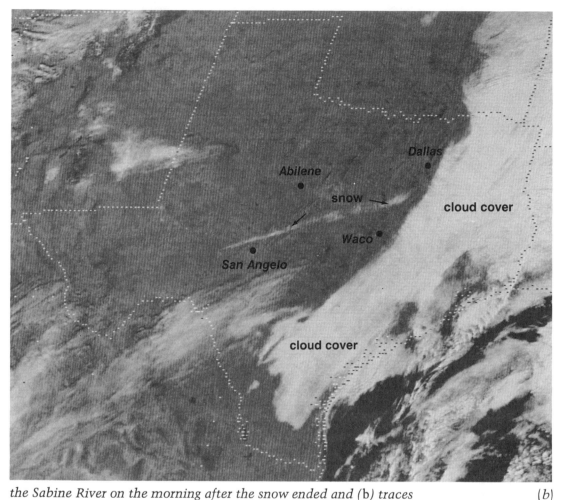

*the Sabine River on the morning after the snow ended and (b) traces
of snow remained on the ground in a narrow band as long as five
days after the snowstorm ended. (National Oceanic and
Atmospheric Administration)*

(b)

thick enough to generate snow, precipitation most often falls in liquid form—at least until it hits the surface, when occasionally the moisture turns into a treacherous glaze that coats everything exposed to the air. One or a few spells of icy weather typically occur every winter in such places as Austin, San Antonio, Houston, and Lufkin, but snowfall accumulations of one inch or more are no more common than once every three to five years.

Snow is even more scarce farther south in Southern Texas. In fact, while flurries that do not stick to the surface may be seen once every three to five years, a snow cover is observed no more often than once or twice every decade. Snow is slightly more probable in northern portions of this region (around Eagle Pass, Cotulla, and Jourdanton) than at points much farther south (at Rio Grande City and Alice). Freezing rain and sleet, with significant accumulations of ice, happen as often as once every other winter.

Nowhere in Texas is snow as close to being extinct as in the Lower Valley. Flurries may be seen once—or possibly twice—in every decade, but if any of the snow sticks to the ground, the accumulation is almost invariably too thin to be measured.

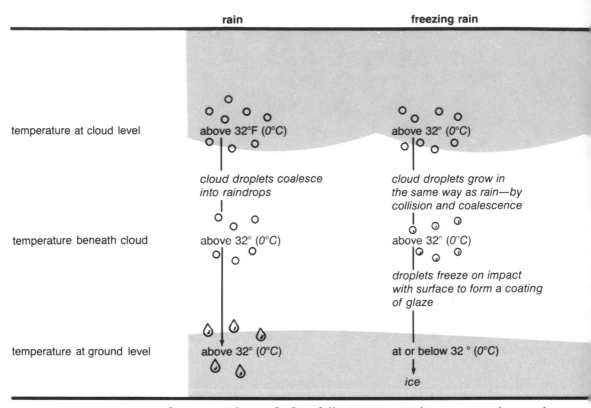

96. *The stages of growth that differentiate rain from various forms of*

Nonetheless, this tip of Texas closest to the Equator has had a couple of noteworthy snowfalls since weather records were begun more than one hundred years ago. The most recent appreciable snowfall at Brownsville, however, came before the turn of the century. Newspaper accounts reveal that a 6-inch snow smothered Texas' southernmost metropolis in mid-February 1895. Those same records show that 4 inches of snow fell in the city in 1866. Sleet or freezing rain is more common than snowfall in the Lower Valley, but even those varieties of frozen precipitation are rare. The heaviest glaze on record in the Lower Valley accompanied a bitter cold wave in late winter of 1951; ice accumulated to as much as 1½ inches in many places during a six-day period ending on February 3.

**Ice Storms—Snow's Ugly Cousin**

Snow is merely one of a number of forms of frozen precipitation that frequently pose problems in winter for motorists and sports-minded persons. In much of Texas, *sleet* is as common as snowfall. Sleet, also called ice pellets, occurs when rain from a layer of relatively warm (above freezing) air aloft falls through a layer of cold air (subfreezing) near the ground (*see fig. 91*). The falling raindrops do not turn into snow but rather into grains or pellets of ice. Closely akin to sleet is *freezing rain*, which is rain that falls through the lower atmosphere in liquid form but freezes upon impact to form a coating of glaze upon the ground and on exposed objects. To have freezing rain, the surface struck by the falling drops initially must be at or below freezing and, furthermore, the drops themselves must be supercooled. It is common, particularly in the northern third of Texas, for freezing rain to be a transient condition between the occurrence of rain and sleet or snow.

For those sections of Texas to the south and east of the Panhandle, icy weather outbreaks are more common than snow-

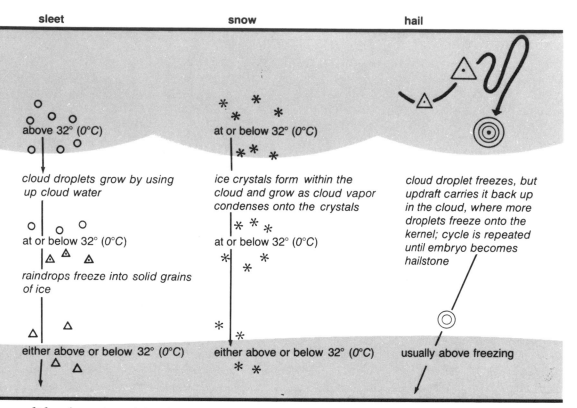

| sleet | snow | hail |
|---|---|---|
| above 32° (0°C) | at or below 32° (0°C) | |
| cloud droplets grow by using up cloud water | ice crystals form within the cloud and grow as cloud vapor condenses onto the crystals | cloud droplet freezes, but updraft carries it back up in the cloud, where more droplets freeze onto the kernel; cycle is repeated until embryo becomes hailstone |
| at or below 32° (0°C) | at or below 32° (0°C) | |
| raindrops freeze into solid grains of ice | | |
| either above or below 32° (0°C) | either above or below 32° (0°C) | usually above freezing |

*solid or frozen precipitation.*

storms. In fact, on many of those infrequent occasions when snow falls, it is accompanied by sleet or freezing rain or a combination of the two. The snow is not scorned nearly as much as a troublemaker as are the other types of frozen precipitation that put motorists in great peril by glazing roadways with sheets of ice. Very little snow fell on New Year's Eve 1978, when the worst ice storm in at least three decades struck North Central Texas. Layers of ice up to 2 inches thick produced heavy damage in a 100-mile-wide swath from Gatesville to Paris, including damage to trees that was the most extensive in thirty years. Some 2,000 residents were treated at area hospitals for injuries sustained in auto accidents, falls on the ice, and frostbite. Heavy accumulations of ice and a brisk wind, which maintained a chill temperature of 10°F (−12°C) or less, snapped power lines, leaving nearly 300,000 residents of Dallas County without electricity for two days. It took the power company ten days to completely restore service at a cost of $3 million. A young lad in Dallas was killed when he contacted a downed electrical wire, and at least five other persons died in ice-related auto accidents. The Great Icestorm of New Year's Eve 1978 was responsible for $14 million worth in damage in Dallas County alone.

## The Numbing Cold of Winter

Often—but not always—the outbreak of very cold Arctic air responsible for winter's worst snow and ice storms also produces the season's coldest temperatures. In the "typical" year, morning low temperatures steadily decline during October, November, and December, eventually bottoming out in mid-January in all sections of Texas (*see fig. 98*). At this point in the season morning minimum temperatures average in the low 20s in the Texas Panhandle (at Amarillo, for instance), whereas in the Lower Valley average low readings are no cooler than

97. *The devastating ice storm that encased northern Texas on New Year's Eve 1978 graphically illustrates the impact that ice can have on society and the environment.*

50°F (10°C). Once every three or four years, however, winter is atypical in that coldest readings occur in December or February, and occasionally as late in the season as early March.

With few exceptions, the coldest temperatures of any winter are felt in the Texas Panhandle, where the names of Dalhart and Lipscomb are as synonomous with cold weather as Presidio is with excessive and prolonged heat in summer. As shown in Appendix B-4, during the period 1960–1982, either Dalhart or Lipscomb garnered the distinction as Texas' coldest spot about half of the time. At other times the state's lowest temperature is gauged elsewhere in the Panhandle, either in the northernmost tier of counties or in one of the counties that form the border with New Mexico. Very occasionally, some locale in the northernmost Trans Pecos near the Texas–New

Mexico border will register the lowest temperature statewide for the year. Rarely are these statewide extremes above 0°F (−18°C); in half of the years, the minimum is −10°F (−23°C) or colder. The High Plains region, week in and week out, is also the most persistently cold of any sector of Texas. Invariably Texas' most poleward region registers the coldest average monthly temperature of any of the state's ten climatic regions.

More evidence of the relative cold that typifies the Texas Panhandle in winter is the occurrence of freeze days. Nighttime temperatures dip to 32°F (0°C) or below on more than 130 days during the seven-month period ending in April in a normal year in the northern fringe of the Panhandle. January traditionally qualifies as the year's coldest month, not only because winter's lowest temperatures usually are

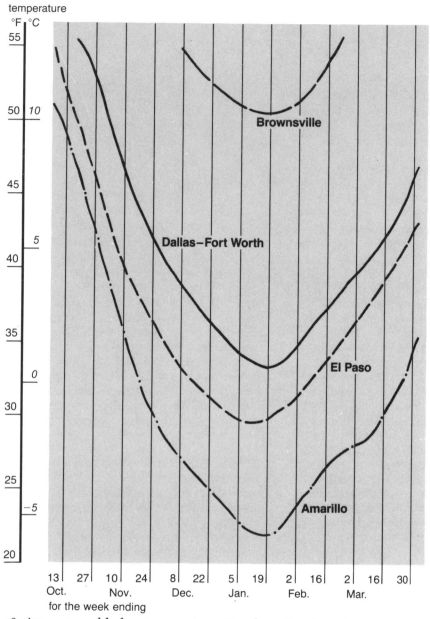

temperature
°F   °C

55
50   10
45
40    5
35
30    0
25
    −5
20

**Brownsville**

**Dallas—Fort Worth**

**El Paso**

**Amarillo**

13  27  10  24  8  22  5  19  2  16  2  16  30
Oct.    Nov.    Dec.    Jan.    Feb.    Mar.
for the week ending

*98. Average weekly low temperature, October—March, at four cities in Texas.*

recorded at that time but also because the month is marked by the greatest number of freeze days of any month of the year (*see Appendix B-2*). Indeed, in much of Texas, the frequency of occurrence of a freezing (or subfreezing) temperature is at least twice as great in January as for any other month.

# 9. The Wind: A Curse and a Blessing

Over a parched expanse of west Texas plain the wind races, lifting precious topsoil, transporting it great distances, and then depositing it in great drifts as "black snow." The pervasive dust finds its way into kitchens, even closets and basements, in spite of gallant efforts to keep it out of the house with cloth stuffed around the edges of doors and windows. Motorists strain to see ahead, then halt their automobiles and sit alongside roadways, waiting restlessly for nature's tirade to end. Street lamps begin glowing in the middle of the day, the light they emit appearing only as a faint, green glow in the distance. Communications are disrupted when static electricity knocks out telephone lines. Tumbleweeds roll along like hollow bounding boulders. Cattle grazing in the fields die of asphyxiation. Some folks try to flee, but it is to no avail. At long last, the fury ends, and stunned residents surface to sift through layers of sand and dirt that cover everything in sight.

That scenario was repeated more than a few times in the High Plains of Texas during the 1930s, a decade remembered by many older Texans as the era of the "Dust Bowl." However, the harsh, relentless wind and one of its most prominent but objectionable manifestations—dust—are not unique to that period. Rather, wind-driven dust and sand continue to be a source of harassment to the plains and prairies of western Texas. The lack of ample rainfall over extended periods of time partially explains the raging dust storms that sometimes blast the western sector of the state. Our insensitive use of a land all too often openly exposed to nature's whims is another, major contributing factor. After all, for decades residents of the High Plains have recognized the unusually high productive potential of the region's farmland, but many have failed to see that nature's bounty cannot be exploited with impunity.

## Essence of the Wind

Plainly, circumstances dictate whether the wind is a curse or a blessing. To the meteorologist, the wind is air in horizontal motion. It courses along at all levels of the atmosphere as invisible streams of energy possessing considerable power. It can suddenly becloud the sky, then just as rapidly clear it again. It channels vast quantities of water inland from the sea and then manipulates the cargo to shed rain on parched hillsides and dusty valleys. It pushes an enshrouding fog landward early at night, then whisks it away not long after the next day's sunrise. It brings masses of air of differing densities together, often with a clash that promptly results in raucous thunderstorms. It refreshes coastal residents beleaguered by an unrelenting summer sun and humidities that suffocate. It ventilates our smog-clogged cities, sweeping away the poisonous exhalations of our machines, both sed-

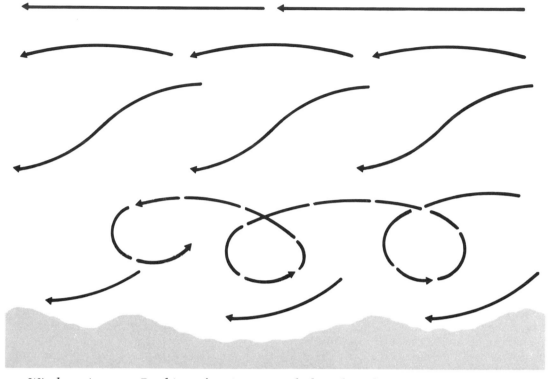

*99. Wind motion near Earth's surface is more turbulent than that
higher in the atmosphere due to the effects of friction.*

entary and mobile. It deposits seeds and
scatters pollen.

### Why the Wind Blows

Uneven air pressure forces the wind to
blow. Differences in atmospheric pressure
stem from variations in temperature
throughout Earth's atmosphere, a result of
the change in the tilt of Earth's axis toward
the sun with the changing of the seasons.
Where differences in air temperature—and
hence pressure—exist, there is a constant
striving by the forces of nature to balance
them. Wind, consequently, is a highly important regulator of the atmosphere in that,
while the equalization of temperature,
pressure, and humidity is never attained, it
does help to maintain an approximate average state for these differences.

The movement of the air at or near
Earth's surface is strongly influenced by the
force of friction that is exerted when the air
"rubs" against that surface. While higher in

the atmosphere the flow of air is predominantly smooth and continuous, in the
lowest layer of the atmosphere adjacent to
Earth's crust, air motion consists mostly of
gusts and lulls because of this frictional
influence. In daytime, intense heating of
Earth's skin by the sun initiates thermal
convection currents that further disturb
the air near the surface. Winds are apt to
be gustier in regions where the terrain is
rougher and where temperature contrasts
between a warm surface and cooler air aloft
are common. This helps explain why, over
large bodies of water, wind speed and wind
direction are more uniform than over land
areas, which are characterized by more rough-
ness and greater temperature variations.

Because of the presence of these wind ed-
dies in the lower atmosphere, the wind is
often highly variable in direction and speed
over short periods of time. Where the wind
shifts in clockwise fashion (i.e., from east
through south to the west), such a change

is called a *veering wind*. The opposite of this (i.e., a wind originally blowing from the west but changing into an east wind through the south direction) is known as a *backing wind*. On the average, however—for intervals of a few days, if not a few hours—the primary flow of air, or the *prevailing wind*, is usually rather steady from one general direction. What is commonly referred to as "wind" is this average air motion over a short span of time (e.g., one or a few minutes) and does not include short-lived gusts and lulls. An NWS report that "the wind is from the south at 12 miles per hour" should be construed as the average wind speed and direction over a one-minute interval at the time the observation was made (usually five or ten minutes before the top of the hour). Even though, during that period of one minute, wind speed may have varied between 8 and 15 mph and wind direction may have modulated between the southwest and southeast, the average value of wind speed and wind direction is cited to describe the nature of the prevailing wind.

Most of the time a wind observation includes only direction and speed, with gusts reported when wind speeds exceed 18 mph. Wind direction is taken to be that sector of the compass *from* which the wind is blowing, and in common parlance the direction is referred to as a particular compass point. Wind speed most often is expressed in miles per hour (mph), although for marine interests in coastal areas and out at sea, it may be given in knots. On most weather maps, wind movement is depicted as in fig. 9; the arrow points in the direction from which the wind is blowing, while the number and length of the tails (or barbs) denote the speed of the wind observed at that point. To express the average wind condition at a given locality for a period of time (say, a month or a season), a wind rose is shown, as in fig. 100.

Before the era of modern weather instruments, wind was estimated with the help of a scale invented by Admiral Beaufort in 1804 as an aid in sailing ships. Beaufort used his own ship as a wind-measuring instrument, and the wind was measured according to the pressure it exerted on the sails of his vessel. Later, the scale was modified for use at stations on land and has since become standard throughout the world (*see Appendix F-3*).

Typical Texas Wind Conditions

For much of any year, a southerly wind—or some component of it (such as a southwesterly or southeasterly wind)—predominates in Texas. This is especially true during summer, when wind shifts necessitated by the invasion of cool fronts are uncommon. In fact, except for the Texas Panhandle, where weak cool fronts often ease southward before dissipating, Texas is under the influence of a southerly wind virtually all the time during July and August. The wind rose for summer (*see fig. 100*) reveals that, on the average, a southerly wind or some derivative of it prevails 90% of the time in the southern half of the state. Farther north, from the High Plains and along the Red River to East Texas, northerly winds are a bit more frequent, though the southerly wind is still the overwhelmingly dominant feature. Southwesterly winds are most common at this time of the year in the High Plains and Trans Pecos, due to the effect of the oscillating dry line (or Marfa front) that almost daily ushers in arid air from the Mexican and New Mexican deserts. Even in July, when the circulation pattern over the state is most prone to stagnate, calm winds—those with speeds of less than 1 mph—are rare, occurring at the most only 5%–7% of the time. The air is seldom still because the intense summer sun heats up the surface layer of air sufficiently to maintain enough pressure differential to keep the air continually on the move.

By contrast, in winter, northerly winds are common in all regions of Texas. The invasion of cold polar or Arctic air is frequent enough (once every 4–5 days on the average) to generate a northerly wind, or some variation of it, about half the time in January. In fact, the persistent influence of these cold-air incursions is such that the average wind direction for January in some locales in South Central Texas and the Upper Coast is northerly (*see Appendix F-2*). In reality, even in these areas southerly and northerly winds are about equal in frequency of occurrence during the year's coldest month.

The wind invariably veers into the north at the time of arrival of a cold air mass, and it usually remains northerly for one, two, or three days afterward. The passage of a cold front—and its concomitant wind shift into the north—merely signals the fact that the edge of the usually massive mound of cold air has penetrated the area; the bulk of the air mass has yet to arrive. Winds persist from the north for one or a few days as the cold air "builds" into the state. Wind speeds, while vigorous or even tempestuous at first, always wane with the passing of time. One almost sure sign of the proximity of the center of the dome of cold air is a very light, if not calm, wind. After the center of the mound of cold air—or the "ridge line"—passes a locale, the wind gradually increases in speed, but from the opposite (southerly) direction. The return of the southerly wind one or more days after the arrival of a cold front signals a warming trend, because air is then being channeled from the relatively warm waters of the Gulf of Mexico. Clearly, it also indicates an increase in moisture in the lower atmosphere.

Without a doubt, the tableland known as the High Plains of Texas is one of the windiest regions on the North American continent during spring. Occasionally, work out of doors becomes impossible, and walking into the wind requires a superhuman exertion. While "average" wind speeds from the Permian Basin in the south to the Canadian River Valley in the north vary between 13 and 17 mph during March, April, and May, sustained wind speeds that often amount to two or three times that modest magnitude may persist over periods of more than a few hours. Gusts of wind exceeding 60 mph are not infrequent. Strongest winds observed in the western third of Texas have occurred during either late winter or spring, with maximum one-minute wind speeds of 70–80 mph egressing from the southwest-to-northwest quadrant.

Spring is the windiest season of the year in other parts of Texas as well. Prevailing wind direction is southerly most of the time, especially in the latter half of the season when the number of cool fronts ushering in northerly winds drops off substantially. As in the High Plains, winds in other sections intensify in advance of the arrival of a cold front and often blow in gusts of as much as 30 or 40 mph. Though on an average basis the wind is most blustery along the Texas coast in spring, in some years the strongest winds occur in late summer or early autumn as a result of tropical weather disturbances. It is not uncommon for winds to be clocked in excess of 100 mph at points along the coast in proximity to the center of an approaching hurricane.

## The Chill of the Wind

The human body can tolerate subfreezing—and even subzero—temperatures as long as there is little or no wind. But at extremely low temperatures, the speed of the wind becomes a far greater determinant of physical well-being than does an additional drop in temperature. During winter in Texas, when "blue northers" often shove the temperature well below the freeze level, it is not so much the cold temperature that makes outdoor activity torturous and numbing as it is the bone-chilling combination of the cold temperature and the strong winds. The cold wind removes the body-warmed air next to one's skin and clothing, so that the body must heat more air. If it is unable to supply heat at the rate it is being lost to the wind, the body then has the sensation of being cool, cold, or even bitterly cold.

A thermometer measures the temperature of the air independent of the speed of the wind. But to the human body, a temperature of 30°F (−1°C) with a calm wind is preferred to the same temperature combined with a 30-mph gale. Neither temperature nor wind alone gives a meaningful index to how cold a person feels under different conditions of wind and temperature. As a result, an empirical formula yielding the *windchill index* was developed after a number of experiments were performed in Antarctica in 1939–40. The index indicates that a temperature of 35°F (2°C) accompanied by a wind of 10 mph will turn one's ears white and numb just as quickly as still

100. Wind roses depicting the frequency of occurrence (%) of winds from various directions at selected points in Texas during the four seasons; based upon data for the period 1961–1980.

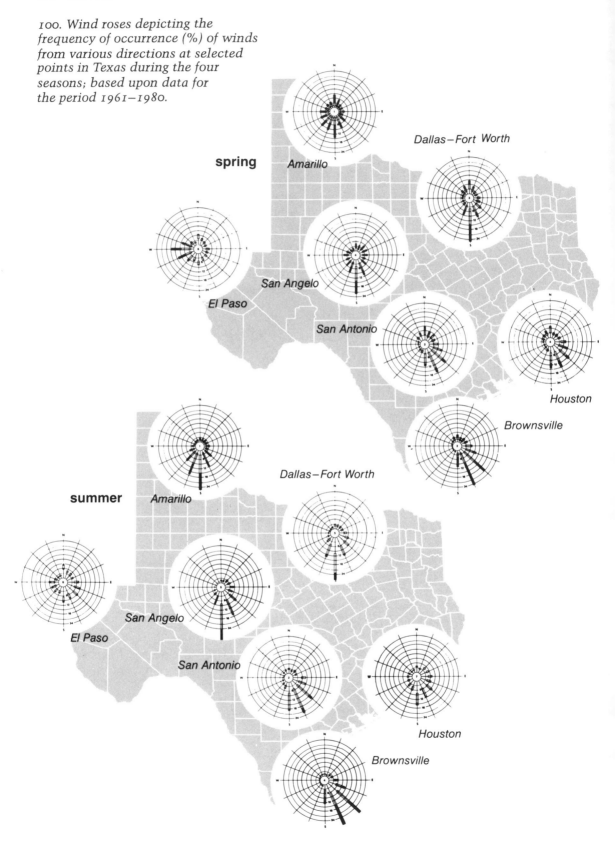

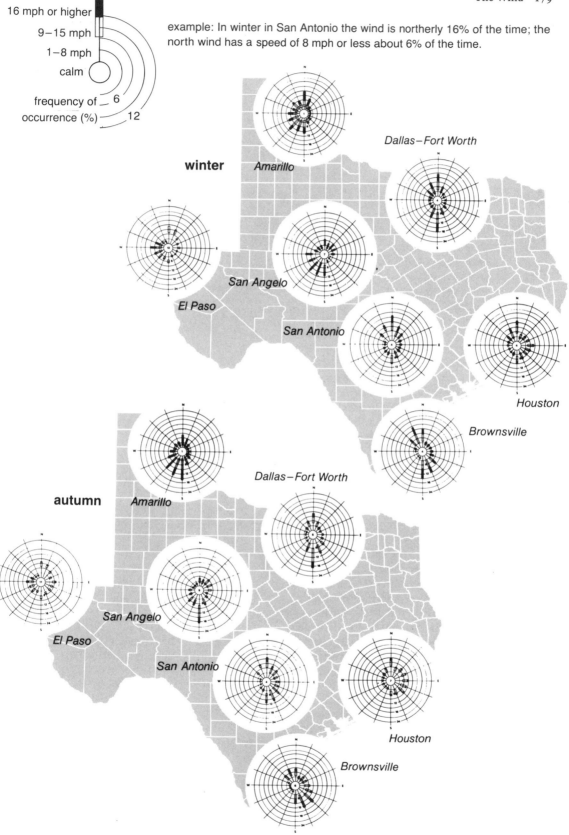

16 mph or higher

9–15 mph

1–8 mph

calm

frequency of — 6
occurrence (%) — 12

example: In winter in San Antonio the wind is northerly 16% of the time; the north wind has a speed of 8 mph or less about 6% of the time.

**winter**

Amarillo

Dallas–Fort Worth

El Paso

San Angelo

San Antonio

Houston

Brownsville

**autumn**

Amarillo

Dallas–Fort Worth

El Paso

San Angelo

San Antonio

Houston

Brownsville

actual thermometer reading (°F)

| estimated wind speed (mph) | 50 | 40 | 30 | 20 | 10 | 0 | −10 | −20 | −30 | −40 | −50 | −60 |
|---|---|---|---|---|---|---|---|---|---|---|---|---|
| | equivalent temperature (°F) | | | | | | | | | | | |
| calm | 50 | 40 | 30 | 20 | 10 | 0 | −10 | −20 | −30 | −40 | −50 | −60 |
| 5 | 48 | 37 | 27 | 16 | 6 | −5 | −15 | −26 | −36 | −47 | −57 | −68 |
| 10 | 40 | 28 | 16 | 4 | −9 | −21 | −33 | −46 | −58 | −70 | −83 | −95 |
| 15 | 36 | 22 | 9 | −5 | −18 | −36 | −45 | −58 | −72 | −85 | −99 | −112 |
| 20 | 32 | 18 | 4 | −10 | −25 | −39 | −53 | −67 | −82 | −96 | −110 | −124 |
| 25 | 30 | 16 | 0 | −15 | −29 | −44 | −59 | −74 | −88 | −104 | −118 | −133 |
| 30 | 28 | 13 | −2 | −18 | −33 | −48 | −63 | −79 | −94 | −109 | −125 | −140 |
| 35 | 27 | 11 | −4 | −20 | −35 | −49 | −67 | −82 | −98 | −113 | −129 | −145 |
| 40* | 26 | 10 | −6 | −21 | −37 | −53 | −69 | −85 | −100 | −116 | −132 | −148 |

| little danger for properly clothed person | increasing danger | great danger |
|---|---|---|

danger from freezing of exposed flesh

*wind speeds greater than 40 mph have little additional effect

*101. The windchill index. (National Science Foundation)*

air at a temperature of 21°F (−6°C); for a wind twice as strong, the effect is the same as that of a calm condition where the temperature is 12°F (−11°C). It is not uncommon for residents of the Texas Panhandle in winter to be subjected to temperatures near zero and wind gusts of 40 mph or more, which translates into a "sensed" temperature of −53°F (−47°C) (*see fig. 101*).

The reader should remember that the windchill index is only an approximation, since how one feels depends not only on temperature of the air and wind speed but also on other variables, such as the amount of exposed flesh, the type of clothing worn, the amount of radiation, the relative humidity, and even the physical condition of the person. Nonetheless, the index is a useful indicator for those planning outdoor activities where proper clothing and protection from the wind are important concerns.

Wind as a Source of Energy

In this age of diminishing energy resources and exorbitant fuel prices, the wind has become an appealing source of power. Wind costs nothing, and it is much more generously distributed over the terrain than any other power resource. Recognition of

this fact is evidenced by the thousands of windmills that jut skyward in much of rural Texas, but many of them stand today purposely inoperable. On the other hand, a growing number of them are being refurbished and put back into operation. It would be difficult to find a cheaper, simpler device for pumping water. What is more, they can be equipped with a small generator to deliver relatively inexpensive electricity. Windmills are especially cost efficient in the plains and prairies of western Texas, where winds of 10 mph or more blow almost incessantly on virtually every day of the year. A windmill in Midland would generate 80% more power in an average month of May than one in Austin because of that western city's characteristically higher wind speed; in a typical September a windmill owner in Amarillo would garner nearly three times as much wind power as an operator of a similar device in Dallas. Still another advantage in wind power is the absence of any adverse environmental effect. While wind will not replace most other sources of power in a matter of months or even a few years from now, it is conceivable that wind power may provide a small but significant fraction of

Texas' power needs by the end of this century. The fuel is ample, inexhaustible, and—most important—free.

A Thief of Water, Too

The movement of the wind not only may stir up and remove vast amounts of rich topsoil but also contributes to the loss of immense quantities of precious water from the soil and from storage facilities, such as reservoirs. This extraction of water results from the process of evaporation and is an important factor in reducing the quantity of water available for domestic supply or irrigation. What is more, the evaporation of water has almost as much influence upon crop production as does rainfall.

Actually, the process of evaporation is an integration of many components of the weather—not just the rate of movement of the air. Rates of evaporation depend on temperature, precipitation, and moisture content of the air (relative humidity), as well as wind speed. Vigorous winds accelerate the rate of evaporation because they carry newly evaporated water vapor away from the water surface, thereby maintaining a gradient in pressure between the air and water. It is considerably more difficult to measure evaporation from land surfaces, since the rate of water loss from the soil depends on the availability of water in the soil as well as on other factors, such as soil type and exposure to sun and wind.

By measuring water loss from exposed pans of water placed at strategic points around the state, and by applying appropriate numerical coefficients to the data to relate them to the actual loss of water from a large water surface (such as a reservoir), it has been possible to derive estimates of the actual amount of water lost through evaporation on large water bodies in Texas. The total amount of water lost through evaporation from a unit area of lake surface is commonly referred to as *gross lake surface evaporation*. The average annual rate of gross lake surface evaporation varies from less than 45 inches in the Sabine River Valley in extreme eastern Texas to nearly 100 inches in the Big Bend area of far southwestern Texas. One would expect such a wide range in gross evaporation rates, for

the Trans Pecos area on the average receives far more sunshine and experiences much lower relative humidities, higher daytime temperatures, and faster winds than the verdant woodlands of East Texas.

A more meaningful measure of the amount of water lost from water bodies due to evaporation is the *net lake surface evaporation* rate. Its value is greater than that of gross lake surface evaporation because it reflects the contribution of rainfall to the sustenance of the water level in the reservoir. The net evaporation rate consists of the gross evaporation rate minus the amount of rainfall that fell over the reservoir watershed (excluding the portion lost as runoff). Because rainfall in the eastern extremity of Texas usually is ample (50 inches or more per year, or some 10 inches more than the average annual gross lake surface evaporation rate), the average annual net lake surface evaporation rate is 10 inches or less. Conversely, in the arid valleys and plateaus of the Big Bend, where yearly rainfall is meager and gross evaporation rates are almost 100 inches per year, the average annual net rate of lake surface evaporation is 90 inches or more. In relatively "wet" years, evaporation rates characteristically are low, but in "dry" years, rates are high and the water supply is correspondingly low. These relationships are particularly meaningful to water engineers, who realize that in critical drought years evaporation losses not only contribute significantly to major water-supply problems but also influence decisions on reservoir design and operation.

The annual net evaporation rate is not evenly distributed throughout the year in Texas. In much of the state, maximum rates of water loss occur in August, while minimum rates are usually sustained in February. Of course, in years of deficient rainfall, net evaporation rates usually are higher in most months, but the monthly distribution of the evaporation tends to be more uniform than in wet years. The evaporation of water is a continuous process in both the desert climate of far western Texas and the humid subtropical climate of the extreme eastern sector of the state. While evaporation is taking place almost all the time, the net rates of evaporation demonstrate in some months that loss of water is partially or totally offset by appreciable rainfall.

## Regional Varieties of the Wind
### The Marfa Front

While the climate of Texas' far western region is best categorized as merely "semi-arid," the Trans Pecos is not without a recurrent, and often searing and desiccating, desertlike wind. Intense heat rising from the solar furnace that makes up a vast expanse of the Mexican desert gives rise to a very dry—and often torrid—wind that

sweeps eastward during spring and summer across parts of the Trans Pecos and even sections of the High Plains as a type of dry line known as the *Marfa front*. This term is frequently used by weather watchers in Texas because the community of Marfa usually lies in the path of this surging desert air mass and because the town is the site of a remote weather station that collects and transmits hourly data that allow trained weather observers to monitor the phenomenon.

With amazing regularity, in the late spring, summer, and early autumn, the arid air mass will push eastward around midday, slow to a virtual stall by midafternoon, and

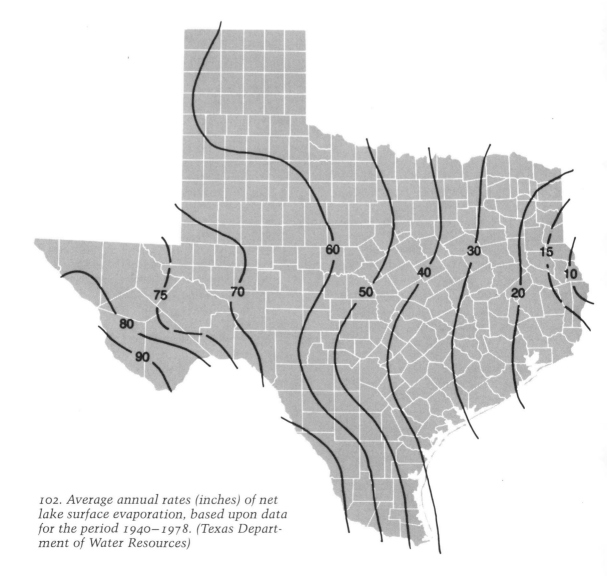

*102. Average annual rates (inches) of net lake surface evaporation, based upon data for the period 1940–1978. (Texas Department of Water Resources)*

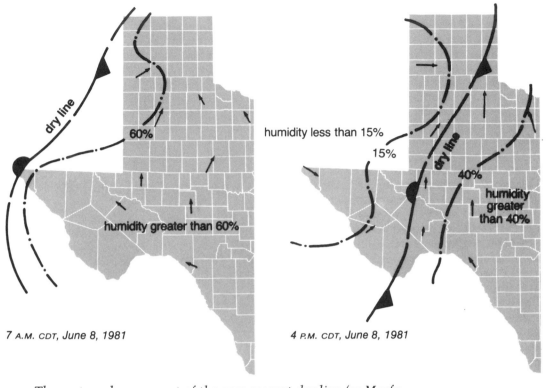

*7 A.M. CDT, June 8, 1981*

*4 P.M. CDT, June 8, 1981*

*103. The eastward movement of the ever-present dry line (or Marfa front) through west Texas on a typical summer day. Arrows denote wind movement by pointing in the direction to which the wind is blowing; length of arrows is proportional to wind speed.*

then slowly recede with nightfall. As a rule, the eastward motion of the dry line in the daytime is faster than the nocturnal retreat westward. Another commonly observed trait of the Marfa front is the tendency for various segments of the system to migrate at appreciably different speeds. On some days the leading edge of the desert air may ram deep into the Trans Pecos, while farther north the arid air mass barely traverses the Texas–New Mexico border. At other times, nearly all of the Panhandle may be engulfed by the intrusion of desert air, while farther south the dry line makes little headway.

The presence of the dry line is easily discernible. Its passage brings a shift in wind direction, most often from southerly to westerly in a few minutes (*see fig. 103*). In addition, with the invasion of the westerly desert breezes, the humidity plummets. On some days in spring, a change in dew-point

temperature of 40°–50°F (22°–28°C) can be observed from one end of the Pecos River Valley to the other. Furthermore, the influx of much drier desert air abets even more of a rise in temperature during the day. The moisture-deficient air sustains temperatures that often reach eye-catching levels well beyond 100°F (38°C) in such renowned locations as Presidio and Lajitas upstream from Big Bend National Park.

The Marfa front is a key element of the climate of far western Texas because of its impact on the formation of thunderstorms. The hot and dry desert wind knifes underneath less dense and more moist air to trigger towering thunderheads that can be seen great distances away and that supply a sizable fraction of the total yearly rainfall in the Trans Pecos and High Plains. It is common for a locale that happens to lie in the path of one of these mammoth thunderstorms to collect an inch or more of rain in one hour or less on a late afternoon or early evening in summer. The dry line can also have a dramatic impact on the temperature and humidity in a locale on many summer days. What customarily takes place when the Marfa front passes through a community is illustrated in fig. 104, which depicts dew-point temperature (a convenient measure of humidity) and wind movement at El Paso on June 7–8, 1981. The Marfa front, or dry line, retreated westward after dusk on June 7, and the result was an increase in moisture accompanied by a southeasterly wind (note the jump in dew point of 16°F (9°C) in nine hours, from 10:00 P.M. to 7:00 A.M. the following morning). When the dry line pushed through from the west around 7:00 A.M. on June 8, the air dried out appreciably. By afternoon of that day, with the wind blowing again from the west, the dew point had plunged to 35°F (2°C), or a relative humidity of 11%, and the temperature of the air stood well above 100°F (38°C). Throughout this whole period, skies remained clear and visibilities varied from 20 to 50 miles. Many other locales in the Trans Pecos and High Plains witness the same sequence of weather events on many days from May through September.

## The Sea Breeze

With humidities the highest anywhere in the state by virtue of its proximity to the Gulf of Mexico, the Texas coastal plain is notorious for its stifling and insufferable summer afternoons. But nature offers compensation to the city dweller and the shoreline visitor most every summer day in the form of a fresh and invigorating *sea breeze.* Although, of all the winds that blow across the state, the sea breeze is the least complex and comparatively autonomous, it is not a simple phenomenon.

The sea breeze and its nocturnal opposite, the *land breeze,* occur with clockwork regularity along the Texas coast, although the land breeze is not nearly as prevalent or well developed as its counterpart. The two phenomena are prevalent in the coastal plain on most days from late spring to mid-autumn. Both are set in motion by differences in the temperature of the air between land and sea. Land areas

are heated much more quickly by a rising sun than are adjacent bodies of water. As a result, this rapid warming of the ground by radiation heats the air above it, which then rises and is replaced by cooler, heavier air flowing in from the sea. This landward influx of not-so-warm air is the sea breeze.

The sea breeze in Texas is the strongest in summer, since the intense heat of the sun causes the air above the land to rise more rapidly than at any other time. The more intense the heat, the faster the vertical wind currents and the greater the pressure difference between land and sea. Since the atmosphere is continually striving to equalize the disparity in pressure from one locale to another (e.g., land to sea), the more the land air is heated, the lighter it becomes and the greater the contrast between air pressure over land versus that over water. The cooler and, hence, heavier air over the sea surges toward the land to restore the equilibrium, in the same way that water always flows from a higher point to a lower one.

Generally, the higher the temperature climbs along the Texas coast, the more intense and more extensive is the sea breeze. With few exceptions, the sea breeze attains its maximum speed at midafternoon, or the time of highest land-surface temperature. In Texas the sea breeze customarily blows at speeds of 8–15 mph and reaches distances of 10 to 15 miles inland. Its ceiling usually is no more than 600–1,000 feet above the land surface. Occasionally, however, the breeze may be of greater depth and may extend up to 50 miles inland during the peak heating period. This means relief to a city like Houston, where the air stagnates and becomes hot and suffocating. It is no overstatement to observe that, when the sea breeze sweeps into a coastal hamlet or inland metropolis, in a matter of a few minutes its soothing effect can transform a person's outlook on life. It is one of the few varieties of wind that folks enjoy having.

Aside from the salubrious effect that a sea breeze can have on sweltering coastal residents, the carefree and often capricious wind often manifests itself in the generation of many cumulus, or "fair weather,"

dew-point temperature

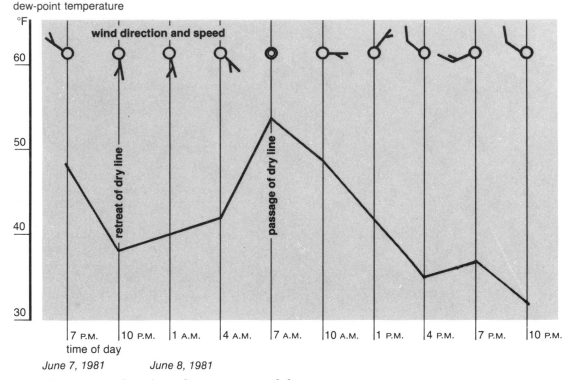

104. Time-series plot of wind movement and dew-point temperature at El Paso on June 7–8, 1981.

clouds that crop up along and inland of the coastline with striking regularity on most summer days. These small convective clouds first appear late in the morning near the shoreline and then migrate inland, where they grow to more appreciable heights. By noontime or early in the afternoon, most of the sky above the whole coastal plain of Texas is bespeckled by these growing cumuli. Whereas the first few batches of puffy white cumuli seen along the coast around midmorning are due to rising land-air currents, later in the day the clouds grow taller and more numerous as some of the moisture-laden air from the Gulf surges inland and is also heated by the hot land surface. Depending upon the availability of enough moisture higher in the atmosphere, quite a few of these sea-breeze-induced convective clouds may blossom into full-fledged thunderstorms capable of generating noteworthy rains onto the coastal plain.

### The Legendary Texas Duster

The origin of the loathsome west Texas dust storm can be traced to land misuse that began early in the nineteenth century when pioneer Americans pushed westward out of the Mississippi River Valley into the vast grass-covered prairies and put plow to the ground for the first time. Suddenly, stalks of wheat, cotton, and grain sorghums sprang forth where nothing but prairie grass had grown previously. But settlers found Earth's crust to be thin and often sandy and observed that (with good rains highly infrequent) blinding, choking, stinging waves of dust became an all too common phenomenon. The scourges reached a peak in the mid-1930s, when strong westerly winds racing down out of the Rocky Mountains scooped up tons of loose topsoil and flung the sediment over monumental distances.

Although the cultivation of marginal farming areas—with their thin, sandy

small fair-weather cumulus clouds

*105. As seen from an orbiting weather satellite, hundreds of small "fair weather" cumulus clouds dot the Texas coastal plain on a typical late morning in summer. (National Oceanic and Atmospheric Administration)*

soils—was a leading contributor to the onset of the devastating dust storms of the "Dust Bowl" era, repeated paucities of rainfall and an almost incessant wind also played important roles. Indeed, even with more modern farming techniques, such as contour farming and strip cropping, employed in much of the Texas High Plains since World War II, dust storms continue to plague the area. Inhabitants of the southern High Plains of Texas—a broad, flat tract of treeless terrain stretching from Plainview

in the north to Midland–Odessa in the south—have come to know that an uneasy co-existence with nettlesome dust is the price to pay for residing on the Llano Estacado. One consolation to these hardy plains people is that dust usually is not a year-round misery. In reality, the frequency of visits by the black pall varies considerably within the elongated High and Low Rolling Plains regions of Texas. Moreover, other regions of Texas are not immune to occasional bombardments of dust, most of

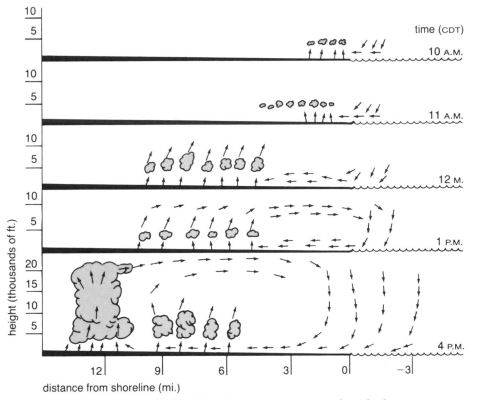

*106. Motion of the air and cloud development associated with the daily landward penetration of the sea breeze. (National Science Foundation)*

which emanate from the semiarid plains of western Texas.

### Ingredients of a Dust Storm

In the meteorological sense, dust consists of solid materials suspended in the atmosphere in the form of small irregular particles, most of which are microscopic in size. Dust cannot be a stable component of the atmosphere because it must eventually fall back to Earth's surface when winds and turbulence become too weak to bear it aloft. While some dust found in the Texas atmosphere has as its source plant pollen and bacteria, salt spray from the Gulf of Mexico, and smoke and ashes from forest fires and industrial combustion processes, much of the material in a Texas dust storm consists of solid soil particles removed by winds raking the surface of Earth. For such a dislocation to occur, a period of drought over what is normally arable land is needed

to loosen the material in the outermost layer of Earth's surface. Most often a dust storm arrives suddenly in the form of an advancing dust wall that may be many miles long and thousands of feet high.

Standard observing practices of the NWS require enough blowing dust to reduce visibility to between 5/8 and 5/16 of a mile to have a bona fide *dust storm*; if the visibility plunges to or below 5/16 of a mile, the event is designated as a *severe dust storm*. Dust storms are also referred to as "dusters" and "black blizzards." Dust storms in Texas are differentiated from "sandstorms" in several ways: Blowing sand particles are confined mostly to the lowest 10 feet of the atmosphere, rarely rising to much above 50 feet; dust storms invariably grow much, much larger. Sandstorms are confined to select areas of Texas where loose sand, often in sand dunes and without much admixture of dust, is prevalent. Sandstorms usually are restricted to parts of the Trans Pecos and select portions of the High and Low Rolling Plains. They are also known as "haboobs" and "desert winds."

Development of a Dust Blowout

The ingredients for a typical blowout of dust usually mesh together early in the year, or at the end of the customary "dry season" in western Texas and just when the racing winds that hint of spring become an almost everyday feature. The main surface and upper-atmospheric components responsible for a large-scale dust storm are represented in fig. 107, which depicts weather conditions about daybreak on

107. *Weather map showing conditions during an intense dust storm raging over the High and Low Rolling Plains of Texas: numbers denote visibilities (miles) at selected observation points; arrows (⊾) indicate speed (10 mph per flag) and direction (stem points toward direction wind is blowing from) of surface winds at 6 A.M. CST, March 11, 1977.*

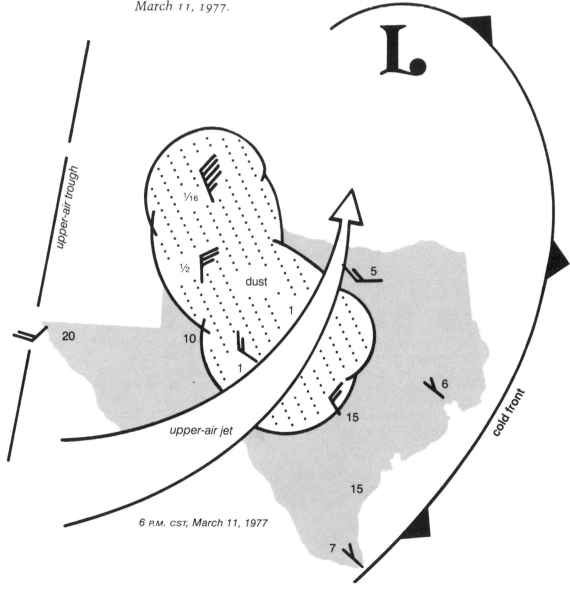

March 11, 1977, when a massive dust storm moved southeastward out of the northwestern portion of Texas. On this occasion—as with most other instances of widespread dust—a cold front had earlier traversed the state, ushering in an increasingly vigorous flow of winds from the central Great Plains area. These strong surface winds scoured up loose material in the High Plains of Texas, as well as in adjacent areas of New Mexico, Colorado, Kansas, and Oklahoma, and transported the particles south and eastward deeper into Texas. As shown in the figure, the visibility at Amarillo was no better than 1/16 of a mile at dawn, while visibilities at Lubbock and Abilene were 1/2 and 1 mile, respectively.

The strong northerly winds pouring down through the High and Low Rolling Plains resulted from a storm system (centered in northern Kansas and designated by an "L") that had sustained considerable intensification on the previous night. Supporting this strengthening was a larger storm system (called a "trough") oriented north-to-south and positioned over the Rocky Mountains (shown as a dashed line in fig. 107). A feature of this intense upper-air storm over the Rockies was a very potent stream of winds high in the atmosphere (commonly referred to as a "jet") moving northward out of Mexico into Texas. As the upper-air trough and its associated jet propagated eastward out of the Rockies, that movement stimulated intensification of the surface low in northern Kansas. As the low was energized further, winds circling it in a counterclockwise fashion increased in speed, and the end product was the stirring up of more dust and the transfer of much of that vast quantity into the eastern and southern quarters of Texas. In short, the upper-air jet aided the development of the dust storm by causing the surface storm center to wind tighter; hence, surface winds pouring out of Kansas into the Texas Panhandle increased in speed, which in turn exacerbated the stirring of greater amounts of dust.

Many of the drastic changes that occur when a dust-producing cold front surges through Texas are illustrated in fig. 108. Strong southwesterly winds gusting to 20–25 mph late on the morning of March

10, 1977, signaled the impending arrival of much drier air from the Rocky Mountains. When the dry air mass arrived around noon, winds veered into the west, the relative humidity plunged quickly to 10%, and the dust that was whipped into the air dropped visibilities to only one mile. The barometric pressure continued to decline because a second front—of polar origin and bringing colder, northwesterly winds—had not yet arrived from the north. That second cold front passed Lubbock shortly before 3:00 P.M., when pressures suddenly began rising and the temperature dipped. Skies continued to be obscured by dust until well after dusk.

Dust Storms of the Past

Aside from the enormous and inveterate spells of choking dust that highlighted the Dust Bowl era of the mid-1930s, perhaps the most injurious dust storm in more recent years came to western Texas in February 1977. A very intense surface low-pressure center that moved out of the Colorado Rockies early on February 22 sent extremely strong winds into the northwestern quarter of Texas, where tons of soil were whirled into the atmosphere, so beclouding the air at Lubbock that visibilities were restricted to one mile or less for thirty-four consecutive hours. Winds blasted through Guadalupe Pass at speeds clocked as high as 114 mph, blowing several tractor-trailer rigs off the highway. Winds of near-hurricane force in El Paso caused extensive damage to roofs of homes and businesses, while signs and utility poles were felled, and twenty persons were injured; property damage amounted to $650,000.

Just two weeks later, the awesome dust storm of March 11, 1977, struck the Panhandle, destroying one-fourth of the total winter wheat crop. Winds pounding the northern High Plains at speeds as high as 75–80 mph raked up enough dust to hold visibilities at Amarillo to one-half mile for ten hours on March 11. The very intense duster blasted to death a wheat crop valued in excess of $6 million. Still reeling from the destruction wrought by the late-February dust storm, El Paso at the same time sustained more sizable property

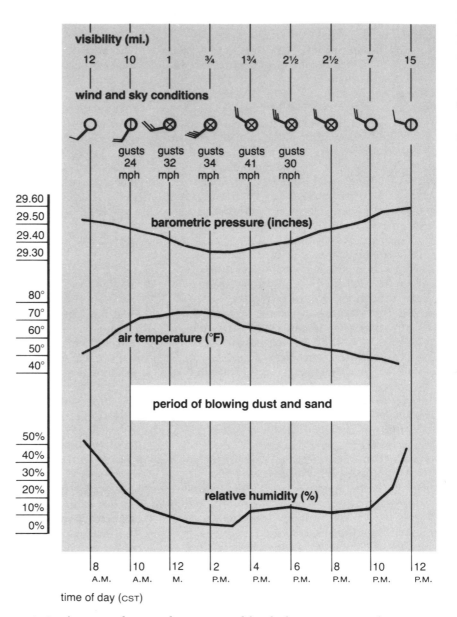

*108. Surface weather conditions at Lubbock during a severe dust storm on March 10, 1977.*

damage ($250,000) from winds measured as high as 84 mph.

A couple of prodigious dust storms struck the western half of the state in the mid-1950s near the end of the twentieth century's worst drought in Texas. More than one-quarter of a million acres of cropland suffered moderate to severe wind erosion from a highly perilous storm that

struck the High Plains with winds of hurricane force on March 2–3, 1956. Dust so clogged the air in Odessa that street lights came on at midafternoon, and in Lubbock visibilities were held to half a city block for hours. The blinding dust caused numerous traffic accidents, several of which led to the deaths of three motorists. Dust residue quickly spread over almost all of the re-

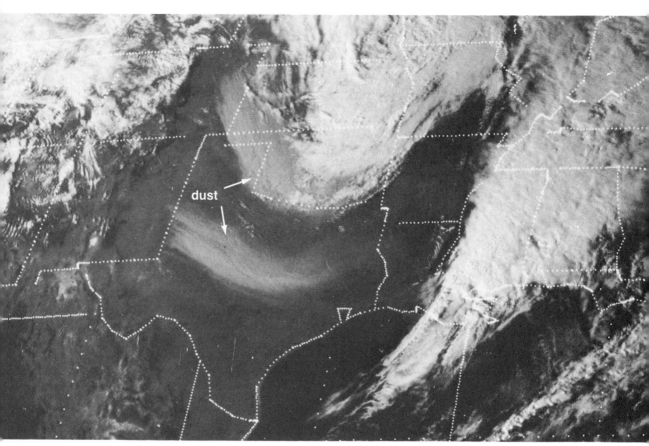

dust

109. Photograph from an orbiting weather satellite of the immense
dust storm that swept across Texas on March 11, 1977. (National
Oceanic and Atmospheric Administration)

mainder of Texas. Later that same year in
November—or just four months before that
devastating mid-1950s drought was abruptly
terminated—so much dust was blown
across the southern High and Low Rolling
Plains that unpicked cotton was buried in
the dust. According to U.S. Weather Bureau
reports, homes and other buildings at Ma-
tador in Motley County were "filled with
dirt." Winds with speeds of 70–75 mph
yanked trees out of the ground and ripped
shingles from the roofs of houses in the
area.

Some dust storms in recent decades were
so pervasive they infiltrated nearly every
corner of Texas, but very few were as far
reaching as the mammoth dust storm of
January 25–26, 1965, which limited visibilities to one or two miles in such dispar-

ate locations as El Paso, San Angelo, Dallas, San Antonio, and Houston. Fourteen
months later a dust storm emanating from
western Texas had similar impact, substantially reducing visibilities from Austin to
Brownsville and Port Arthur. Also ranked
near the top was the colossal dust storm of
March 14–15, 1971; after moving out of
northwestern Texas, it sent visibilities
plunging to less than one mile at Dallas
and just barely more than that as far south
as Brownsville.

Events both tragic and bizarre have been
associated with some of the more notorious
dust storms in Texas in recent years. Dust
that billowed to more than 15,000 feet and
held visibilities to zero was driven by
winds gauged in excess of 100 mph at
Guadalupe Pass in far western Texas in

February 1960; those winds accounted for the death of a small child, who was blown under a bus and killed. Several years later, aircraft pilots reported dust as high as 31,000 feet from the Texas–New Mexico border to as far east as Abilene on January 25, 1965, while motorists traveling on the open highway at maximum legal speeds saw tumbleweeds overtaking them. Dust was whipped along by winds that toppled three 160-foot light towers at Jones Stadium in Lubbock in January 1967; damage to the towers, which were designed to withstand winds of 100 mph, amounted to $100,000. On the same day, residents in the Panhandle were besieged by a combination of dust and snow being driven along by winds of near-hurricane force. Visibilities were reduced by blowing dust at Shamrock in Wheeler County to the extent that a car pileup involving as many as a score of automobiles resulted in thirty-two injured motorists.

When the vicious winds of spring whip dust into a frenzy, the quantity of soil displaced often is astounding. One dust storm that lasted half a day left drifts of sand as high as 10 feet in Lamb County near Littlefield on March 17, 1977. A weather observer at Reese Air Force Base near Lubbock found a layer of dust 3½ inches thick in his rain gauge after a dust storm raged across the southern High Plains in January 1965. A study made of a series of ten dust storms that hit Lubbock during the severe drought of the mid-1930s revealed that an average of 122 tons of sediment was carried through that High Plains city each hour during those storms.

When and Where Dust Storms
Are Prone to Occur

Folks who have resided in the southern High Plains of Texas for as little as one spring season know that annoying dust storms are as common to the region as thunderheads and mesquite. With much of the terrain of this sector of Texas marked by quarter-section subdivisions with the familiar inscribed circles denoting irrigated cropland, and given the fact that brisk winds blow almost incessantly during spring, numerous dust storms are inevitable. Circumstances are aggravated all the

more early in the year before irrigation water is applied and particularly when the preceding winter has been unusually dry. In these instances the fertile cropland is highly susceptible to wind erosion. Consequently, most spring seasons bring as few as a half-dozen and sometimes as many as a score of notable dust storms. The phenomena are especially prevalent in a twenty-county area centered at Lubbock. Skies beclouded by dust are most common in March, although in many years as many or even more dust storms may come earlier in February or later in April and even May. Most spells of dust prevail for only six hours or less, when visibilities are restricted to ½ mile or less. Summers, by contrast, are rather free of the large-scale, stifling dust storms common in spring. Much of the dust seen in summer in the southern High Plains is highly localized and the result of gusts of cold air plummeting down out of thunderstorms.

Dust storms farther north in the Panhandle usually are not as long lasting or nettlesome as those in the southern High Plains. Nevertheless, it is rare when a year does not provide at least a few dust storms that reduce visibilities to 3 miles or less in this northernmost sector of Texas. More than half of all dust storms in the Panhandle occur during March and April, and dusters are not uncommon in the late winter or autumn. While the event is rare, dust storms in this region have been known to coincide with snowstorms, as happened during the mammoth dust storm of March 11–12, 1977. The combination of blowing dust and blowing snow driven by winds of 40–50 mph stifled visibilities to as little as 1/16 of a mile during that time.

Aside from the irrigated croplands of the southern High Plains, the northern Trans Pecos rates as the dustiest strip anywhere in Texas. More than half of all dust storms come with the very brisk winds of spring, which is not only the most blustery time of the year but also the conclusion of the lengthy "dry season." Occasionally one or two dust storms occur in autumn or early winter after a particularly dry summer. However, few of them are severe enough to lower visibilities to one or two miles or less, and nearly all of them strike during

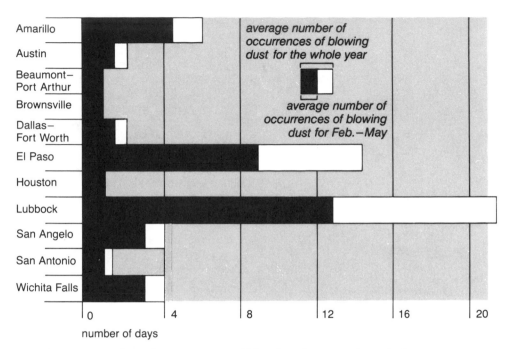

Amarillo
Austin
Beaumont–Port Arthur
Brownsville
Dallas–Fort Worth
El Paso
Houston
Lubbock
San Angelo
San Antonio
Wichita Falls

*average number of occurrences of blowing dust for the whole year*

*average number of occurrences of blowing dust for Feb.–May*

number of days

0    4    8    12    16    20

*110. Average number of occurrences of blowing dust at selected places in Texas. Shown here are periods when the visibility is reduced to 5 miles or less by blowing dust or sand. Data are for the period 1961–1980.*

the daytime between noon and dusk, or when winds are most vigorous.

It is a rare year that fails to furnish at least one or two dust storms in the Low Rolling Plains and Edwards Plateau. Virtually all the dusters that afflict these regions stem from stiff westerly winds that carry dust out of the High Plains and Trans Pecos. Nearly always the dust is not thick enough to drop and maintain visibilities as little as one or two miles for any appreciable length of time. Over half of all dust storms persist in any one locale no longer than six hours. As in the source regions to the north and west, dust, when it occurs, most often develops during the afternoon and early evening in the territory stretching from Del Rio through San Angelo and Abilene to the Red River.

The cross-timber woodlands and grassy prairies of North and South Central Texas lend no support to dust storms migrating out of the higher elevations to the west. Still, the vigorous dusters that materialize over the Trans Pecos and High Plains

sometimes remain intact as far east as the state's blackland belt. With very few exceptions, dust is never a problem in these areas except during the windy late winter and early spring. During that time enough dust is blown eastward to litter the atmosphere to the extent that visibilities are reduced to only a few miles, but such modest dusters seldom last more than six hours. If a dust storm occurs early in the year, it is an even bet that a second or third will follow before spring has ended. Dust in the air sufficient to restrict visibilities substantially in autumn or early winter is seen only once every three to four years.

Though they are far removed from the breeding grounds of most dust storms, Southern Texas and the Lower Valley nonetheless are touched every year or two by a modest-sized dust storm or sandstorm. Ninety percent of the time the dust that afflicts these regions is transported by northerly winds from the central portion of the state, while, occasionally, dust may emanate from northern Mexico. Usually the dust is not so thick as to lessen visibilities to one or two miles or less, and dust most often limits visibilities for periods no longer than six to nine hours. With the lush cover of native grasslands interspersed with numerous thickets of piney woods and post oaks, East Texas and the Upper Coast regions are the least likely of all sectors of Texas to suffer from a dust storm or a sandstorm. It is an extreme rarity for dust to congest the atmosphere there such that visibilities dip to one or a few miles. That happens only when the residue of a massive and vigorous dust storm that formed in western Texas sweeps the breadth of the state. Severely restricted visibilities due to dust are observed only once every three to five years, on the average.

## Dust Devils

A far more frequent weather disturbance involving dust or sand in semiarid regions of Texas is the *dust devil*, a whirling column of air resembling a miniature tornado. Dust devils are particularly prominent in the western half of Texas on days with a blazing sun and hot, gusty winds. Unlike tornadoes, dust devils develop on Earth's surface and seldom, if ever, attain the violence of funnels that drop from clouds. Sometimes these rapidly spinning columns of dust mingled with small twigs, leaves, and other lightweight debris attain heights of several hundred feet. Most of them occur over dusty roads or plowed fields, extend only a few feet in diameter and less than 50 feet in height, and persist for no more than a few minutes.

A flat, hot surface, such as a hardened, dusty road or a barren stretch of desert, and a burst of wind are the ingredients necessary for a dust devil. Air next to the bare, dusty surface is heated rapidly by the intense rays of the sun and, as this warmed air expands and rises, cooler air nearby swarms in to replace the rising hot air. As little as a gentle breeze, and more often a gust of wind, provides this rising air column with a twist, such that a spiraling eddy of air—a dust devil—is formed in only a matter of seconds.

Dust devils are common occurrences during spring and summer in the state's most arid regions—the Trans Pecos and High Plains. Since many dust devils are never seen, and an even larger number never actually are reported by weather observers, the usual number of dust devils on a characteristically hot, windy summer day is not known. Likely, tens—if not hundreds—of the small spinning dust demons form every summer afternoon. What is more, dust devils may be seen on hot, breezy days in any region of the state, although, naturally, the frequency lessens greatly in heavily wooded areas of eastern Texas.

Most dust devils in Texas advance at such slow speeds that people rarely are injured, and damage to property, if sustained at all, usually is insignificant. Being struck by a dust devil usually means no more severe damage to a person than an eyeful, earful, or noseful of dust and small debris. There are a few notable exceptions to this rule, however, for a few dust devils may grow large enough and persist long enough to cause meaningful damage to property unfortunately positioned in the devil's erratic path. For example, on a hot, sultry afternoon in the middle of the summer heatwave of 1980, a dust devil ripped off the roof of a ranch barn and downed power lines near Granite Mountain in Burnet County. The owner of the barn described the noise made by the strong whirlwind as similar to that made by a jet aircraft. Earlier in the same summer, under the same set of weather conditions, a dust devil tore into an auto repair shop in the Falcon Heights community of Starr County and removed its doors, windows, and roof.

# 10. *Doing Something about the Weather*

The adage that nature's elements are so overbearing that we can only endure them has been expressed innumerable times by a people grown weary in submitting to weather's vagarious nature. Nevertheless, the statement that we remain at the mercy of our environment is not quite as true as in the time when a melting glacier forced hordes of people to vacate their seaside dwellings. Today a rapidly improving technology allows us not only to recognize developing weather conditions and to project with reasonable accuracy their imminent behavior but also to intervene in modest ways to lessen the impact of harsh winters and stifling summers, disruptive rains and pernicious droughts. Indeed, we can do more than merely heat and cool our homes and dam up our rivers and streams. We can also exercise prudence in dealing with those facets of weather that inflict injury to a disastrous degree on the citizenry, and we can act to mitigate the impact of those inevitable weather tragedies through the timely application of personal and communal first aid.

## Coping with a Hurricane

Few people who witness the fury of a hurricane ever forget the experience. Because of the great probability that these tempests will produce death and widespread destruction, it is imperative that residents in the Texas coastal plain—especially those who live within a few miles of the shoreline—take proper precautions. The key to hurricane protection is preparation. By taking adequate and sensible measures before, during, and after a hurricane strikes, many lives can be saved and considerable property damage averted. In the final analysis, the responsibility for hurricane preparedness rests with the individual. At the time a cyclone is churning in the open Gulf, local officials are usually preoccupied with implementing plans for the imminent emergency and are unable to assist individuals with planning or other background information. It behooves coastal residents to familiarize themselves with the dangers posed by a hurricane and to be ready for them well in advance.

Advance Preparations

The first bit of advice is to ascertain the degree to which your area is vulnerable to the forces of wind and water. Determine the elevation of your home (or business establishment) above sea level and its distance from open water. Also find out if your neighborhood is susceptible to freshwater flooding. Learn what maximum storm-surge height you could expect in your locale. The NWS office nearest you (in Brownsville, Corpus Christi, Victoria, Houston, Galveston, or Port Arthur) can supply estimates of the highest potential storm surge and the expected extent of inland flooding. Generally, if your home or business is less than 25 feet above sea level and no more than 20 miles from the shoreline, it could be highly vulnerable to storm-surge flooding.

A second precaution to observe prior to the arrival of a hurricane is to secure adequate insurance for your home and other possessions from a reputable company. Be sure to determine whether your policies do or do not cover damage to your property caused by rising or wind-driven water. There are other measures to take as well. When—if not before—a *hurricane watch* is issued for your area, you may want to store drinking water in jugs, ensure that flashlights and other battery-operated equipment are adequately charged, obtain and store food that needs no refrigeration, and fill your car with gasoline. At the time—if not sooner—a *hurricane warning* is issued, abandon low-lying beaches that may be swept by the storm surge; make sure young or helpless people and livestock are moved to higher, safer ground; secure all boats and other items on piers or boathouses. If you live in an area that is out of danger from high tides and is well built, plan to stay home. If you decide to leave, keep in mind that if you delay your departure roads to safer areas may become flooded before the main portion of the storm arrives. Board up windows or otherwise protect them with shutters and/or a sturdy tape. Tie down any item (such as garbage cans, porch furniture, and toys) that might be picked up by the winds and hurled through the air.

When the Storm Is Raging

When the brunt of the storm has arrived, it is important that someone in your group stay awake at all times to keep track of the storm's progress through NWS advisories and bulletins and to guard against such hazards as fire and snakes. Remember to stay indoors during a hurricane, for it is treacherous to travel or move about when winds and tides are lashing your vicinity. Remember that, if the calm "eye" of the storm passes overhead, there will be a sudden lull in the winds. Do not be deceived into thinking that the storm has passed. Such a respite from the storm's fury usually lasts up to a half-hour or even longer, but the second half of the cyclone inevitably will come, with winds rising very rapidly to hurricane force or greater in a matter

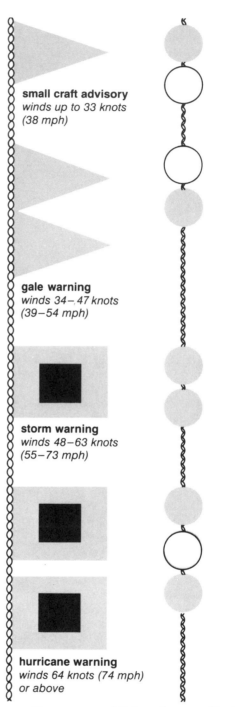

**small craft advisory**
*winds up to 33 knots
(38 mph)*

**gale warning**
*winds 34–47 knots
(39–54 mph)*

**storm warning**
*winds 48–63 knots
(55–73 mph)*

**hurricane warning**
*winds 64 knots (74 mph)
or above*

*111. Pennants and lights that are displayed in some ports and in other areas of the Texas coast to warn of hazardous sea conditions. (Henry et al.,* Hurricanes on the Texas Coast)

of minutes. To help eliminate the risk that windows or doors in your home will be blown out due to great pressure variations set up by the storm, keep a window partially open on the downwind side of your home (i.e., opposite the direction from which the storm is approaching). Be sure, though, not to situate yourself near that open window or door. By following these instructions and using common sense, most of those residents who experience a hurricane will remember it as little more than a dramatic event.

In the Storm's Wake

After the hurricane or tropical storm has passed and winds have receded, it is important that people stay out of disaster areas. The presence of large numbers of unqualified people hampers rescue and first-aid operations. It is after such a storm that the extent of property damage, as well as injuries and deaths, can cause widespread shock. Seek medical care for the injured at a disaster station or nearby hospital. Remember that health hazards are in abundance. Avoid loose or dangling wires and report them to a law-enforcement officer or power company. Also report to the proper authorities any broken sewer or water mains. Be careful what water you drink; private wells and municipal water supplies may be contaminated. Check for food that may have been spoiled due to lack of refrigeration. If you travel, drive carefully along those streets and roads that are strewn with debris, and remember that some roads, undermined by floodwaters, may collapse under the weight of your car. Stay away from the banks of rivers and streams, for they often become unsafe after a deluge.

## Reacting to a Violent Thunderstorm
Taking Refuge from a Tornado

What you should do when a tornado is approaching is dictated, naturally, by the circumstances in which you find yourself at that precise moment. There are different measures to take, and some actions to avoid, depending upon where you happen to be at the time you are made aware of a tornado hazard. The following is a list of tornado safety rules intended to assist you

in securing the maximum amount of protection possible when a tornado threatens:

If you are outdoors in the open, move away from the oncoming tornado at a right angle (i.e., if the tornado is moving west to east, move toward either the north or the south). If there is no time to escape, lie flat in the nearest low spot, such as a ditch or ravine.

If you are in a home with a basement, seek refuge next to the wall in the most sheltered and deepest portion of the basement. For added protection, get under heavy furniture or a workbench.

If you are in a home without a basement, go to the lowest floor and to the smallest room having stout walls. Get under some heavy furniture (e.g., a tipped-over couch or chair). Stay away from doors and windows, so that you are less apt to be struck by flying glass and other debris.

If you are in an office building, go to the basement or an interior hallway on one of the lower floors. If you have no time to descend to a lower area, get in a closet or other small room having stout walls. Otherwise, crawl under heavy furniture. Be sure to stay away from windows.

If you are in a mobile home (a structure that is particularly vulnerable to overturning and destruction during strong winds), vacate it and seek shelter in another, more fortified preselected structure or get in a ditch out in the open.

An automobile is not a safe haven from a tornado, unless of course the driver can be certain of getting out of the twister's path. Often, it can be a tragic mistake to get into your auto and attempt to outrun a twister. One lesson gleaned from the investigation of the death and destruction wrought by the April 10, 1979, tornado in Wichita Falls was that an auto may become nothing other than a sealed tomb when a tornado strikes an area. About half of the forty-five persons who died from the Wichita Falls tornado were occupants of autos; many of them had been trying to run away from the storm and were caught in massive traffic jams from which they could not extricate themselves. An automobile may be useful when a ravine or ditch is not available, in that you may obtain some shelter from flying debris by crawling under one.

Schoolteachers and students should be familiar with designated shelter areas and proper measures to be taken when a tornado threat is made known. School administrators ought to ensure that a special alarm system is available to indicate to school personnel that a tornado has been sighted and is approaching. School supervisors should be assigned to round up children who may be on playgrounds or in other outdoor areas. Students should be escorted from structures of weak fabrication (e.g., portable or temporary buildings) to sturdier ones. When the danger of a tornado is imminent, students should be assembled in school basements or interior hallways and instructed to crouch on elbows and knees facing the wall. Buses should not be operated, but, if a bus is caught out in the open when a tornado is approaching, the occupants should leave the bus and go to a nearby ditch or ravine to lie down with their hands over their heads. It is wise for school administrators to drill their students periodically, so that they will know precisely what to do to protect themselves.

Individuals, families, and employees can take numerous steps long before the tornado season gets underway to ensure their safety in the event a tornado should enter their neighborhood. Some homeowners prefer that their houses be equipped with either basements or storm cellars. If you have such a facility available to you, it is a good idea to place a battery-operated radio in it, so that you and other occupants can remain apprised of weather developments even though the home's electrical power supply is severed. Also, a thoughtful step is to store other necessities—such as drinking water, supplemental clothing, even food—in these protected areas. This will ensure that the family members have sufficient provisions should they be confined for a lengthy period of time, or if it should be their misfortune to have a tornado remove all their other, unpreserved belongings. If you do not have access to a basement or cellar, and if you think that other areas of your home are unsafe, you should preselect a nearby culvert or deep ditch

to which you can flee should the need arise. An employer should see that shelter areas are selected and marked for use by all employees. Someone should be trained to direct the building's occupants to these shelters when the situation warrants it. Those who live in mobile home parks would do well to select a leader responsible for ensuring constant radio monitoring and storm spotting and for sounding an alarm if a storm cloud approaches.

## Avoiding the Lightning Bolt

The more intense a thunderstorm, the more frequent are the lightning strokes generated by it. Consequently, a person should develop the habit of keeping an eye to the sky. When a thunderstorm approaches, you should seek shelter inside a home or a large building or, if neither of those is available, get into an all-metal vehicle (such as an automobile). Inside a home or office, use the telephone and television sparingly, if at all.

Not everyone can reach a safe haven when lightning begins to strike. For the camper, hunter, fisher, soldier on maneuvers, or farmer in the field, there may not be enough time to reach a safe building or vehicle. In that case, remember and abide by the following safety rules:

Avoid getting underneath a tall object, such as a tree, that stands alone in an open area; it is a natural lightning rod. If you are in a forest, find cover in a low area under a thick growth of smaller trees.

In wide-open areas, seek out a low place, such as a ravine or valley. However, avoid any low-lying area where surface water or saturated soil is likely to be a better conductor of electricity than the surrounding landscape. If you find yourself hopelessly stranded in a level field and you sense that lightning is about to strike (an event often presaged by hair standing on end), squat on the balls of your feet with your arms wrapped around under your knees. The important thing to remember is to keep as small an area of the body as possible from making contact with the ground. Do not lie flat on the ground. Try not to project yourself above the sur-

*112. When towering cumuliform clouds are observed, be alert to the possibility of lightning; frequent strokes of lightning are common with these budding thunderstorms. (Texas Department of Highways and Public Transportation)*

rounding landscape, as you would if you stood on the top of a hill, in an open field, on a beach, or in a boat.

Get out of and as far away from water as possible.

Stay away from objects like tractors and other farm equipment, motorcycles, golf carts, and bicycles.

Get away from metallic objects (such as wire fences, rails, clotheslines, metal pipes, steel towers, even golf clubs) that could carry lightning to you from some distance away.

Spread out from others if you are in a group (such as a party of hikers or mountain climbers) so that, if lightning should hit, a fewer number of people will be affected.

Lightning is also a potential hazard on water. Damage to pleasure craft is not uncommon and often is expensive. The masts of sailboats are highly susceptible, even when taller boats or other objects are nearby. One incident illustrative of the severe beating that lightning can administer to pleasure craft on water occurred on Eagle Mountain Lake in 1954. A blaze of lightning hit the mast of a small sailboat and surged down the headstay, putting the stem out of commission. It continued on down the

shrouds, putting a large hole in the starboard side of the craft and numerous other holes the size of a fist in the port side. Even the sail, which was stowed below deck, sustained five burn holes. Fortunately, since all the holes in the hull were above the waterline, the boat did not sink. The little craft was not grounded, that is, none of the metal rigging was in contact with the water. Though the percentage of boat fires attributed to lightning is quite small, it still behooves a boat owner to take precautionary measures to limit the damage that might be done by a chance lightning hit. If caught in a thunderstorm and unable to reach shore, the seafarer should stay below deck, if possible, and keep away from large metal objects in the boat (such as the motor) if they are not grounded to the boat's lightning-protection system.

## Taking Refuge from a Flood

When a *flood warning* is given, and even when you receive no warning but suspect that weather conditions have deteriorated to the extent that you are threatened, move to a safe area immediately. Be sure to scout your neighborhood and vicinity well enough to know where you should go when the need to flee arises. Do not procrastinate in getting away from a flood-threatened area; to delay may mean you are denied access to safety because roads have been cut off by rising water. When you seek a safe place, or if you are out in the middle of a flash-flooding rainstorm, remember the following:

Do not attempt to cross on foot a flowing stream where water is above your knees.

If you are in an automobile, do not try to drive over a flooded road. Your car could stall out, leaving you stranded and trapped. If your auto does stall in water, abandon it immediately and get to higher ground. Rapidly rising water will likely sweep the vehicle—with its occupants—away quickly. Of course, if you encounter a water-filled dip in the road ahead and you do not know its depth, do not try to ford it. (Many deaths are caused by attempts to move stalled-out vehicles.)

Be exceedingly cautious at all times when you are near high water, especially at night,

*113. If stranded in an open field during an electrical storm, squat on the balls of your feet.*

when it is more difficult to recognize flood dangers. Once you reach a place of safety, stay there. Do not venture out to areas that may be subject to sudden flooding.

If you are forced to evacuate your home or other property because rising floodwaters threaten, certain steps can be taken to help limit the loss of personal belongings. If you determine that you may not have enough time to take these measures before floodwaters reach you, forget about them. The safety of you and your family is the most important consideration of all. Remember that floodwaters usually rise very rapidly. Long before you find yourself in a flood situation, take time to make an itemized list of your personal property. Include on the list all your clothing, furnishings, and valuables. It helps the insurance adjuster to settle claims expeditiously if you are thoughtful enough to take photographs of your belongings and of the interior and exterior of your home. This precaution is worth taking also because you are better equipped in proving uninsured losses, which are tax deductible. Of course, do not leave this itemized list, with photographs, lying around the house somewhere. Put it, along with your insurance policy, in a place of safe keeping, such as a safe-deposit box. At the time you are alerted to flee your property because rising water threatens you, do the following if you have time:

Shut off all your utilities—electricity, gas, and water—by turning the main switches. Avoid contacting any electrical equipment unless you and the device are in a dry area. It is wise to wear rubber boots and gloves when you are dealing with an electrical circuit switch.

Transport valuable belongings—such as jewelry, clothing, papers, and furs—to upper portions of the property or take them to higher elevations.

If your property has a basement with windows, open them to allow equalization of water pressure on the walls and foundations of the structure.

After floodwaters subside and you have had an opportunity to ascertain damages done to your home, or business, you should follow certain steps to aid in recovery efforts. If you sustain losses that are covered by a flood-insurance policy, you should immediately notify the insurance agent who sold the coverage. Before entering your property, you should look for structural damage and, even if you do not suspect the presence of gas, you should avoid using an open flame as a source of light; a battery-operated flashlight will suffice. You should not switch on any lights or electrical appliances. Be sure not to touch any electrical wires that are loose or dangling in water. Opening windows and doors will facilitate the drying-out process. If perishable items are to be discarded, you should make a note of them and, if possible, get photographs before throwing the items away. If water is to be pumped out of a basement, the job should be done in stages to lessen likelihood of structural damage caused by rapid changes in water pressure.

## Withstanding the Chill of Winter

The key to withstanding the chill of a wintry wind is in wearing clothing that is impervious to the wind. In fact, one's ability to withstand cold temperatures depends more upon the maintenance of an adequate amount of air space between skin and clothing than upon the thickness and weight of the clothing. The capacity of clothing to protect the body has been found to be proportional to the amount of "dead air" that exists between the outer layer of clothing and the surface of one's skin. A vigorous wind diminishes this insulating capacity of the dead air by penetrating the clothing and pressing it against the body. A string vest or some other soft, but heavy, garment can be worn to prevent the wind from plastering the inner clothing against the body. Care should be taken, however, when wearing clothing that is impenetrable by the wind; garments so impermeable as to impede the evaporation of perspiration from the body should be avoided.

Of course, the extremities of the body—the hands, ears, face, and feet—should be protected to prevent chapped skin, chilblains, and frostbite. The greater the surface area of the skin exposed to the air, the more rapidly that part of the body loses heat. Possibly the greatest loss of body heat is through the head area. Studies have indicated that, with an air temperature of 5°F (−15°C), about three-fourths of the body's heat is lost from the head. Insulated hats with earmuffs, thick socks, gloves, and possibly a covering over the face are essential at very cold temperatures to stem undue heat loss from the body. Without adequate covering, extremities of the body may quickly turn numb and pale, and in extreme cases, with the body's blood supply cut off, the danger of gangrene becomes a real threat.

An increase in wind speed also means that additional energy is needed to maintain a level of warmth in the home in winter or sufficient cooling in summer. The degree of heat loss in a home during a cold spell is approximately proportional to the square root of the wind speed. This means, for instance, that a poorly insulated home stands to lose up to five times as much heat when winds of 20–30 mph are blowing than if the air outside is still. As with clothing, the key to proper insulation is the amount of "dead air" between the inner wall of the house and the building's exterior. Even the type of material used on the exterior of the home contributes significantly to heating efficiency (e.g., a substance like sandstone, with high porosity, would be preferable to cement or wood).

## Contending with Summer's Stifling Heat
### Heat's Impact on the Human Body

Prolonged, excessive heat like that of the 1980 killer heat wave exacts a gruesome toll by adversely affecting the human body in a number of ways. More than any other organ in the body, the heart is subjected to a great amount of stress during heat waves. When the weather is hot, the heart must work harder, pumping more than the usual amount of blood to tiny capillaries in the outermost layer of skin, through which the blood sheds some of its heat to the atmosphere. Usually, in this way, the temperature of the body can be maintained in the neighborhood of 98.6°F (37°C). However, if the air temperature climbs beyond that level, the blood—and hence the body—cannot give off some of its heat. One of the most basic laws of thermodynamics states that heat energy flows from a body having a higher temperature to one with a lower temperature; the reverse cannot happen. Since the atmosphere is warmer than the blood of the human body during a heat wave, it is prone not to accept the heat that the blood strives to rid itself of. The only other means by which the body can control its temperature is to lose some of its heat through perspiration.

The evaporation of sweat cools the skin and, subsequently, the blood. When the air is relatively dry (say, when the relative humidity is less than 30% or 40%), the atmosphere accepts a great deal of moisture from the skin; as a result, a person feels tolerably well even though the temperature may be as high as 100°F (38°C). But if the humidity is high (say, more than 60%), the moisture-rich atmosphere will extract much less water from the skin. Thus, cooling of the skin and blood is greatly diminished, and an individual feels quite uncomfortable. To compensate for excessive heat, the body's respiration rate may increase to the extent that an abnormal amount of carbon dioxide is lost from the blood and hyperventilation ensues. There are limits of temperature and humidity that the human body can tolerate. When these limits are exceeded for a lengthy period of time or to a great degree, unconsciousness and even death may become real threats.

Sieges of extreme heat pose a host of other problems in large, heavily industrialized metropolitan areas. Air pollution becomes more pronounced when the atmosphere does a poor job of ventilating itself. This is most apparent during summer in the Upper Coast, where air pollution advisories issued by the NWS frequently warn residents with respiratory problems living in Houston and other nearby locales to confine themselves indoors. The large dome of high pressure that often settles over Texas at this time of the year entraps hot and humid air near the ground, so that abnormally heavy concentrations of carbon monoxide, ozone, sulfur dioxide, and other particulate matter accumulate. Those folks most affected by the deadened and polluted air are the elderly, especially those with troubling heart conditions. The poor, unable to afford air conditioning, also suffer to an alarming degree. Even the ability to afford air conditioning may mean little in those situations where severely hot weather strains the abilities of electric utilities to generate enough energy to keep cooling systems in operation. In recent years, some urbanites in Texas have come to know firsthand what life is like during a "brown-out" or a "blackout."

How to Combat the Heat

The hot and humid weather that consistently typifies summer in Texas puts major emphasis on finding ways to provide shade from the sun and to encourage evaporation of excessive moisture. Combatting the debilitating effects of the summer heat is partially accomplished by locating homes in the shade of trees to lessen the impact of solar insolation. In recent years we have gone one step further by using air conditioners, evaporative coolers, and dehumidifiers to create artificially a climate that satisfies our need for comfort. Yet, there are numerous other considerations that can have a pronounced impact on our ability to stay cool and comfortable in the midst of the stifling heat outdoors. The need for shade suggests the elimination of radiation on the east and west walls of the home, for example, and this can be attained in part by constructing houses with marked east-west elongation with small east and west walls. The use of attic and ceiling fans facilitates airflow within a dwelling that promotes cooling by evaporation as well as a feeling of comfort associated with air movement. In addition to the shade they provide, trees—along with shrubs, walls, and fences—can create pressure and suction zones in the wind flow and thereby influence the movement of air over and around low structures, such as houses and small office buildings. Heat-absorbing glass and double panes of glass in windows are effective in soaking up over 40% of the impinging radiant energy, although in winter such energy-conservation fixtures might work against the homeowner who desires to have the increased transmission of insolation. Consequently, shading devices—especially those that are movable—might be preferable. Since light colors reflect more sunlight than dark colors, white window blinds or shades would provide more protection than dark ones. Of more crucial significance, however, is the location of such shading devices; experiments have demonstrated that an average of about 35% in increased protection is afforded when outside, rather than inside, shading devices are used.

There are precautionary measures that, if observed during a wave of torrid temperatures, will help ensure that the health of the human body is not put in jeopardy. These heat-wave safety rules are as follows:

Pay attention to your body's early signals that heat syndrome is imminent; then slow down and get to a cooler environment. Drink an ample amount of fluids, particularly water.

Wear lightweight, light-colored clothing to reflect much of the sun's rays. This includes a covering for the top of the head.

Consume only a modest amount of the foods (like proteins) that accelerate your metabolic heat production and enhance the loss of water from the body.

Avoid sunburn, since ultraviolet radiation burns significantly retard the ability of the skin to shed excess heat.

If you must be out in the heat, limit your exposure time. Try to get inside a cool house, store, restaurant, or theater for short spells several times during the day.

If you discover one or more symptoms of heat syndrome, administer appropriate first aid immediately.

Likely the least threatening symptom is "heat asthenia," a condition characterized by easy fatigue, headaches, a poor appetite, insomnia, heavy sweating, and shallow breathing. Heat cramps are nearly as common as heat asthenia in those individuals who physically exert themselves outdoors in the heat. Ironically, while it is prudent to drink lots of water when the weather is hot and sticky, it may be the intake of large quantities of water without salt that brings on heat cramps.

Caring for the Heat Victim

When a person's skin becomes cold, pale, and clammy with sweat and the body sweats profusely, becomes weak, and sustains heat cramps, very likely *heat exhaustion* has oc-

curred. It is imperative that the victim be
placed in a cooler environment immediately. Though the person may be nauseated
at first, he or she usually can—and should—
take fluids after a period of rest. Certainly,
if the heat exhaustion is severe, medical
help should be sought as quickly as possible. If intense heat and high humidity lead
to the failure of the body's thermoregula-
tory and cardiovascular systems, a *heat
stroke* takes place. The following are among
the indications that the ultimate form of
heat syndrome has occurred: body temperature rises sharply, often to 106°F (41°C) or
more, the pulse is bounding and full, and
blood pressure is elevated; weakness, ver-
tigo, nausea, headache, muscle cramps, and
excessive sweating (although sweating
stops just prior to the onset of the heat
stroke) are felt; and the skin, flushed and
pink at first, becomes ashen and purplish.
Delirium and even coma are common during heat stroke as well. It is emphasized
that, if some or all of these symptoms are
discerned, the victim has incurred a severe
medical disorder and immediate emergency
care by a physician is essential.

# Appendixes

TABLE A-1.
*Extremes in Weather*

| Category | Record | Location | Date |
|---|---|---|---|
| **Temperature** (°F) | | | |
| Coldest | −23° | Tulia | Feb. 12, 1899 |
| | | Seminole | Feb. 8, 1933 |
| Hottest | 120° | Seymour | Aug. 12, 1936 |
| **Rainfall** (inches) | | | |
| Greatest in 24-hour period[a] | 29.05 | Albany | Aug. 4, 1978 |
| Greatest in one month[b] | 35.70 | Alvin | Jul. 1979 |
| Greatest in one year | 109.38 | Clarksville | 1873 |
| Least in one year[c] | 1.64 | Presidio | 1956 |
| Greatest excess in one year | 57.30 | Freeport | 1979 |
| Greatest deficit in one year | 39.10 | Orange | 1896 |
| **Snowfall** (inches) | | | |
| Greatest in 24-hour period | 24.0 | Plainview | Feb. 3–4, 1956 |
| Greatest maximum depth at time of observation | 33.0 | Hale Center | Feb. 5, 1956 |
| | | Vega | Feb. 7, 1956 |
| Greatest in a single storm | 61.0 | Vega | Feb. 1–8, 1956 |
| Greatest in one month | 61.0 | Vega | Feb. 1956 |
| Greatest in one season | 65.0 | Romero | 1923–24 |
| **Wind** (mph)[d] | | | |
| Highest sustained speed | SE 145 | Matagorda | Sep. 11, 1961 |
| | NE 145 | Port Lavaca | Sep. 11, 1961 |
| Highest peak gust[e] | SW 180 | Aransas Pass | Aug. 3, 1970 |
| | WSW 180 | Robstown | Aug. 3, 1970 |

*Source:* Based upon data for the period 1880–1982.

[a] An unofficial 24-hour total of 38.20 inches (and a 12-hour total of 32.0 inches) occurred 2 miles north of Thrall (in Williamson County) on September 9–10, 1921.
[b] An amount of 45 inches, judged by NWS investigators to be of reasonable accuracy, was reported 3 miles northwest of Alvin (in Brazoria County) in July 1979 as a result of Tropical Storm Claudette.

[c] An estimated amount, since some daily data were missing.
[d] These extremes in wind speed were associated with hurricanes (Carla, 1961; Celia, 1970). Higher wind speeds have occurred in tornadoes, but measurements have not been made by the conventional NWS network.
[e] Estimates.

TABLE B-1.
*Average Dates of Last and First Freezes*

| Location | Last Freeze | First Freeze |
|---|---|---|
| Abilene | Mar. 24 | Nov. 13 |
| Alpine | Apr. 5 | Nov. 3 |
| Amarillo | Apr. 14 | Oct. 29 |
| Austin | Mar. 1 | Nov. 30 |
| Beaumont–Port Arthur | Feb. 27 | Nov. 28 |
| Beeville | Feb. 20 | Dec. 2 |
| Big Spring | Mar. 24 | Nov. 13 |
| Brownsville | Jan. 29 | * |
| Brownwood | Mar. 21 | Nov. 16 |
| Childress | Mar. 29 | Nov. 7 |
| Chisos Basin | Mar. 17 | Nov. 14 |
| College Station | Mar. 6 | Nov. 21 |
| Corpus Christi | Feb. 14 | Dec. 8 |
| Corsicana | Mar. 14 | Nov. 18 |
| Dalhart | Apr. 25 | Oct. 22 |
| Dallas–Fort Worth | Mar. 17 | Nov. 20 |
| Del Rio | Feb. 26 | Nov. 30 |
| Eagle Pass | Feb. 22 | Nov. 28 |
| El Paso | Mar. 20 | Nov. 10 |
| Fort Stockton | Mar. 28 | Nov. 10 |
| Galveston | Feb. 4 | * |
| Hereford | Apr. 21 | Oct. 24 |
| Houston (Hobby) | Feb. 15 | Dec. 6 |
| Huntsville | Mar. 5 | Nov. 24 |
| Kerrville | Mar. 29 | Oct. 31 |
| Laredo | Feb. 10 | Dec. 2 |
| Llano | Mar. 23 | Nov. 8 |
| Longview | Mar. 19 | Nov. 10 |
| Lubbock | Apr. 8 | Nov. 1 |
| Lufkin | Mar. 15 | Nov. 9 |
| McAllen | Feb. 2 | * |
| McKinney | Mar. 24 | Nov. 10 |
| Midland–Odessa | Mar. 26 | Nov. 13 |
| Mineral Wells | Mar. 19 | Nov. 13 |
| Mount Locke | Apr. 17 | Oct. 27 |
| Paris | Mar. 21 | Nov. 11 |
| Pecos | Mar. 30 | Nov. 4 |
| Presidio | Mar. 10 | Nov. 14 |
| San Angelo | Mar. 24 | Nov. 13 |
| San Antonio | Mar. 4 | Nov. 24 |
| Sherman | Mar. 21 | Nov. 10 |
| Sonora | Mar. 28 | Nov. 8 |
| Spearman | Apr. 19 | Oct. 23 |
| Temple | Mar. 3 | Nov. 24 |
| Tyler | Mar. 27 | Nov. 8 |
| Uvalde | Mar. 6 | Nov. 17 |
| Van Horn | Mar. 30 | Nov. 11 |
| Victoria | Feb. 14 | Dec. 2 |
| Waco | Mar. 13 | Nov. 24 |
| Wichita Falls | Mar. 29 | Nov. 9 |

*Source:* Based upon data for the period 1951–1980.

* A freeze occurs before January 1 in one of every two years at McAllen, one of every three years at Galveston, and one of every five years at Brownsville; when a freeze does occur before January 1, the average date of occurrence is December 6 for McAllen, December 10 for Galveston, and December 18 for Brownsville.

TABLE B-2.
*Average and Extreme Numbers of Freeze Occurrences*

| Location | Jan. | Feb. | Mar. | Apr. | Oct. | Nov. | Dec. | Total | Most in Any Season No. | Most in Any Season Years |
|---|---|---|---|---|---|---|---|---|---|---|
| Abilene | 17 | 11 | 5 | 0 | 0 | 5 | 14 | 54 | 73 | 1976–77 |
| Alpine | 15 | 13 | 7 | 2 | 1 | 7 | 15 | 60 | 97 | 1979–80 |
| Amarillo | 27 | 22 | 16 | 4 | 2 | 15 | 26 | 111 | 130 | 1979–80 |
| Austin | 9 | 5 | 1 | 0 | 0 | 1 | 5 | 21 | 38 | 1977–78 |
| Beaumont–Port Arthur | 6 | 3 | 1 | 0 | 0 | 1 | 4 | 16 | 24 | 1976–77[a] |
| Beeville | 6 | 3 | 1 | 0 | 0 | 1 | 4 | 16 | 30 | 1959–60 |
| Big Spring | 19 | 12 | 6 | * | 0 | 6 | 16 | 60 | 88 | 1972–73 |
| Brownsville | 1 | 0 | 0 | 0 | 0 | 0 | 0 | 2 | 8 | 1972–73 |
| Brownwood | 17 | 11 | 5 | 0 | 0 | 5 | 14 | 51 | 79 | 1976–77 |
| Childress | 22 | 17 | 9 | 1 | 0 | 9 | 20 | 78 | 92 | 1963–64[a] |
| Chisos Basin | 9 | 6 | 3 | * | * | 4 | 8 | 32 | 52 | 1976–77 |
| College Station | 9 | 5 | 2 | 0 | 0 | 2 | 6 | 23 | 37 | 1976–77[a] |
| Corpus Christi | 3 | 2 | 0 | 0 | 0 | 0 | 1 | 6 | 15 | 1962–63 |
| Corsicana | 16 | 9 | 4 | 0 | 0 | 4 | 11 | 43 | 65 | 1966–67 |
| Dalhart | 29 | 25 | 21 | 7 | 4 | 21 | 29 | 136 | 153 | 1969–70 |
| Dallas–Fort Worth | 15 | 9 | 4 | 0 | 0 | 3 | 10 | 41 | 62 | 1977–78 |
| Del Rio | 7 | 3 | * | 0 | 0 | 1 | 5 | 17 | 28 | 1966–67 |
| Eagle Pass | 9 | 3 | * | 0 | 0 | 2 | 6 | 20 | 32 | 1963–64 |
| El Paso | 17 | 11 | 5 | 0 | 0 | 8 | 18 | 60 | 84 | 1976–77 |
| Fort Stockton | 18 | 13 | 7 | 1 | * | 7 | 17 | 64 | 83 | 1963–64 |
| Galveston | 2 | * | 0 | 0 | 0 | 0 | 0 | 4 | 12 | 1962–63 |
| Hereford | 28 | 24 | 19 | 5 | 3 | 19 | 28 | 127 | 148 | 1972–73 |
| Houston (Hobby) | 5 | 2 | * | 0 | 0 | 1 | 3 | 11 | 42 | 1977–78 |
| Huntsville | 11 | 6 | 2 | 0 | 0 | 2 | 6 | 27 | 52 | 1977–78 |
| Kerrville | 19 | 13 | 6 | 1 | 1 | 8 | 17 | 65 | 85 | 1963–64 |
| Laredo | 3 | 2 | 0 | 0 | 0 | * | 1 | 7 | 16 | 1977–78 |
| Llano | 17 | 11 | 4 | 0 | 0 | 6 | 16 | 56 | 74 | 1976–77 |
| Longview | 14 | 10 | 4 | 0 | 0 | 4 | 11 | 44 | 64 | 1959–60 |
| Lubbock | 26 | 20 | 11 | 2 | 1 | 12 | 24 | 96 | 110 | 1963–64[a] |
| Lufkin | 12 | 8 | 3 | 0 | 0 | 4 | 10 | 37 | 63 | 1977–78 |
| McAllen | 2 | * | 0 | 0 | 0 | 0 | * | 3 | 8 | 1972–73[a] |
| McKinney | 16 | 10 | 5 | 0 | 0 | 5 | 13 | 49 | 67 | 1977–78[a] |
| Midland–Odessa | 19 | 13 | 6 | * | 0 | 7 | 17 | 62 | 76 | 1968–69[a] |
| Mineral Wells | 17 | 10 | 4 | 0 | 0 | 5 | 13 | 50 | 67 | 1976–77 |
| Mount Locke | 15 | 12 | 9 | 3 | 1 | 9 | 14 | 63 | 82 | 1963–64 |
| Paris | 19 | 13 | 5 | 0 | 0 | 6 | 15 | 58 | 77 | 1959–60 |
| Pecos | 23 | 17 | 8 | * | * | 10 | 22 | 82 | 109 | 1959–60 |
| Presidio | 15 | 8 | 2 | 0 | 0 | 4 | 14 | 44 | 57 | 1955–56 |
| San Angelo | 16 | 10 | 4 | 0 | 0 | 5 | 13 | 50 | 68 | 1963–64 |
| San Antonio | 9 | 5 | 2 | 0 | 0 | 2 | 6 | 23 | 40 | 1966–67 |
| Sherman | 19 | 13 | 5 | 0 | 0 | 6 | 15 | 58 | 83 | 1977–78 |
| Sonora | 17 | 11 | 5 | * | 0 | 6 | 14 | 55 | 75 | 1963–64 |
| Spearman | 28 | 23 | 17 | 5 | 3 | 18 | 27 | 121 | 141 | 1955–56 |
| Temple | 12 | 7 | 2 | 0 | 0 | 2 | 7 | 30 | 47 | 1977–78 |
| Tyler | 16 | 11 | 5 | 0 | 0 | 6 | 14 | 53 | 83 | 1976–77 |
| Uvalde | 11 | 6 | 2 | 0 | 0 | 3 | 8 | 30 | 56 | 1963–64 |
| Van Horn | 19 | 14 | 7 | 1 | 0 | 8 | 19 | 68 | 92 | 1954–55 |
| Victoria | 5 | 2 | 0 | 0 | 0 | 1 | 3 | 11 | 21 | 1978–79 |
| Waco | 13 | 7 | 2 | 0 | 0 | 3 | 8 | 33 | 50 | 1976–77 |
| Wichita Falls | 21 | 14 | 7 | 1 | 0 | 7 | 17 | 67 | 89 | 1970–71 |

*Source:* Based upon data for the period 1951–1980.

* Less than one occurrence (but more than 0.5); in some instances the sum of the monthly amounts does not equal the total because of rounding.
[a] Also on earlier dates.

TABLE B-3.
*Greatest Number of 32° and Below Days during Most Severe Cold Winters*

| Location | Season | Total | Consecutive No. | Dates |
|----------|--------|-------|-----------------|-------|
| Amarillo | 1979–80 | 130 | 31 | Jan. 19–Feb. 18 |
|          | 1976–77 | 129 | 57 | Nov. 26–Jan. 21 |
|          | 1963–64 | 128 | 41 | Nov. 22–Jan. 1 |
|          | 1955–56 | 128 | 23 | Jan. 29–Feb. 20 |
| Brownsville | 1950 | 8 | 6 | Jan. 29–Feb. 3 |
|          | 1972–73 | 8 | 3 | Feb. 9–11[a] |
|          | 1962–63 | 7 | 2 | Jan. 23–24[a] |
| Dallas–Fort Worth | 1977–78 | 62 | 18 | Jan. 25–Feb. 11 |
|          | 1963–64 | 61 | 9 | Jan. 9–18 |
|          | 1959–60 | 59 | 8 | Jan. 18–25 |
|          | 1964–65 | 58 | 8 | Jan. 27–Feb. 3 |
| El Paso | 1976–77 | 84 | 13 | Nov. 27–Dec. 9 |
|          | 1960–61 | 78 | 25 | Dec. 5–29 |
| Houston | 1978–79 | 23 | 3 | Feb. 8–10 |
|          | 1977–78 | 20 | 4 | Feb. 19–22 |
|          | 1962–63 | 19 | 5 | Jan. 12–16 |
| San Antonio | 1966–67 | 40 | 5 | Dec. 10–14 |
|          | 1977–78 | 39 | 6 | Feb. 17–22 |
|          | 1976–77 | 36 | 5 | Jan. 17–21 |

*Source:* Based upon data for the period 1951–1980.
[a] Also on earlier dates.

TABLE B-4.
*Coldest Temperatures, 1960–1982*

| Year | °F | Location | County | Region | Date |
|------|------|----------|--------|--------|------|
| 1960 | −9 | Dalhart | Dallam | High Plains | Jan. 19 |
| 1961 | −12 | Dalhart | Dallam | High Plains | Dec. 12 |
| 1962 | −14 | Dalhart | Dallam | High Plains | Jan. 10 |
|  | −14 | Red Bluff Dam | Reeves | Trans Pecos | Jan. 11 |
|  | −14 | Wink | Winkler | Trans Pecos | Jan. 11 |
| 1963 | −18 | Dalhart | Dallam | High Plains | Jan. 13 |
| 1964 | −5 | Spearman | Hansford | High Plains | Jan. 13 |
|  | −5 | Tahoka | Lynn | High Plains | Jan. 13 |
| 1965 | 0 | Bravo | Hartley | High Plains | Feb. 24 |
| 1966 | −12 | Lipscomb | Lipscomb | High Plains | Jan. 23 |
| 1967 | −1 | Lipscomb | Lipscomb | High Plains | Mar. 8 |
| 1968 | −3 | Gruver | Hansford | High Plains | Jan. 7 |
| 1969 | −1 | Perryton | Ochiltree | High Plains | Jan. 1 |
| 1970 | −3 | Vega | Oldham | High Plains | Jan. 6 |
| 1971 | −12 | Sunray | Moore | High Plains | Jan. 6 |
| 1972 | −12 | Perryton | Ochiltree | High Plains | Dec. 6 |
| 1973 | −10 | Lipscomb | Lipscomb | High Plains | Jan. 12 |
| 1974 | −18 | Lipscomb | Lipscomb | High Plains | Jan. 4, 5 |
| 1975 | −7 | Dalhart | Dallam | High Plains | Jan. 12 |
|  | −7 | Perryton | Ochiltree | High Plains | Jan. 12 |
| 1976 | −11 | Lipscomb | Lipscomb | High Plains | Jan. 8 |
| 1977 | −8 | Lipscomb | Lipscomb | High Plains | Jan. 10, 12 |
| 1978 | −10 | Bootleg Corner | Deaf Smith | High Plains | Dec. 9 |
| 1979 | −12 | Bravo | Hartley | High Plains | Jan. 2 |
|  | −12 | Lipscomb | Lipscomb | High Plains | Jan. 2 |
| 1980 | −2 | Boys Ranch | Oldham | High Plains | Nov. 26 |
|  | −2 | Lipscomb | Lipscomb | High Plains | Mar. 2 |
| 1981 | −10 | Gruver | Hansford | High Plains | Feb. 11 |
|  | −10 | Vega | Oldham | High Plains | Feb. 11 |
| 1982 | −15 | Perryton | Ochiltree | High Plains | Feb. 6 |

TABLE B-5.
*Average Monthly Low Temperatures (°F)*

| Location | Jan. | Feb. | Mar. | Apr. | May | Jun. | Jul. | Aug. | Sep. | Oct. | Nov. | Dec. | Annual |
|---|---|---|---|---|---|---|---|---|---|---|---|---|---|
| Abilene | 31.2 | 35.5 | 42.6 | 52.8 | 60.8 | 69.0 | 72.7 | 71.7 | 64.9 | 54.1 | 42.0 | 34.3 | 52.6 |
| Alice | 43.0 | 46.3 | 53.1 | 61.9 | 67.8 | 72.4 | 73.8 | 73.4 | 70.2 | 61.2 | 52.4 | 45.8 | 60.1 |
| Alpine | 32.0 | 34.3 | 39.5 | 47.4 | 54.8 | 62.6 | 64.1 | 62.8 | 57.9 | 48.5 | 38.1 | 33.3 | 47.9 |
| Amarillo | 21.7 | 26.1 | 32.0 | 42.0 | 51.9 | 61.5 | 66.2 | 64.5 | 56.9 | 45.5 | 32.1 | 24.8 | 43.8 |
| Angleton | 41.9 | 44.4 | 50.8 | 59.2 | 65.2 | 70.4 | 72.3 | 71.8 | 68.4 | 58.0 | 50.0 | 44.2 | 58.1 |
| Austin | 38.8 | 42.2 | 49.3 | 58.3 | 65.1 | 71.5 | 73.9 | 73.7 | 69.1 | 58.7 | 48.1 | 41.4 | 57.5 |
| Ballinger | 29.3 | 33.3 | 40.8 | 51.7 | 60.1 | 67.6 | 70.2 | 69.1 | 63.0 | 51.2 | 39.5 | 32.0 | 50.7 |
| Bay City | 42.4 | 45.0 | 51.6 | 60.5 | 66.8 | 71.9 | 74.0 | 73.1 | 69.3 | 59.8 | 51.1 | 44.7 | 59.2 |
| Beaumont– Port Arthur | 42.1 | 44.3 | 51.0 | 59.4 | 66.1 | 71.8 | 73.7 | 73.3 | 69.8 | 58.9 | 49.8 | 44.4 | 58.7 |
| Beeville | 41.4 | 44.3 | 51.5 | 60.1 | 66.0 | 70.9 | 72.5 | 72.4 | 68.9 | 59.5 | 50.5 | 44.0 | 58.5 |
| Big Spring | 29.7 | 33.6 | 40.6 | 50.5 | 59.4 | 67.4 | 71.0 | 69.7 | 63.0 | 52.4 | 40.0 | 32.9 | 50.9 |
| Blanco | 33.3 | 36.7 | 43.4 | 53.2 | 60.5 | 67.6 | 69.8 | 69.0 | 64.0 | 53.2 | 42.4 | 35.8 | 52.4 |
| Bonham | 31.4 | 35.4 | 42.5 | 52.4 | 60.3 | 68.0 | 71.9 | 70.6 | 63.8 | 52.7 | 41.8 | 34.6 | 52.1 |
| Borger | 23.6 | 27.9 | 33.8 | 43.9 | 53.2 | 62.4 | 67.1 | 65.4 | 57.9 | 46.8 | 34.0 | 27.0 | 45.3 |
| Brady | 30.4 | 34.1 | 40.9 | 51.2 | 59.1 | 66.4 | 69.6 | 68.5 | 63.0 | 51.9 | 40.1 | 33.2 | 50.7 |
| Brenham | 38.5 | 41.4 | 48.1 | 57.2 | 64.2 | 70.1 | 72.6 | 72.2 | 67.7 | 57.3 | 47.5 | 41.3 | 56.5 |
| Brownsville | 50.8 | 53.0 | 59.5 | 66.6 | 71.3 | 74.7 | 75.6 | 75.4 | 73.1 | 66.1 | 58.3 | 52.6 | 64.8 |
| Brownwood | 31.4 | 35.2 | 42.9 | 53.6 | 61.5 | 69.1 | 72.5 | 71.3 | 65.1 | 53.9 | 41.8 | 34.1 | 52.7 |
| Cameron | 38.4 | 41.5 | 48.3 | 57.0 | 63.9 | 70.1 | 72.8 | 72.3 | 67.5 | 57.3 | 47.2 | 40.8 | 56.4 |
| Canyon | 22.7 | 26.8 | 32.7 | 42.4 | 51.6 | 60.8 | 65.2 | 63.6 | 56.7 | 44.7 | 32.6 | 25.8 | 43.8 |
| Center | 34.0 | 36.5 | 43.3 | 52.6 | 60.4 | 67.1 | 70.0 | 68.8 | 63.6 | 51.2 | 41.6 | 35.6 | 52.1 |
| Childress | 26.2 | 30.4 | 37.3 | 48.3 | 57.7 | 66.9 | 71.0 | 69.4 | 61.8 | 50.1 | 37.5 | 29.7 | 48.9 |
| Chisos Basin | 37.0 | 38.7 | 43.9 | 51.6 | 57.6 | 62.7 | 63.3 | 62.1 | 58.2 | 51.2 | 42.4 | 37.9 | 50.6 |
| Clarksville | 30.1 | 34.0 | 41.2 | 51.0 | 59.5 | 67.0 | 70.5 | 69.1 | 62.9 | 50.9 | 40.4 | 32.9 | 50.8 |
| Cleburne | 33.4 | 37.1 | 44.3 | 54.0 | 61.5 | 68.7 | 72.1 | 71.3 | 65.6 | 54.4 | 43.6 | 36.4 | 53.5 |
| College Station | 39.3 | 42.5 | 49.1 | 57.9 | 64.8 | 71.0 | 73.5 | 73.0 | 68.5 | 57.9 | 47.9 | 41.6 | 57.3 |
| Conroe | 38.3 | 40.9 | 47.9 | 56.9 | 63.7 | 69.6 | 71.9 | 71.4 | 67.2 | 55.9 | 46.5 | 40.4 | 55.9 |
| Corpus Christi | 46.1 | 48.7 | 55.7 | 63.9 | 69.5 | 74.1 | 75.6 | 75.8 | 72.8 | 64.1 | 54.9 | 48.8 | 62.5 |
| Corsicana | 33.6 | 37.1 | 44.0 | 54.0 | 61.9 | 69.4 | 73.0 | 72.1 | 66.1 | 54.6 | 43.3 | 36.5 | 53.8 |
| Crockett | 34.8 | 37.5 | 44.2 | 53.5 | 61.0 | 67.6 | 70.6 | 69.8 | 65.2 | 53.6 | 43.3 | 36.9 | 53.2 |
| Dalhart | 18.4 | 22.8 | 28.3 | 38.7 | 48.8 | 58.9 | 63.8 | 62.0 | 53.6 | 41.2 | 28.2 | 21.3 | 40.5 |
| Dallas– Fort Worth | 33.9 | 37.8 | 44.9 | 55.0 | 62.9 | 70.8 | 74.7 | 73.7 | 67.5 | 56.3 | 44.9 | 37.4 | 55.0 |
| Del Rio | 38.3 | 42.5 | 50.2 | 59.4 | 66.1 | 72.1 | 74.2 | 73.6 | 68.9 | 59.2 | 47.5 | 40.1 | 57.7 |
| Denton | 31.8 | 35.9 | 43.2 | 53.4 | 61.1 | 69.1 | 73.0 | 71.8 | 65.3 | 54.5 | 43.0 | 35.5 | 53.1 |
| Eagle Pass | 38.1 | 42.5 | 50.1 | 59.8 | 66.5 | 72.3 | 74.4 | 73.7 | 69.3 | 59.2 | 47.4 | 39.8 | 57.8 |
| El Paso | 30.4 | 34.1 | 40.5 | 48.5 | 56.6 | 65.7 | 69.6 | 67.5 | 60.6 | 48.7 | 37.0 | 30.6 | 49.2 |
| Fort Stockton | 30.2 | 33.5 | 39.0 | 48.2 | 57.0 | 65.2 | 67.9 | 66.2 | 60.5 | 49.7 | 38.2 | 32.1 | 49.0 |
| Fredericksburg | 35.5 | 38.8 | 45.7 | 54.6 | 61.1 | 67.2 | 69.1 | 68.4 | 64.1 | 54.6 | 44.2 | 37.9 | 53.4 |
| Gainesville | 28.2 | 32.3 | 39.9 | 50.7 | 59.2 | 68.0 | 71.9 | 70.6 | 63.4 | 51.5 | 39.6 | 31.7 | 50.6 |
| Galveston | 47.9 | 50.2 | 56.5 | 64.9 | 71.6 | 77.2 | 79.1 | 78.8 | 75.4 | 67.7 | 57.6 | 51.2 | 64.8 |
| Graham | 28.0 | 32.3 | 40.0 | 50.9 | 59.5 | 68.2 | 72.0 | 70.6 | 63.4 | 51.0 | 39.1 | 30.9 | 50.5 |
| Harlingen | 48.0 | 50.4 | 57.1 | 64.6 | 69.3 | 72.5 | 73.4 | 73.3 | 70.9 | 63.3 | 55.8 | 50.2 | 62.4 |
| Hereford | 20.5 | 24.0 | 29.5 | 39.6 | 49.5 | 59.1 | 63.4 | 61.5 | 53.9 | 42.1 | 29.9 | 23.3 | 41.4 |
| Hillsboro | 32.8 | 36.7 | 44.0 | 54.3 | 62.1 | 69.6 | 73.3 | 72.4 | 66.4 | 54.9 | 43.2 | 36.2 | 53.8 |
| Houston | 40.8 | 43.2 | 49.8 | 58.3 | 64.7 | 70.2 | 72.5 | 72.1 | 68.1 | 57.5 | 48.6 | 42.7 | 57.4 |
| Huntsville | 38.3 | 41.1 | 48.1 | 57.2 | 63.9 | 69.8 | 72.1 | 71.4 | 66.6 | 56.8 | 47.2 | 40.8 | 56.1 |
| Junction | 31.7 | 35.4 | 43.2 | 52.7 | 59.9 | 67.0 | 69.1 | 68.0 | 62.6 | 51.5 | 40.2 | 33.2 | 51.2 |
| Kingsville | 45.6 | 48.0 | 55.1 | 62.8 | 68.4 | 72.7 | 74.1 | 73.8 | 70.4 | 61.8 | 53.4 | 47.6 | 61.1 |
| Laredo | 44.5 | 48.3 | 56.1 | 64.2 | 69.6 | 73.9 | 75.5 | 75.0 | 71.6 | 63.4 | 53.1 | 46.6 | 61.8 |
| Levelland | 23.4 | 26.6 | 32.6 | 43.0 | 52.7 | 61.7 | 64.7 | 63.1 | 56.3 | 44.4 | 32.6 | 26.3 | 44.0 |

| Location | Jan. | Feb. | Mar. | Apr. | May | Jun. | Jul. | Aug. | Sep. | Oct. | Nov. | Dec. | Annual |
|---|---|---|---|---|---|---|---|---|---|---|---|---|---|
| Liberty | 39.8 | 42.3 | 49.2 | 57.7 | 64.1 | 69.5 | 71.8 | 71.2 | 67.0 | 55.7 | 47.3 | 41.5 | 56.4 |
| Llano | 31.9 | 35.8 | 43.8 | 54.6 | 62.2 | 69.6 | 72.1 | 70.8 | 65.1 | 53.4 | 41.8 | 33.9 | 52.9 |
| Lubbock | 24.3 | 27.9 | 35.2 | 45.8 | 55.2 | 64.3 | 67.6 | 65.7 | 58.7 | 47.3 | 34.8 | 27.4 | 46.2 |
| Lufkin | 38.0 | 40.5 | 47.5 | 56.5 | 63.8 | 70.0 | 72.8 | 71.9 | 67.1 | 55.4 | 45.4 | 39.5 | 55.7 |
| Marshall | 33.3 | 36.7 | 43.4 | 53.0 | 60.8 | 67.9 | 71.5 | 70.4 | 64.8 | 52.5 | 42.3 | 35.6 | 52.7 |
| McAllen | 47.6 | 50.1 | 57.1 | 64.6 | 69.4 | 72.7 | 73.6 | 73.8 | 71.1 | 63.5 | 55.3 | 49.5 | 62.4 |
| McKinney | 32.6 | 36.6 | 43.8 | 53.6 | 61.4 | 69.0 | 72.6 | 71.3 | 65.2 | 54.3 | 43.2 | 35.8 | 53.3 |
| Midland–Odessa | 29.7 | 33.3 | 40.2 | 49.4 | 58.2 | 66.6 | 69.2 | 68.0 | 61.9 | 51.1 | 39.0 | 32.2 | 49.9 |
| Mineral Wells | 32.1 | 35.9 | 43.5 | 53.6 | 61.4 | 69.2 | 72.9 | 71.9 | 65.5 | 54.4 | 42.6 | 35.1 | 53.2 |
| Mount Locke | 32.0 | 33.0 | 37.5 | 45.1 | 51.7 | 57.8 | 58.6 | 57.7 | 54.2 | 47.3 | 37.7 | 33.2 | 45.5 |
| New Braunfels | 37.8 | 41.5 | 48.1 | 57.1 | 64.1 | 70.7 | 72.7 | 72.2 | 68.0 | 57.6 | 47.1 | 40.2 | 56.4 |
| Palestine | 35.1 | 38.2 | 45.5 | 55.0 | 62.2 | 68.8 | 71.4 | 70.5 | 65.4 | 55.1 | 44.7 | 37.9 | 54.2 |
| Paris | 30.2 | 34.3 | 41.8 | 52.5 | 61.0 | 68.8 | 72.8 | 71.4 | 64.7 | 53.2 | 41.8 | 33.8 | 52.2 |
| Pecos | 27.4 | 30.9 | 37.7 | 47.7 | 56.5 | 66.4 | 69.4 | 67.3 | 61.1 | 48.5 | 35.8 | 28.9 | 48.1 |
| Plainview | 22.8 | 26.4 | 32.5 | 42.9 | 52.5 | 62.0 | 65.7 | 63.8 | 56.8 | 45.1 | 33.1 | 26.2 | 44.2 |
| Port Lavaca | 44.1 | 47.6 | 54.4 | 62.7 | 68.9 | 74.1 | 75.9 | 75.5 | 71.9 | 63.2 | 53.9 | 47.0 | 61.6 |
| Presidio | 33.1 | 36.7 | 43.8 | 53.0 | 61.6 | 70.9 | 72.8 | 70.9 | 65.6 | 54.1 | 41.0 | 33.8 | 53.1 |
| Rio Grande City | 43.5 | 46.5 | 54.3 | 63.4 | 68.9 | 72.8 | 73.8 | 73.6 | 70.1 | 61.1 | 51.9 | 45.6 | 60.5 |
| San Angelo | 32.2 | 36.1 | 43.4 | 53.4 | 61.5 | 69.3 | 72.0 | 71.2 | 65.0 | 54.0 | 42.0 | 34.7 | 52.9 |
| San Antonio | 39.0 | 42.4 | 49.8 | 58.8 | 65.5 | 72.0 | 74.3 | 73.7 | 69.4 | 58.9 | 48.2 | 41.4 | 57.8 |
| San Marcos | 35.9 | 39.3 | 46.0 | 55.6 | 62.9 | 69.3 | 71.3 | 70.8 | 66.4 | 55.7 | 44.9 | 38.0 | 54.7 |
| Seminole | 25.3 | 28.5 | 35.0 | 44.8 | 53.4 | 62.3 | 65.3 | 63.8 | 57.6 | 46.1 | 34.3 | 27.6 | 45.3 |
| Sherman | 29.7 | 34.1 | 41.6 | 52.2 | 60.3 | 68.6 | 72.6 | 71.2 | 64.4 | 52.8 | 41.5 | 33.5 | 51.9 |
| Snyder | 25.8 | 29.9 | 36.7 | 47.9 | 57.1 | 65.7 | 69.0 | 67.5 | 60.5 | 48.8 | 36.4 | 28.9 | 47.9 |
| Spearman | 19.5 | 24.2 | 30.7 | 41.4 | 51.2 | 60.9 | 65.5 | 63.6 | 55.8 | 43.4 | 30.4 | 23.0 | 42.5 |
| Sulphur Springs | 30.7 | 34.6 | 41.3 | 50.9 | 59.5 | 67.1 | 70.7 | 69.3 | 62.9 | 51.3 | 40.7 | 34.1 | 51.1 |
| Temple | 35.7 | 39.0 | 45.9 | 55.7 | 63.3 | 70.2 | 73.5 | 73.0 | 67.8 | 57.2 | 45.7 | 38.4 | 55.5 |
| Uvalde | 37.0 | 40.9 | 48.4 | 57.6 | 64.4 | 70.0 | 71.2 | 70.3 | 66.7 | 57.1 | 46.4 | 39.2 | 55.8 |
| Victoria | 43.1 | 45.9 | 52.8 | 61.5 | 67.7 | 73.1 | 75.2 | 74.7 | 70.9 | 61.0 | 51.5 | 45.4 | 60.2 |
| Waco | 35.7 | 39.4 | 46.6 | 56.5 | 64.2 | 71.5 | 75.2 | 74.5 | 68.6 | 57.2 | 46.1 | 38.5 | 56.2 |
| Waxahachie | 33.1 | 36.8 | 43.8 | 53.9 | 61.9 | 69.1 | 72.9 | 71.6 | 65.6 | 54.6 | 43.4 | 36.4 | 53.6 |
| Weatherford | 29.9 | 33.7 | 40.6 | 51.2 | 59.4 | 67.6 | 71.7 | 70.4 | 63.8 | 52.3 | 40.6 | 33.2 | 51.2 |
| Weslaco | 49.7 | 52.2 | 59.1 | 66.0 | 70.2 | 73.4 | 74.1 | 74.1 | 71.7 | 64.8 | 57.3 | 51.8 | 63.7 |
| Wichita Falls | 28.2 | 32.6 | 39.9 | 50.6 | 59.4 | 68.2 | 72.5 | 71.2 | 63.7 | 52.0 | 39.6 | 31.6 | 50.8 |

*Source:* Based upon data for the period 1951–1980.

TABLE B-6.
*Temperature Extremes*

| Location | Coldest Ever Observed °F | Date | Hottest Ever Observed °F | Date |
|---|---|---|---|---|
| Abilene | − 9 | Jan. 4, 1947 | 111 | Aug. 3, 1943 |
| Alpine | − 2 | Feb. 8, 1933 | 106 | Jul. 20, 1936[a] |
| Amarillo | −16 | Feb. 12, 1899 | 108 | Jun. 24, 1953 |
| Austin | − 2 | Jan. 31, 1949 | 109 | Jul. 26, 1954 |
| Beaumont–Port Arthur | 11 | Jan. 18, 1930 | 107 | Aug. 10, 1962 |
| Beeville | 5 | Feb. 12, 1899 | 111 | Jul. 9, 1939 |
| Big Spring | − 7 | Feb. 8, 1933 | 110 | Jun. 20, 1951 |
| Brownsville | 12 | Feb. 13, 1899 | 104 | Sep. 20, 1947 |
| Brownwood | − 2 | Jan. 23, 1940 | 111 | Aug. 7, 1964 |
| Childress | − 7 | Jan. 4, 1947 | 115 | Aug. 2, 1944[a] |
| Chisos Basin | − 3 | Jan. 30, 1949 | 103 | Jun. 28, 1972 |
| College Station | − 3 | Jan. 31, 1949 | 109 | Jul. 12, 1954 |
| Corpus Christi | 11 | Feb. 12, 1899 | 105 | Jul. 24, 1934 |
| Corsicana | − 7 | Feb. 12, 1899 | 113 | Jul. 26, 1954 |
| Dalhart | −21 | Jan. 4, 1959 | 110 | Sep. 7, 1907 |
| Dallas–Fort Worth | − 8 | Feb. 12, 1899 | 113 | Jun. 27, 1980[a] |
| Del Rio | 11 | Feb. 2, 1951 | 111 | Jul. 30, 1960 |
| Eagle Pass | 7 | Feb. 12, 1899 | 115 | Jul. 25, 1944[a] |
| El Paso | − 8 | Jan. 11, 1962 | 112 | Jul. 10, 1979 |
| Fort Stockton | − 7 | Jan. 3, 1911 | 114 | Jun. 17, 1924 |
| Galveston | 8 | Feb. 12, 1899 | 101 | Jul. 16, 1932 |
| Hereford | −17 | Feb. 1, 1951 | 111 | Jun. 8, 1910 |
| Houston | 5 | Jan. 23, 1940 | 107 | Aug. 23, 1980 |
| Huntsville | − 2 | Feb. 13, 1899[a] | 107 | Aug. 23, 1980[a] |
| Kerrville | − 7 | Jan. 31, 1949 | 110 | Jul. 27, 1954 |
| Laredo | 16 | Dec. 21, 1973 | 115 | Jun. 11, 1942 |
| Llano | − 7 | Dec. 22, 1929 | 115 | Jul. 14, 1933 |
| Longview | − 7 | Feb. 12, 1899 | 113 | Aug. 10, 1936 |
| Lubbock | −17 | Feb. 8, 1933 | 109 | Jul. 10, 1940 |
| Lufkin | − 2 | Feb. 2, 1951[a] | 110 | Aug. 19, 1909 |
| McAllen | 13 | Jan. 12, 1962 | 107 | Jun. 27, 1980 |
| McKinney | − 7 | Jan. 18, 1930 | 118 | Aug. 10, 1936 |
| Midland–Odessa | −11 | Feb. 8, 1933 | 109 | Jun. 20, 1951 |
| Mineral Wells | 3 | Feb. 2, 1951 | 114 | Jun. 28, 1980[a] |
| Mount Locke | −10 | Jan. 11, 1962 | 104 | Aug. 17, 1969 |
| Paris | −13 | Feb. 12, 1899 | 115 | Aug. 10, 1936 |
| Pecos | − 9 | Jan. 11, 1962 | 118 | Jun. 29, 1968 |
| Presidio | 4 | Jan. 12, 1962 | 117 | Jun. 18, 1960[a] |
| San Angelo | 1 | Feb. 2, 1951[a] | 111 | Jul. 29, 1960 |
| San Antonio | 0 | Jan. 31, 1949 | 107 | Aug. 20, 1909 |
| Sherman | − 2 | Jan. 31, 1949 | 113 | Aug. 10, 1936 |
| Sonora | − 8 | Feb. 2, 1951 | 109 | Jun. 28, 1980 |
| Spearman | −22 | Jan. 4, 1959 | 111 | Aug. 13, 1936 |
| Temple | − 4 | Feb. 12, 1899 | 112 | Aug. 11, 1947 |
| Tyler | − 8 | Feb. 12, 1899 | 108 | Jul. 25, 1954 |
| Uvalde | 6 | Feb. 2, 1951 | 114 | Jun. 9, 1910 |
| Van Horn | − 7 | Jan. 11, 1962 | 112 | Jun. 25, 1969 |
| Victoria | 9 | Jan. 18, 1930 | 110 | Jul. 9, 1939[a] |
| Waco | − 5 | Jan. 31, 1949 | 112 | Aug. 11, 1969 |
| Wichita Falls | −12 | Jan. 4, 1947 | 117 | Jun. 28, 1980 |

*Source:* Based upon data for periods that vary widely; some began as early as 1878, some as late as 1927.

[a] Also on earlier dates.

TABLE B-7.
*Average Monthly High Temperatures (°F)*

| Location | Jan. | Feb. | Mar. | Apr. | May | Jun. | Jul. | Aug. | Sep. | Oct. | Nov. | Dec. | Annual |
|---|---|---|---|---|---|---|---|---|---|---|---|---|---|
| Abilene | 55.5 | 60.3 | 68.6 | 77.6 | 84.1 | 91.8 | 95.4 | 94.5 | 87.1 | 77.6 | 64.8 | 58.4 | 76.3 |
| Alice | 67.2 | 71.1 | 78.2 | 84.7 | 88.7 | 93.8 | 96.6 | 97.0 | 92.2 | 85.4 | 76.2 | 69.5 | 83.4 |
| Alpine | 60.6 | 63.6 | 70.4 | 78.4 | 85.0 | 90.8 | 89.5 | 88.3 | 83.5 | 77.3 | 67.3 | 61.5 | 76.4 |
| Amarillo | 49.1 | 53.1 | 60.8 | 71.0 | 79.1 | 88.2 | 91.4 | 89.6 | 82.4 | 72.7 | 58.7 | 51.8 | 70.7 |
| Angleton | 63.2 | 65.7 | 71.7 | 77.8 | 83.6 | 89.1 | 91.9 | 91.5 | 88.1 | 81.7 | 72.6 | 66.6 | 78.6 |
| Austin | 59.4 | 64.1 | 71.7 | 79.0 | 84.7 | 91.6 | 95.4 | 95.3 | 89.3 | 80.8 | 69.2 | 62.8 | 78.6 |
| Ballinger | 59.0 | 63.9 | 72.1 | 80.8 | 86.2 | 93.3 | 96.4 | 95.7 | 88.8 | 80.0 | 67.4 | 61.6 | 78.8 |
| Bay City | 63.4 | 66.0 | 72.2 | 79.0 | 84.8 | 90.2 | 92.8 | 93.0 | 89.5 | 83.2 | 73.1 | 66.6 | 79.5 |
| Beaumont–<br>Port Arthur | 61.7 | 65.4 | 71.8 | 78.5 | 85.0 | 90.5 | 92.5 | 92.2 | 88.6 | 81.5 | 71.4 | 65.0 | 78.7 |
| Beeville | 64.5 | 67.9 | 75.3 | 81.6 | 86.3 | 91.8 | 95.2 | 95.4 | 90.4 | 83.4 | 73.7 | 67.4 | 81.1 |
| Big Spring | 57.1 | 61.3 | 69.8 | 79.0 | 85.7 | 93.0 | 94.6 | 93.6 | 86.9 | 78.3 | 65.8 | 59.7 | 77.1 |
| Blanco | 59.5 | 63.2 | 70.9 | 78.5 | 84.1 | 91.3 | 95.2 | 95.1 | 88.9 | 80.5 | 68.8 | 62.6 | 78.2 |
| Bonham | 53.6 | 58.7 | 66.9 | 75.8 | 82.7 | 90.1 | 95.2 | 95.2 | 87.8 | 77.9 | 65.3 | 57.0 | 75.5 |
| Borger | 51.2 | 56.1 | 63.9 | 73.7 | 81.3 | 90.0 | 93.3 | 91.5 | 84.7 | 75.4 | 60.9 | 53.8 | 73.0 |
| Brady | 58.4 | 62.2 | 70.4 | 79.0 | 84.2 | 91.7 | 95.7 | 95.1 | 88.1 | 79.5 | 67.4 | 60.9 | 77.7 |
| Brenham | 60.0 | 64.1 | 71.9 | 79.4 | 85.9 | 92.5 | 96.4 | 96.6 | 90.7 | 82.7 | 71.1 | 63.6 | 79.6 |
| Brownsville | 69.7 | 72.5 | 77.5 | 83.2 | 87.0 | 90.5 | 92.6 | 92.8 | 89.8 | 84.4 | 77.0 | 71.9 | 82.4 |
| Brownwood | 56.6 | 61.6 | 69.7 | 79.0 | 85.1 | 92.9 | 96.9 | 96.4 | 89.2 | 79.6 | 67.0 | 60.2 | 77.9 |
| Cameron | 60.6 | 65.2 | 72.9 | 79.9 | 85.3 | 92.0 | 96.2 | 96.4 | 90.1 | 81.8 | 70.5 | 64.0 | 79.6 |
| Canyon | 53.7 | 57.8 | 65.1 | 74.2 | 81.6 | 89.8 | 92.0 | 90.2 | 84.3 | 75.7 | 62.4 | 55.9 | 73.6 |
| Center | 57.2 | 61.9 | 69.3 | 77.4 | 83.9 | 90.3 | 94.4 | 94.4 | 88.9 | 80.6 | 68.4 | 60.9 | 77.3 |
| Childress | 51.8 | 56.9 | 65.0 | 75.5 | 82.9 | 91.7 | 95.6 | 94.3 | 86.2 | 76.3 | 62.7 | 54.9 | 74.5 |
| Chisos Basin | 59.6 | 62.6 | 69.6 | 77.4 | 83.5 | 87.7 | 85.9 | 84.5 | 80.2 | 74.5 | 65.6 | 60.2 | 74.3 |
| Clarksville | 52.8 | 57.6 | 65.4 | 74.8 | 81.4 | 88.7 | 93.4 | 93.2 | 87.1 | 78.1 | 65.3 | 56.6 | 74.5 |
| Cleburne | 57.6 | 63.0 | 70.7 | 78.8 | 84.8 | 92.9 | 97.6 | 97.6 | 90.2 | 80.7 | 67.9 | 61.0 | 78.6 |
| College Station | 59.0 | 63.4 | 70.7 | 78.1 | 84.5 | 91.1 | 94.8 | 95.0 | 89.0 | 80.6 | 69.1 | 62.1 | 78.1 |
| Conroe | 60.5 | 64.5 | 72.0 | 79.1 | 85.3 | 91.3 | 94.6 | 94.5 | 89.5 | 81.9 | 70.9 | 63.7 | 79.0 |
| Corpus Christi | 66.5 | 69.9 | 76.1 | 82.1 | 86.7 | 91.2 | 94.2 | 94.1 | 90.1 | 83.9 | 75.1 | 69.3 | 81.6 |
| Corsicana | 55.3 | 59.9 | 67.8 | 76.5 | 83.2 | 91.1 | 96.3 | 96.5 | 89.6 | 80.1 | 67.6 | 59.4 | 76.9 |
| Crockett | 57.6 | 61.8 | 69.6 | 77.5 | 84.1 | 90.5 | 94.5 | 94.8 | 89.0 | 80.7 | 68.8 | 61.3 | 77.5 |
| Dalhart | 48.8 | 53.3 | 59.8 | 69.8 | 78.2 | 88.2 | 91.4 | 89.2 | 81.7 | 72.1 | 58.2 | 51.5 | 70.2 |
| Dallas–<br>Fort Worth | 54.0 | 59.1 | 67.2 | 76.8 | 84.4 | 93.2 | 97.8 | 97.3 | 89.7 | 79.5 | 66.2 | 58.1 | 76.9 |
| Del Rio | 63.2 | 68.6 | 76.5 | 84.2 | 89.1 | 95.1 | 97.7 | 97.0 | 91.7 | 82.4 | 71.2 | 64.8 | 81.8 |
| Denton | 56.0 | 61.1 | 69.0 | 77.4 | 83.6 | 91.3 | 95.9 | 95.8 | 89.0 | 79.6 | 66.9 | 59.3 | 77.1 |
| Eagle Pass | 63.9 | 69.3 | 77.5 | 85.6 | 90.2 | 96.3 | 99.1 | 98.7 | 92.9 | 83.9 | 72.8 | 66.0 | 83.0 |
| El Paso | 57.9 | 62.7 | 69.6 | 78.7 | 87.1 | 95.9 | 95.3 | 93.0 | 87.5 | 78.5 | 65.7 | 58.2 | 77.5 |
| Fort Stockton | 60.6 | 64.5 | 72.2 | 81.5 | 87.9 | 94.5 | 95.0 | 94.2 | 88.2 | 80.2 | 68.8 | 62.6 | 79.2 |
| Fredericksburg | 60.6 | 64.6 | 72.4 | 79.2 | 84.2 | 90.7 | 94.5 | 94.2 | 88.2 | 79.7 | 68.5 | 62.9 | 78.3 |
| Gainesville | 52.2 | 57.6 | 65.8 | 75.5 | 82.6 | 90.9 | 96.1 | 95.9 | 88.2 | 78.1 | 64.8 | 56.5 | 75.4 |
| Galveston | 59.2 | 60.9 | 66.4 | 73.3 | 79.8 | 85.1 | 87.3 | 87.5 | 84.6 | 77.6 | 68.3 | 62.3 | 74.4 |
| Graham | 54.8 | 60.0 | 68.3 | 77.8 | 83.8 | 92.0 | 97.2 | 97.0 | 89.6 | 79.8 | 66.6 | 59.1 | 77.2 |
| Harlingen | 69.4 | 72.7 | 79.0 | 85.2 | 89.1 | 92.9 | 95.1 | 95.8 | 91.6 | 85.7 | 77.2 | 71.8 | 83.8 |
| Hereford | 50.3 | 54.0 | 61.3 | 71.2 | 79.1 | 88.1 | 90.4 | 88.6 | 82.2 | 73.4 | 59.8 | 52.9 | 70.9 |
| Hillsboro | 55.6 | 60.5 | 68.8 | 77.2 | 83.7 | 91.6 | 96.0 | 96.2 | 89.5 | 80.0 | 67.6 | 60.3 | 77.3 |
| Houston<br>(Hobby) | 61.9 | 65.7 | 72.1 | 79.0 | 85.1 | 90.9 | 93.6 | 93.1 | 88.7 | 81.9 | 71.6 | 65.2 | 79.1 |
| Huntsville | 58.4 | 62.9 | 70.4 | 78.2 | 84.8 | 91.5 | 95.1 | 94.7 | 89.0 | 81.0 | 69.3 | 62.0 | 78.1 |
| Junction | 62.1 | 66.4 | 74.2 | 81.3 | 86.9 | 93.3 | 96.7 | 95.9 | 90.0 | 81.3 | 70.2 | 64.1 | 80.2 |
| Kingsville | 69.2 | 72.7 | 79.4 | 84.5 | 88.5 | 92.9 | 95.4 | 95.4 | 91.5 | 85.6 | 77.0 | 71.8 | 83.7 |
| Laredo | 67.7 | 72.5 | 80.4 | 88.0 | 92.5 | 97.1 | 99.3 | 98.8 | 93.6 | 85.9 | 75.9 | 69.3 | 85.1 |
| Levelland | 54.6 | 58.4 | 66.4 | 76.0 | 84.0 | 91.8 | 92.2 | 90.6 | 83.9 | 75.7 | 63.4 | 56.8 | 74.5 |

## TABLE B-7
*cont.*

| Location | Jan. | Feb. | Mar. | Apr. | May | Jun. | Jul. | Aug. | Sep. | Oct. | Nov. | Dec. | Annual |
|---|---|---|---|---|---|---|---|---|---|---|---|---|---|
| Liberty | 61.5 | 65.1 | 71.7 | 78.7 | 85.5 | 91.1 | 93.9 | 94.0 | 89.9 | 82.6 | 72.0 | 64.9 | 79.2 |
| Llano | 59.9 | 63.7 | 71.6 | 79.8 | 85.0 | 93.0 | 97.2 | 96.7 | 90.1 | 81.1 | 69.3 | 63.0 | 79.2 |
| Lubbock | 53.3 | 57.3 | 65.1 | 74.8 | 82.8 | 90.8 | 91.9 | 90.1 | 83.6 | 74.7 | 62.1 | 55.5 | 73.5 |
| Lufkin | 59.2 | 63.5 | 71.1 | 78.6 | 84.8 | 90.6 | 93.8 | 93.8 | 88.8 | 81.0 | 69.4 | 62.4 | 78.1 |
| Marshall | 54.8 | 59.7 | 67.3 | 76.4 | 83.4 | 90.1 | 94.1 | 94.0 | 88.1 | 79.0 | 66.5 | 58.5 | 76.0 |
| McAllen | 69.9 | 73.5 | 80.1 | 86.2 | 89.6 | 93.2 | 95.1 | 96.2 | 92.6 | 86.3 | 77.6 | 71.9 | 84.4 |
| McKinney | 54.9 | 60.2 | 68.2 | 76.7 | 83.1 | 90.8 | 95.9 | 95.9 | 88.7 | 79.2 | 66.3 | 58.5 | 76.5 |
| Midland–Odessa | 57.6 | 62.1 | 69.8 | 78.8 | 86.0 | 93.0 | 94.2 | 93.1 | 86.4 | 77.7 | 65.5 | 59.7 | 77.0 |
| Mineral Wells | 55.6 | 60.9 | 69.0 | 77.9 | 84.5 | 92.8 | 97.3 | 96.6 | 89.2 | 79.3 | 66.4 | 59.5 | 77.4 |
| Mount Locke | 53.5 | 56.0 | 63.0 | 71.3 | 78.2 | 84.5 | 82.6 | 80.8 | 76.1 | 69.9 | 60.6 | 54.6 | 69.3 |
| New Braunfels | 61.9 | 66.6 | 74.4 | 81.2 | 86.5 | 93.1 | 96.4 | 96.5 | 90.7 | 82.5 | 71.3 | 64.9 | 80.5 |
| Palestine | 55.8 | 60.6 | 68.2 | 76.5 | 83.2 | 90.1 | 94.6 | 94.9 | 88.6 | 80.0 | 67.4 | 59.7 | 76.6 |
| Paris | 51.0 | 56.2 | 64.3 | 74.1 | 81.6 | 89.3 | 94.3 | 94.1 | 86.9 | 77.5 | 64.2 | 55.3 | 74.1 |
| Pecos | 60.9 | 66.2 | 74.5 | 84.4 | 91.9 | 99.3 | 99.7 | 98.5 | 91.7 | 82.5 | 69.6 | 63.0 | 81.9 |
| Plainview | 51.9 | 56.2 | 63.6 | 73.7 | 81.6 | 90.1 | 92.4 | 90.5 | 83.9 | 74.9 | 61.6 | 54.8 | 72.9 |
| Port Lavaca | 63.5 | 66.7 | 72.9 | 79.6 | 85.0 | 90.6 | 93.3 | 93.4 | 89.4 | 82.8 | 73.5 | 67.0 | 79.8 |
| Presidio | 67.6 | 73.3 | 82.0 | 90.8 | 97.8 | 103.3 | 102.0 | 100.2 | 95.1 | 87.0 | 74.9 | 67.5 | 86.8 |
| Rio Grande City | 69.5 | 73.9 | 81.9 | 89.4 | 92.5 | 97.0 | 99.3 | 99.8 | 94.3 | 86.7 | 77.4 | 71.0 | 86.1 |
| San Angelo | 58.7 | 63.3 | 71.5 | 80.2 | 86.3 | 93.4 | 96.5 | 95.4 | 88.0 | 79.2 | 67.2 | 61.2 | 78.4 |
| San Antonio | 61.7 | 66.3 | 73.7 | 80.3 | 85.5 | 91.8 | 94.9 | 94.6 | 89.3 | 81.5 | 70.7 | 64.6 | 79.6 |
| San Marcos | 60.8 | 65.2 | 72.8 | 79.8 | 85.1 | 91.4 | 95.0 | 95.5 | 89.8 | 81.8 | 70.8 | 64.0 | 79.3 |
| Seminole | 56.9 | 61.4 | 69.0 | 78.7 | 85.8 | 93.6 | 94.5 | 93.2 | 86.9 | 78.1 | 65.6 | 59.1 | 76.9 |
| Sherman | 51.6 | 56.7 | 64.7 | 74.6 | 81.8 | 90.1 | 95.3 | 95.2 | 87.7 | 77.6 | 64.3 | 55.7 | 74.6 |
| Snyder | 55.0 | 59.5 | 67.9 | 77.8 | 84.7 | 92.6 | 95.2 | 94.2 | 86.9 | 77.7 | 64.7 | 58.1 | 76.2 |
| Spearman | 49.1 | 54.5 | 61.9 | 72.9 | 81.1 | 90.4 | 94.6 | 92.8 | 85.4 | 75.1 | 59.4 | 52.1 | 72.4 |
| Sulphur Springs | 53.1 | 57.7 | 65.2 | 74.7 | 81.6 | 89.0 | 94.3 | 94.6 | 88.1 | 78.5 | 65.5 | 57.3 | 75.0 |
| Temple | 57.4 | 61.6 | 69.5 | 77.6 | 83.7 | 91.3 | 95.7 | 95.9 | 89.5 | 80.6 | 68.1 | 60.9 | 77.8 |
| Uvalde | 64.4 | 69.0 | 76.9 | 83.9 | 88.2 | 94.2 | 97.1 | 96.8 | 91.8 | 83.6 | 72.9 | 66.3 | 82.1 |
| Victoria | 63.6 | 67.1 | 73.8 | 80.2 | 85.6 | 90.8 | 93.7 | 93.7 | 89.3 | 82.8 | 73.0 | 66.7 | 80.0 |
| Waco | 56.6 | 61.6 | 69.5 | 77.6 | 84.2 | 92.1 | 96.5 | 96.7 | 89.7 | 80.3 | 67.9 | 60.3 | 77.8 |
| Waxahachie | 56.3 | 61.3 | 69.4 | 77.8 | 84.5 | 92.4 | 96.9 | 96.8 | 89.8 | 80.7 | 67.7 | 60.1 | 77.8 |
| Weatherford | 53.2 | 57.9 | 66.2 | 75.6 | 82.5 | 91.6 | 96.7 | 96.2 | 88.3 | 77.9 | 64.8 | 57.2 | 75.7 |
| Weslaco | 70.6 | 74.2 | 80.6 | 86.1 | 89.3 | 92.5 | 94.9 | 95.7 | 92.0 | 86.1 | 77.7 | 72.6 | 84.4 |
| Wichita Falls | 52.3 | 58.0 | 66.7 | 76.8 | 84.1 | 93.2 | 98.5 | 97.3 | 88.7 | 78.2 | 64.4 | 56.2 | 76.2 |

*Source:* Based upon data for the period 1951–1980.

TABLE B-8.
*Hottest Temperatures, 1960–1982*

| Year | °F | Location | County | Region | Date |
|------|-----|----------|--------|--------|------|
| 1960 | 117 | Presidio | Presidio | Trans Pecos | Jun. 18 |
| 1961 | 111 | Presidio | Presidio | Trans Pecos | May 12 |
| 1962 | 114 | Gonzales | Gonzales | South Central | Aug. 10 |
| 1963 | 113 | Falcon Dam | Starr | Southern | Apr. 10 |
| 1964 | 114 | Albany | Shackelford | North Central | Aug. 7 |
|      | 114 | Guthrie | King | Rolling Plains | Aug. 7 |
|      | 114 | Henrietta | Clay | North Central | Aug. 7 |
|      | 114 | Seymour | Baylor | Rolling Plains | Aug. 7 |
| 1965 | 110 | Candelaria | Presidio | Trans Pecos | Jun. 14 |
|      | 110 | La Pryor | Zavala | Southern | Jul. 10 |
|      | 110 | Presidio | Presidio | Trans Pecos | Jul. 9 |
| 1966 | 110 | Candelaria | Presidio | Trans Pecos | Jul. 22 |
|      | 110 | Pecos | Reeves | Trans Pecos | Aug. 2 |
|      | 110 | Presidio | Presidio | Trans Pecos | Aug. 1 |
| 1967 | 110 | Boquillas | Brewster | Trans Pecos | Jun. 14 |
|      | 110 | Presidio . | Presidio | Trans Pecos | Jun. 23 |
|      | 110 | Dryden | Terrell | Trans Pecos | May 12 |
|      | 110 | Encinal | LaSalle | Southern | May 13 |
| 1968 | 118 | Pecos | Reeves | Trans Pecos | Jun. 29 |
| 1969 | 114 | Presidio | Presidio | Trans Pecos | Jun. 21 |
| 1970 | 112 | Candelaria | Presidio | Trans Pecos | Jun. 17 |
|      | 112 | Wichita Falls | Wichita | Rolling Plains | Aug. 9 |
| 1971 | 111 | Matador | Motley | Rolling Plains | Jul. 5 |
| 1972 | 115 | Albany | Shackelford | North Central | Jun. 27 |
| 1973 | 112 | Candelaria | Presidio | Trans Pecos | Jun. 30 |
|      | 112 | Pecos | Reeves | Trans Pecos | Jun. 30 |
| 1974 | 113 | Bridgeport | Wise | North Central | Jul. 24 |
| 1975 | 111 | Falcon Dam | Starr | Southern | Aug. 1 |
| 1976 | 115 | Pecos | Reeves | Trans Pecos | Jun. 23 |
| 1977 | 114 | Pecos | Reeves | Trans Pecos | Jun. 18 |
| 1978 | 116 | Munday | Knox | Rolling Plains | Jul. 15 |
|      | 116 | Olney | Young | Rolling Plains | Jul. 15 |
| 1979 | 112 | El Paso | El Paso | Trans Pecos | Jul. 10 |
| 1980 | 119 | Weatherford | Parker | North Central | Jun. 26 |
| 1981 | 114 | Boquillas | Brewster | Trans Pecos | Jun. 10 |
|      | 114 | Castolon | Brewster | Trans Pecos | Jun. 10 |
| 1982 | 114 | Lajitas | Brewster | Trans Pecos | Jun. 28 |

TABLE B-9.
*Average and Extreme Numbers of Days of 100° and Above*

| Location | May | Jun. | Average Jul. | Aug. | Sep. | Year | Most in Any Season No. | Year |
|---|---|---|---|---|---|---|---|---|
| Abilene | * | 3 | 7 | 6 | 1 | 17 | 43 | 1952 |
| Alpine | 0 | 2 | 0 | 0 | 0 | 2 | 12 | 1951 |
| Amarillo | 0 | 2 | 2 | 1 | 0 | 5 | 26 | 1953 |
| Austin | 0 | 1 | 4 | 6 | * | 12 | 40 | 1963 |
| Beaumont–Port Arthur | 0 | 0 | 0 | * | 0 | 1 | 12 | 1962 |
| Beeville | 0 | 1 | 4 | 7 | 1 | 14 | 57 | 1953 |
| Big Spring | 1 | 5 | 5 | 4 | * | 15 | 43 | 1969 |
| Brownsville | 0 | 0 | 0 | 0 | 0 | 0 | 3 | 1980 |
| Brownwood | * | 3 | 10 | 10 | 2 | 25 | 53 | 1963 + 1964 |
| Childress | 1 | 4 | 9 | 7 | 1 | 23 | 49 | 1954 |
| Chisos Basin | 0 | 0 | 0 | 0 | 0 | 0 | 3 | 1972 |
| College Station | 0 | 1 | 5 | 6 | * | 12 | 43 | 1980 |
| Corpus Christi | 0 | 0 | 0 | * | 0 | 1 | 7 | 1954 |
| Corsicana | 0 | 2 | 10 | 10 | 2 | 23 | 82 | 1954 |
| Dalhart | 0 | 2 | 2 | 1 | 0 | 5 | 20 | 1980 |
| Dallas–Fort Worth | 0 | 2 | 9 | 8 | 1 | 21 | 69 | 1980 |
| Del Rio | 1 | 5 | 11 | 9 | 2 | 27 | 78 | 1953 |
| Eagle Pass | 2 | 9 | 17 | 16 | 5 | 51 | 100 | 1953 |
| El Paso | 0 | 8 | 7 | 3 | 0 | 18 | 55 | 1980 |
| Fort Stockton | 2 | 6 | 6 | 4 | 1 | 20 | 45 | 1980 |
| Galveston | 0 | 0 | 0 | 0 | 0 | 0 | 0 | |
| Hereford | 0 | 1 | 1 | 0 | 0 | 3 | 13 | 1980 |
| Houston (Hobby) | 0 | 0 | 0 | 1 | 0 | 1 | 9 | 1962 |
| Huntsville | 0 | * | 4 | 5 | * | 11 | 43 | 1980 |
| Kerrville | 0 | * | 2 | 3 | * | 6 | 6 | 1951 + 1962 |
| Laredo | 4 | 9 | 18 | 16 | 4 | 53 | 93 | 1953 |
| Llano | 0 | 3 | 12 | 10 | 2 | 28 | 83 | 1951 |
| Longview | 0 | * | 4 | 4 | * | 9 | 44 | 1954 |
| Lubbock | 0 | 4 | 2 | 1 | 0 | 7 | 20 | 1953 + 1980 |
| Lufkin | 0 | 0 | 3 | 4 | 0 | 7 | 32 | 1951 |
| McAllen | 0 | 1 | 3 | 4 | 1 | 10 | 54 | 1980 |
| McKinney | 0 | 1 | 9 | 8 | 1 | 20 | 56 | 1954 + 1956 |
| Midland–Odessa | 1 | 5 | 4 | 3 | * | 13 | 52 | 1964 |
| Mineral Wells | 0 | 3 | 12 | 11 | 2 | 28 | 63 | 1956 |
| Mount Locke | 0 | 0 | 0 | 0 | 0 | 0 | 2 | 1969 |
| Paris | 0 | 1 | 6 | 5 | 1 | 13 | 55 | 1954 |
| Pecos | 6 | 16 | 18 | 15 | 6 | 63 | 119 | 1951 |
| Presidio | 13 | 23 | 22 | 18 | 8 | 88[a] | 121 | 1962 |
| San Angelo | 1 | 4 | 8 | 7 | 1 | 21 | 60 | 1969 |
| San Antonio | 0 | 1 | 3 | 4 | 0 | 8 | 32 | 1951 |
| Sherman | 0 | 2 | 8 | 8 | 1 | 19 | 56 | 1956 |
| Sonora | 1 | 3 | 6 | 4 | * | 14 | 52 | 1980 |
| Spearman | 0 | 4 | 7 | 4 | 0 | 16 | 34 | 1980 |
| Temple | 0 | 1 | 7 | 8 | 1 | 17 | 41 | 1951 |
| Tyler | 0 | 0 | 4 | 4 | 0 | 9 | 34 | 1956 |
| Uvalde | 1 | 5 | 12 | 10 | 3 | 31 | 82 | 1953 |
| Van Horn | * | 6 | 6 | 4 | 0 | 17 | 71 | 1969 |
| Victoria | 0 | 0 | 1 | 2 | 0 | 3 | 20 | 1951 |
| Waco | 0 | 2 | 9 | 10 | 1 | 22 | 63 | 1980 |
| Wichita Falls | * | 5 | 15 | 12 | 2 | 36 | 79 | 1980 |

*Source:* Based upon data for the period 1951–1980.

* Less than one occurrence (but more than 0.5); in some instances the sum of the monthly amounts does not equal the total because of rounding.
[a] Includes two days in April and one in October.

TABLE B-10.
*Greatest Number of Days of 100° and Above during Most Notable Heat Waves*

| Location | Year | Total | Consecutive No. | Dates |
|---|---|---|---|---|
| Amarillo | 1953 | 26 | 5 | Jul. 1–5 |
| | 1952 | 17 | 6 | Jun. 20–25 |
| | 1980 | 14 | 7 | Jun. 23–29 |
| Dallas–Fort Worth | 1980 | 69 | 42 | Jun. 23–Aug. 3 |
| | 1954 | 52 | 20 | Jul. 9–28 |
| | 1956 | 48 | 17 | Aug. 2–18 |
| El Paso | 1980 | 55 | 21 | Jun. 10–30 |
| | 1969 | 36 | 9 | Jun. 27–Jul. 5 |
| | 1979 | 35 | 11 | Jun. 21–Jul. 1 |
| Houston (Hobby) | 1962 | 9 | 9 | Aug. 6–14 |
| | 1980 | 8 | 4 | Jul. 15–18 |
| | 1951 | 7 | 3 | Aug. 4–6 |
| Midland–Odessa | 1964 | 52 | 12 | Jun. 30–Jul. 11 |
| | 1977 | 43 | 7 | Sep. 24–30 |
| | 1953 | 32 | 8 | Jun. 20–27 |
| San Antonio | 1951 | 32 | 8 | Aug. 11–18 |
| | 1980 | 31 | 11 | Jun. 23–Jul. 3 |
| | 1962 | 26 | 21 | Jul. 24–Aug. 13 |

*Source:* Based upon data for the period 1951–1980.

TABLE C-1.
*Average Precipitation (inches) by Climatic Region*

| Region | Total | Driest Months | | Wettest Months | |
|---|---|---|---|---|---|
| High Plains | 17.73 | Jan. | 0.44 | May | 2.65 |
| | | Dec. | 0.45 | Jun. | 2.61 |
| | | Feb. | 0.60 | Jul. | 2.51 |
| Low Rolling Plains | 22.80 | Jan. | 0.75 | May | 3.60 |
| | | Dec. | 0.82 | Sep. | 3.07 |
| | | Feb. | 0.88 | Jun. | 2.73 |
| North Central | 32.15 | Jan. | 1.74 | May | 4.43 |
| | | Dec. | 1.90 | Apr. | 3.75 |
| | | Feb. | 2.00 | Sep. | 3.67 |
| East | 44.72 | Aug. | 2.73 | Apr. | 4.83 |
| | | Jul. | 3.04 | May | 4.80 |
| | | Mar. | 3.37 | Sept. | 4.25 |
| Trans Pecos | 11.65 | Dec. | 0.38 | Sep. | 2.05 |
| | | Jan. | 0.40 | Aug. | 1.84 |
| | | Feb. & Apr. | 0.42 | Jul. | 1.81 |
| | | Mar. | 0.45 | | |
| Edwards Plateau | 23.35 | Jan. | 0.94 | Sep. | 3.19 |
| | | Dec. | 1.02 | May | 3.10 |
| | | Mar. | 1.16 | Oct. | 2.63 |
| South Central | 34.02 | Mar. | 1.55 | Sep. | 5.07 |
| | | Jul. | 1.93 | May | 4.05 |
| | | Jan. | 2.01 | Jun. | 3.43 |
| | | | | Oct. | 3.38 |
| Upper Coast | 45.93 | Mar. | 2.27 | Sep. | 6.19 |
| | | Feb. | 3.10 | Jun. | 4.56 |
| | | Jan. | 3.26 | May | 4.37 |
| | | | | Aug. | 4.22 |
| Southern | 22.91 | Mar. | 0.71 | Sep. | 4.05 |
| | | Dec. | 0.95 | May | 3.04 |
| | | Jan. | 1.02 | Jun. | 2.55 |
| Lower Valley | 24.73 | Mar. | 0.64 | Sep. | 4.96 |
| | | Dec. | 1.08 | Oct. | 3.04 |
| | | Feb. | 1.26 | Jun. | 2.88 |

*Source:* Based upon data for the period 1951–1980.

# TABLE C-2.
## *Average Precipitation (inches) in Selected Locations*

| Location | Jan. | Feb. | Mar. | Apr. | May | Jun. | Jul. | Aug. | Sep. | Oct. | Nov. | Dec. | Annual |
|---|---|---|---|---|---|---|---|---|---|---|---|---|---|
| Abilene | .97 | .96 | 1.08 | 2.35 | 3.25 | 2.52 | 2.11 | 2.47 | 3.06 | 2.32 | 1.32 | .85 | 23.26 |
| Albany | 1.30 | 1.35 | 1.27 | 2.87 | 3.74 | 2.34 | 2.36 | 3.05 | 3.87 | 2.44 | 1.50 | 1.01 | 27.10 |
| Alice | 1.31 | 1.52 | .75 | 1.66 | 2.95 | 3.33 | 1.96 | 2.81 | 6.35 | 3.14 | 1.59 | 1.14 | 28.51 |
| Alpine | .47 | .43 | .35 | .35 | 1.03 | 1.95 | 2.91 | 2.58 | 2.61 | 1.16 | .57 | .42 | 14.83 |
| Alvin | 3.49 | 3.23 | 2.77 | 3.15 | 3.98 | 5.02 | 4.97 | 3.80 | 6.51 | 3.59 | 3.88 | 3.60 | 47.99 |
| Amarillo | .46 | .57 | .87 | 1.08 | 2.79 | 3.50 | 2.70 | 2.95 | 1.72 | 1.39 | .58 | .49 | 19.10 |
| Angleton | 3.79 | 3.72 | 2.90 | 3.15 | 4.63 | 5.49 | 4.54 | 4.99 | 6.97 | 3.51 | 4.25 | 4.34 | 52.28 |
| Arlington | 1.78 | 2.13 | 2.68 | 4.39 | 4.58 | 2.54 | 2.07 | 1.94 | 3.73 | 3.21 | 2.24 | 1.95 | 33.24 |
| Athens | 2.57 | 3.01 | 2.97 | 4.35 | 5.11 | 3.01 | 1.57 | 2.14 | 4.21 | 3.81 | 3.36 | 3.32 | 39.43 |
| Atlanta | 3.55 | 3.49 | 4.21 | 5.73 | 4.22 | 3.65 | 3.04 | 2.84 | 3.68 | 2.91 | 4.25 | 4.07 | 45.64 |
| Austin | 1.60 | 2.49 | 1.68 | 3.11 | 4.19 | 3.06 | 1.89 | 2.24 | 3.60 | 3.38 | 2.20 | 2.06 | 31.50 |
| Ballinger | .97 | 1.06 | .88 | 2.20 | 3.36 | 2.24 | 1.45 | 2.27 | 3.26 | 2.30 | 1.23 | .89 | 22.11 |
| Bay City | 2.83 | 2.90 | 2.02 | 2.97 | 4.33 | 4.43 | 3.77 | 4.33 | 6.62 | 3.41 | 3.18 | 3.10 | 43.89 |
| Beaumont– Port Arthur | 4.18 | 3.71 | 2.93 | 4.05 | 4.50 | 3.96 | 5.37 | 5.45 | 6.13 | 3.63 | 4.33 | 4.55 | 52.79 |
| Beeville | 1.78 | 2.11 | 1.05 | 2.33 | 3.61 | 3.25 | 1.94 | 2.95 | 5.47 | 2.97 | 1.95 | 1.70 | 31.11 |
| Big Spring | .58 | .64 | .75 | 1.42 | 2.77 | 1.88 | 1.85 | 1.88 | 2.87 | 1.74 | .73 | .61 | 17.72 |
| Blanco | 1.88 | 2.54 | 2.05 | 3.37 | 4.24 | 3.14 | 1.98 | 2.61 | 4.75 | 3.93 | 2.19 | 1.97 | 34.65 |
| Bonham | 2.06 | 2.67 | 3.58 | 4.77 | 5.07 | 4.08 | 2.90 | 2.33 | 4.83 | 3.50 | 2.99 | 2.77 | 41.55 |
| Borger | .47 | .76 | 1.01 | 1.34 | 3.19 | 2.91 | 3.12 | 2.48 | 1.57 | 1.27 | .68 | .53 | 19.33 |
| Brady | 1.06 | 1.36 | 1.19 | 2.30 | 3.60 | 2.17 | 2.07 | 2.35 | 3.67 | 2.48 | 1.35 | 1.06 | 24.66 |
| Brenham | 2.71 | 2.99 | 2.09 | 4.02 | 4.57 | 3.63 | 1.84 | 2.60 | 4.72 | 3.52 | 3.81 | 3.22 | 39.72 |
| Brownfield | .43 | .55 | .83 | .95 | 2.34 | 2.45 | 2.34 | 2.10 | 2.31 | 1.77 | .57 | .33 | 16.97 |
| Brownsville | 1.25 | 1.55 | .50 | 1.57 | 2.15 | 2.70 | 1.51 | 2.83 | 5.24 | 3.54 | 1.44 | 1.16 | 25.44 |
| Brownwood | 1.38 | 1.44 | 1.52 | 2.69 | 3.88 | 2.89 | 1.66 | 1.95 | 3.00 | 2.96 | 1.50 | 1.23 | 26.10 |
| Cameron | 2.15 | 2.74 | 2.23 | 3.93 | 3.94 | 2.65 | 1.68 | 1.78 | 4.22 | 3.43 | 2.94 | 2.57 | 34.26 |
| Canyon | .45 | .56 | .81 | 1.17 | 2.67 | 3.14 | 2.50 | 2.89 | 1.53 | 1.58 | .64 | .43 | 18.37 |
| Center | 4.38 | 3.80 | 4.26 | 4.89 | 5.01 | 3.94 | 3.60 | 3.96 | 4.47 | 2.93 | 4.04 | 4.44 | 49.72 |
| Childress | .57 | .71 | 1.07 | 1.76 | 3.50 | 2.60 | 1.95 | 1.91 | 2.32 | 2.04 | .76 | .70 | 19.89 |
| Chisos Basin | .54 | .46 | .37 | .53 | 1.47 | 1.85 | 3.16 | 3.04 | 2.74 | 1.45 | .58 | .52 | 16.71 |
| Clarksville | 2.67 | 3.13 | 4.01 | 5.22 | 4.96 | 3.52 | 3.03 | 2.25 | 4.07 | 3.93 | 3.91 | 3.41 | 44.11 |
| Cleburne | 1.79 | 1.96 | 2.36 | 4.18 | 4.73 | 2.79 | 1.90 | 2.35 | 3.03 | 3.24 | 2.22 | 1.82 | 32.37 |
| College Station | 2.48 | 2.97 | 2.39 | 4.34 | 4.35 | 3.21 | 2.39 | 2.30 | 4.93 | 3.42 | 3.33 | 2.97 | 39.08 |
| Columbus | 2.86 | 3.05 | 2.29 | 4.02 | 5.29 | 4.10 | 2.48 | 2.74 | 5.09 | 3.35 | 3.17 | 3.01 | 41.45 |
| Conroe | 3.47 | 3.39 | 2.52 | 4.85 | 4.77 | 3.89 | 3.40 | 3.58 | 4.92 | 3.77 | 3.84 | 4.20 | 46.60 |
| Corpus Christi | 1.63 | 1.55 | .84 | 1.99 | 3.05 | 3.36 | 1.96 | 3.51 | 6.15 | 3.19 | 1.55 | 1.40 | 30.18 |
| Corsicana | 2.32 | 2.47 | 2.55 | 4.59 | 5.09 | 2.83 | 1.61 | 2.22 | 3.49 | 3.73 | 2.79 | 2.94 | 36.63 |
| Crockett | 3.43 | 3.05 | 2.79 | 4.72 | 4.24 | 3.69 | 2.45 | 2.83 | 4.35 | 3.33 | 3.84 | 3.51 | 42.23 |
| Dalhart | .38 | .45 | .73 | 1.18 | 2.61 | 2.11 | 2.98 | 2.60 | 1.33 | 1.12 | .58 | .38 | 16.45 |
| Dallas–Fort Worth | 1.65 | 1.93 | 2.42 | 3.63 | 4.27 | 2.59 | 2.00 | 1.76 | 3.31 | 2.47 | 1.76 | 1.67 | 29.45 |
| Del Rio | .51 | .89 | .63 | 1.85 | 1.99 | 1.72 | 1.69 | 1.60 | 2.73 | 2.24 | .80 | .55 | 17.19 |
| Denton | 1.55 | 2.00 | 2.63 | 4.13 | 4.71 | 3.04 | 2.12 | 2.04 | 4.14 | 3.03 | 2.21 | 1.93 | 33.53 |
| Eagle Pass | .62 | .81 | .67 | 1.83 | 3.24 | 2.37 | 2.02 | 2.42 | 2.92 | 2.47 | .88 | .70 | 20.95 |
| El Paso | .38 | .45 | .32 | .19 | .24 | .56 | 1.60 | 1.21 | 1.42 | .73 | .33 | .39 | 7.82 |
| Ennis | 2.19 | 2.47 | 2.45 | 4.72 | 5.38 | 3.31 | 2.10 | 2.17 | 3.89 | 3.65 | 2.77 | 2.56 | 37.66 |
| Fort Stockton | .42 | .60 | .43 | .52 | 1.61 | 1.43 | 1.30 | 1.46 | 2.14 | 1.09 | .77 | .44 | 12.21 |
| Fredericksburg | 1.16 | 1.66 | 1.35 | 2.66 | 3.76 | 3.23 | 1.78 | 2.89 | 4.02 | 3.18 | 1.72 | 1.26 | 28.67 |
| Freeport | 3.61 | 3.19 | 1.89 | 2.91 | 3.92 | 4.95 | 5.56 | 5.20 | 7.98 | 4.02 | 3.83 | 3.95 | 51.01 |
| Gainesville | 1.65 | 1.95 | 2.74 | 3.40 | 4.34 | 3.24 | 2.15 | 2.23 | 4.25 | 3.18 | 2.19 | 1.67 | 32.99 |
| Galveston | 2.96 | 2.34 | 2.10 | 2.62 | 3.30 | 3.48 | 3.77 | 4.40 | 5.82 | 2.60 | 3.23 | 3.62 | 40.24 |
| Gonzales | 1.86 | 2.38 | 1.61 | 3.68 | 4.27 | 3.58 | 1.35 | 2.10 | 4.54 | 3.26 | 2.43 | 2.09 | 33.15 |
| Graham | 1.30 | 1.36 | 1.47 | 3.09 | 4.11 | 2.84 | 2.04 | 2.33 | 3.94 | 2.59 | 1.71 | 1.23 | 28.01 |
| Harlingen | 1.38 | 1.46 | .68 | 1.77 | 2.27 | 2.79 | 1.83 | 3.26 | 5.25 | 2.98 | 1.61 | 1.20 | 26.48 |
| Hereford | .38 | .52 | .70 | .98 | 1.85 | 2.92 | 2.15 | 2.37 | 1.58 | 1.52 | .64 | .40 | 16.01 |

# TABLE C-2
*cont.*

| Location | Jan. | Feb. | Mar. | Apr. | May | Jun. | Jul. | Aug. | Sep. | Oct. | Nov. | Dec. | Annual |
|---|---|---|---|---|---|---|---|---|---|---|---|---|---|
| Hillsboro | 1.88 | 2.32 | 2.37 | 3.99 | 4.83 | 3.04 | 2.21 | 1.83 | 3.25 | 3.61 | 2.57 | 2.32 | 34.22 |
| Houston | 3.21 | 3.25 | 2.68 | 4.24 | 4.69 | 4.06 | 3.33 | 3.66 | 4.93 | 3.67 | 3.38 | 3.66 | 44.77 |
| Huntsville | 3.20 | 3.36 | 2.83 | 4.60 | 4.73 | 3.86 | 2.69 | 3.06 | 4.96 | 3.42 | 3.50 | 3.99 | 44.20 |
| Jasper | 4.39 | 4.12 | 3.80 | 4.59 | 5.05 | 4.30 | 4.28 | 3.50 | 4.10 | 3.22 | 4.35 | 4.89 | 50.59 |
| Junction | .95 | 1.36 | 1.03 | 2.13 | 3.28 | 2.26 | 1.54 | 2.88 | 2.66 | 2.36 | 1.12 | .95 | 22.52 |
| Kerrville | 1.33 | 1.89 | 1.83 | 2.95 | 3.91 | 2.64 | 1.90 | 2.95 | 3.64 | 3.68 | 1.75 | 1.50 | 29.97 |
| Kingsville | 1.51 | 1.57 | .94 | 1.53 | 3.07 | 3.22 | 1.98 | 3.16 | 5.04 | 2.96 | 1.38 | 1.14 | 27.50 |
| Laredo | .73 | 1.05 | .40 | 1.22 | 2.54 | 2.74 | 1.06 | 2.67 | 3.45 | 2.18 | 1.23 | .87 | 20.14 |
| Levelland | .43 | .59 | .71 | 1.03 | 2.20 | 2.46 | 2.66 | 2.59 | 2.53 | 1.98 | .53 | .43 | 18.14 |
| Liberty | 4.00 | 3.93 | 2.66 | 4.35 | 4.31 | 4.46 | 4.19 | 3.84 | 5.34 | 4.22 | 4.59 | 4.76 | 50.65 |
| Llano | 1.13 | 1.72 | 1.35 | 3.06 | 3.92 | 2.34 | 1.54 | 2.35 | 3.49 | 2.77 | 1.61 | 1.28 | 26.56 |
| Longview | 3.79 | 3.69 | 3.73 | 5.19 | 4.95 | 4.18 | 2.97 | 2.54 | 4.45 | 3.02 | 3.88 | 4.08 | 46.47 |
| Lubbock | .38 | .57 | .90 | 1.08 | 2.59 | 2.81 | 2.34 | 2.20 | 2.06 | 1.81 | .59 | .43 | 17.76 |
| Lufkin | 3.55 | 3.05 | 3.38 | 4.27 | 4.31 | 3.39 | 2.81 | 2.46 | 3.72 | 2.98 | 3.59 | 3.97 | 41.48 |
| Marshall | 4.14 | 3.72 | 3.99 | 5.10 | 4.86 | 3.73 | 3.31 | 2.42 | 4.11 | 3.18 | 3.74 | 4.11 | 46.41 |
| McAllen | 1.25 | 1.10 | .64 | 1.56 | 2.08 | 2.79 | 1.59 | 2.38 | 4.29 | 3.20 | 1.15 | 1.01 | 23.04 |
| McKinney | 1.88 | 2.33 | 3.03 | 4.46 | 5.02 | 3.20 | 2.61 | 2.19 | 4.52 | 2.88 | 2.64 | 2.12 | 36.88 |
| Midland-Odessa | .42 | .58 | .51 | .84 | 2.05 | 1.44 | 1.72 | 1.60 | 2.08 | 1.41 | .60 | .45 | 13.70 |
| Mineral Wells | 1.61 | 1.68 | 1.99 | 3.41 | 4.06 | 2.59 | 2.27 | 2.18 | 3.30 | 2.93 | 1.82 | 1.43 | 29.27 |
| Mount Locke | .60 | .52 | .43 | .35 | 1.36 | 2.12 | 3.90 | 4.05 | 3.02 | 1.46 | .65 | .48 | 18.94 |
| New Braunfels | 1.86 | 2.33 | 1.60 | 3.08 | 4.50 | 3.20 | 1.69 | 2.77 | 4.27 | 3.50 | 2.80 | 2.01 | 33.61 |
| Orange | 4.84 | 4.49 | 3.29 | 4.27 | 5.09 | 4.96 | 6.32 | 5.66 | 6.44 | 3.75 | 4.50 | 5.59 | 59.20 |
| Ozona | .63 | .86 | .73 | 1.63 | 2.29 | 2.05 | 1.75 | 2.01 | 2.78 | 2.01 | .97 | .52 | 18.23 |
| Palestine | 3.13 | 2.98 | 3.62 | 4.52 | 4.74 | 3.75 | 2.07 | 2.76 | 3.76 | 3.56 | 3.45 | 3.38 | 41.72 |
| Pampa | .45 | .84 | 1.05 | 1.31 | 3.09 | 2.92 | 2.62 | 2.47 | 2.02 | 1.47 | .84 | .49 | 19.57 |
| Paris | 2.31 | 2.96 | 3.71 | 4.80 | 5.14 | 3.87 | 3.54 | 2.94 | 5.01 | 3.95 | 3.39 | 3.35 | 44.97 |
| Pecos | .33 | .39 | .28 | .43 | .95 | 1.02 | 1.34 | 1.19 | 1.79 | 1.08 | .46 | .31 | 9.57 |
| Perryton | .45 | .59 | 1.21 | 1.17 | 3.50 | 2.89 | 3.01 | 2.59 | 1.57 | 1.18 | .97 | .49 | 19.62 |
| Plainview | .49 | .66 | .79 | 1.24 | 3.27 | 3.07 | 2.57 | 2.03 | 2.03 | 1.68 | .64 | .50 | 18.97 |
| Port Lavaca | 2.35 | 2.65 | 1.57 | 3.01 | 3.81 | 5.50 | 2.61 | 4.03 | 7.24 | 4.13 | 2.75 | 2.56 | 42.21 |
| Presidio | .26 | .31 | .18 | .26 | .49 | 1.35 | 1.62 | 1.62 | 1.65 | .78 | .40 | .28 | 9.20 |
| Richardson | 1.93 | 2.18 | 2.87 | 4.58 | 4.69 | 2.88 | 2.29 | 2.20 | 3.91 | 3.28 | 2.45 | 2.22 | 35.48 |
| Rio Grande City | .81 | .90 | .57 | 1.50 | 2.28 | 2.02 | 1.30 | 2.16 | 5.11 | 2.28 | .95 | .69 | 20.57 |
| San Angelo | .64 | .84 | .79 | 1.75 | 2.52 | 1.88 | 1.22 | 1.85 | 3.04 | 2.05 | .97 | .64 | 18.19 |
| San Antonio | 1.55 | 1.86 | 1.33 | 2.73 | 3.67 | 3.03 | 1.92 | 2.69 | 3.75 | 2.88 | 2.34 | 1.38 | 29.13 |
| San Marcos | 1.86 | 2.69 | 1.60 | 3.37 | 4.42 | 3.38 | 1.81 | 2.39 | 4.39 | 3.46 | 2.81 | 2.13 | 34.31 |
| Sanderson | .38 | .52 | .32 | .92 | 1.37 | 1.59 | 1.24 | 1.58 | 2.56 | 1.37 | .61 | .37 | 12.83 |
| Seminole | .47 | .61 | .66 | 1.02 | 2.04 | 1.81 | 2.14 | 2.18 | 2.28 | 1.58 | .64 | .37 | 15.80 |
| Sherman | 1.72 | 2.49 | 3.13 | 4.55 | 5.33 | 3.46 | 2.28 | 2.24 | 4.84 | 3.21 | 2.81 | 2.12 | 38.18 |
| Snyder | .50 | .64 | .78 | 1.67 | 3.05 | 2.45 | 2.03 | 2.40 | 3.05 | 2.22 | .91 | .59 | 20.29 |
| Spearman | .52 | .64 | 1.18 | 1.12 | 3.23 | 2.59 | 3.01 | 2.50 | 1.71 | 1.23 | .92 | .50 | 19.15 |
| Sulphur Springs | 2.65 | 2.93 | 3.57 | 5.37 | 5.01 | 3.89 | 2.80 | 2.55 | 4.53 | 3.97 | 3.47 | 3.42 | 44.16 |
| Temple | 1.94 | 2.57 | 1.97 | 3.56 | 4.65 | 2.94 | 1.75 | 2.50 | 3.51 | 3.30 | 2.69 | 2.37 | 33.75 |
| Tyler | 2.95 | 3.35 | 3.51 | 4.91 | 4.88 | 3.29 | 2.82 | 2.46 | 4.08 | 3.27 | 3.81 | 3.68 | 43.01 |
| Uvalde | 1.00 | 1.26 | .95 | 2.17 | 3.26 | 2.81 | 1.74 | 2.94 | 2.73 | 3.08 | 1.19 | .97 | 24.10 |
| Van Horn | .48 | .31 | .20 | .23 | .60 | .93 | 1.84 | 2.06 | 2.30 | 1.08 | .62 | .41 | 11.06 |
| Victoria | 1.87 | 2.24 | 1.34 | 2.61 | 4.47 | 4.53 | 2.58 | 3.33 | 6.24 | 3.31 | 2.24 | 2.14 | 36.90 |
| Waco | 1.69 | 2.04 | 1.99 | 3.79 | 4.73 | 2.58 | 1.78 | 1.95 | 3.18 | 3.06 | 2.24 | 1.92 | 30.95 |
| Waxahachie | 1.94 | 2.43 | 2.68 | 4.56 | 5.02 | 2.77 | 1.78 | 2.26 | 4.21 | 3.52 | 2.57 | 2.51 | 36.25 |
| Weatherford | 1.65 | 1.83 | 2.18 | 3.91 | 4.24 | 2.70 | 2.18 | 2.32 | 3.44 | 3.16 | 1.82 | 1.63 | 31.06 |
| Weslaco | 1.21 | 1.14 | .64 | 1.39 | 2.22 | 2.85 | 1.88 | 2.33 | 4.58 | 2.65 | 1.41 | .95 | 23.25 |
| Wichita Falls | .93 | 1.00 | 1.82 | 2.99 | 4.34 | 2.85 | 2.00 | 2.14 | 3.41 | 2.61 | 1.42 | 1.22 | 26.73 |

*Source:* Based upon data for the period 1951–1980.

TABLE C-3.
*Driest and Wettest Points, 1951–1981*

| Year | Least Annual Total | | Greatest Annual Total | |
| | Location (County) | Inches | Location (County) | Inches |
|---|---|---|---|---|
| 1951 | Salt Flat (Hudspeth) | 3.32 | Neuville (Shelby) | 53.85 |
| 1952 | Boquillas (Brewster) | 4.09 | Evadale (Jasper) | 61.65 |
| 1953 | Imperial (Pecos) | 1.95 | Bon Wier (Newton) | 72.99 |
| 1954 | McCamey (Upton) | 5.77 | Marshall (Harrison) | 45.96 |
| 1955 | Fabens (El Paso) | 5.74 | Bon Wier (Newton) | 65.34 |
| 1956 | Presidio (Presidio) | 1.64 | Orange (Orange) | 53.46 |
| 1957 | Presidio (Presidio) | 4.38 | Glenfawn (Rusk) | 84.52 |
| 1958 | Fabens (El Paso) | 10.16 | Orange (Orange) | 64.54 |
| 1959 | Fabens (El Paso) | 2.45 | Houston–Independent Heights (Harris) | 86.97 |
| 1960 | La Tuna (El Paso) | 5.19 | Cypress (Harris) | 77.50 |
| 1961 | Fabens (El Paso) | 4.03 | Freeport (Brazoria) | 77.14 |
| 1962 | Grandfalls (Ward) | 6.07 | Bonham (Fannin) | 56.35 |
| 1963 | Imperial (Pecos) | 4.91 | Orange (Orange) | 67.07 |
| 1964 | Fabens (El Paso) | 3.67 | Bon Wier (Newton) | 57.43 |
| 1965 | Toyah (Reeves) | 4.36 | Long Lake (Anderson) | 56.32 |
| 1966 | Imperial (Pecos) | 6.85 | Orange (Orange) | 75.46 |
| 1967 | Mentone (Loving) | 3.60 | Negley (Red River) | 55.21 |
| 1968 | Garcia Lake (Deaf Smith) | 9.37 | Center (Shelby) | 76.49 |
| 1969 | El Paso (El Paso) | 4.34 | Jacksonville (Cherokee) | 56.33 |
| 1970 | Ysleta (El Paso) | 5.96 | Beaumont (Jefferson) | 69.49 |
| 1971 | Imperial (Pecos) | 6.22 | Robstown (Nueces) | 55.62 |
| 1972 | El Paso (El Paso) | 9.00 | Hemphill (Sabine) | 63.87 |
| 1973 | La Tuna (El Paso) | 6.20 | Angleton (Brazoria) | 100.21 |
| 1974 | Boquillas (Brewster) | 9.43 | Buffalo (Leon) | 77.49 |
| 1975 | La Tuna (El Paso) | 3.71 | Orange (Orange) | 72.35 |
| 1976 | Cornudas Station (Hudspeth) | 6.64 | Buffalo (Leon) | 83.81 |
| 1977 | Boquillas (Brewster) | 4.90 | Liberty (Liberty) | 61.11 |
| 1978 | Ysleta (El Paso) | 10.71 | Center (Shelby) | 57.11 |
| 1979 | El Paso (El Paso) | 5.84 | Freeport (Brazoria) | 106.44 |
| 1980 | Salt Flat (Hudspeth) | 7.08 | Evadale (Jasper) | 63.48 |
| 1981 | Dell City (Hudspeth) | 7.50 | Houston: Hobby Airport (Harris) | 82.14 |

TABLE C-4.
*Extremes in Yearly Precipitation Amounts, 1873–1982*

| Inches | Location (County) | Year |
|---|---|---|
| **Greatest Annual Totals** | | |
| 109.38 | Clarksville (Red River) | 1873 |
| 106.44 | Freeport (Brazoria) | 1979 |
| 102.58 | Alvin (Brazoria) | 1979 |
| 100.21 | Angleton (Brazoria)[a] | 1973 |
| 98.08 | Anahuac (Chambers) | 1946 |
| 95.28 | Beaumont (Jefferson) | 1923 |
| 92.62 | Cypress (Harris) | 1973 |
| 89.50 | Port Arthur (Jefferson) | 1979 |
| 89.38 | Cleveland (Liberty) | 1973 |
| 88.63 | Houston Deer Park (Harris) | 1979 |
| **Least Annual Totals**[b] | | |
| 1.64 | Presidio (Presidio)[c] | 1956 |
| 1.76 | Wink (Winkler) | 1956 |
| 1.95 | Imperial (Pecos)[d] | 1953 |
| 2.00 | Fowlerton (La Salle)[d] | 1917 |
| 2.36 | Pecos (Reeves) | 1956 |
| 2.45 | Fabens (El Paso) | 1959 |
| 2.55 | Bakersfield (Pecos) | 1956 |
| 2.58 | Toyah (Reeves) | 1956 |
| 2.94 | Fabens (El Paso) | 1956 |

[a] An estimate was made for the February amount that year.
[b] Several weather stations operated by the Galveston, Harrisburg, and San Antonio Railroad registered less than 1 inch of rainfall in 1910 (including Ysleta, 0.7 inch; Marathon, 0.8; and Maxon, 0.9) and little more than 1 inch in 1909 (including Watkins in Terrell County, 1.1 inches); two other stations in El Paso County in 1909 and six stations elsewhere in the Trans Pecos in 1910 recorded less than 2 inches.
[c] An estimate was made for the April amount that year.
[d] An estimate was made for three months in that year.

TABLE C-5.
*Greatest One-Day and One-Month Precipitation Totals, 1880–1982*

| Inches | Location (County) | Date |
|---|---|---|
| **One-Day Totals**[a] | | |
| 29.05 | Albany (Shackelford) | Aug. 4, 1978 |
| 25.75 | Alvin (Brazoria) | Jul. 26, 1979 |
| 23.11 | Taylor (Williamson) | Sept. 9–10, 1921 |
| 21.02 | Kaffie Ranch (Jim Hogg) | Sep. 12, 1971 |
| 20.70 | Hye (Blanco) | Sep. 11, 1952 |
| 20.60 | Montell (Uvalde) | Jun. 27, 1913 |
| | Deweyville (Orange) | Sep. 18, 1963 |
| 19.29 | Danevang (Wharton) | Aug. 27–28, 1945 |
| 19.20 | Benavides (Duval) | Sep. 11, 1971 |
| 19.03 | Austin (Travis) | Sep. 9–10, 1921 |
| 18.00 | Fort Clark (Kinney) | Jun. 14–15, 1899 |
| 17.83 | Taylor Ranch (San Saba) | Jul. 3, 1976 |
| 17.76 | Port Arthur (Jefferson) | Jul. 27–28, 1943 |
| 17.47 | Blanco (Blanco) | Sep. 11, 1952 |
| 16.72 | Freeport (Brazoria) | Jul. 26, 1979 |
| 16.31 | Gonzales (Gonzales) | Aug. 31, 1981 |
| 16.05 | Smithville (Bastrop) | Jun. 30, 1940 |
| 16.02 | Hills Ranch (Travis) | Sep. 10, 1921 |
| | Pandale (Val Verde) | Jun. 27, 1954 |
| 16.00 | Hempstead (Waller) | Nov. 24, 1940 |
| **One-Month Totals** | | |
| 35.70 | Alvin (Brazoria) | Jul. 1979 |
| 32.78 | Falfurrias (Brooks) | Sep. 1967 |
| 31.61 | Freeport (Brazoria) | Sep. 1979 |
| 31.19 | Albany (Shackelford) | Aug. 1978 |
| 30.95 | Freeport (Brazoria) | Jul. 1979 |
| 30.57 | Brownsville (Cameron) | Sep. 1886 |
| 30.30 | Pilot Point (Denton) | May 1982 |
| 29.76 | Port Lavaca (Calhoun) | Jun. 1960 |
| 29.22 | Aransas Pass (San Patricio) | Sep. 1967 |
| 29.19 | Whitsett (Live Oak) | Sep. 1967 |
| 28.96 | Deweyville (Orange) | Oct. 1970 |
| 27.94 | Weatherford (Parker) | May 1884 |
| 27.89 | Kaffie Ranch (Jim Hogg) | Sep. 1971 |
| 27.65 | San Angelo (Tom Green) | Sep. 1936 |
| 27.47 | Boyd (Wise) | Oct. 1981 |
| 26.86 | Port Arthur (Jefferson) | Jul. 1979 |

[a]Does not include the unofficial total of more than 38 inches reported at a point near Thrall on September 9–10, 1921.

TABLE D-1.
Statistics on Hurricanes That Hit the Texas Coast in the Twentieth Century

| Year | Name[a] | Date[b] | Location[b] | Maximum Winds[b] (mph) | Approach Speed[b] (mph) | Surge Height (ft.) | Pressure in the Eye[b] (in.) | Pressure in the Eye[b] (mb) | Damage ($ million) | Deaths |
|---|---|---|---|---|---|---|---|---|---|---|
| 1900 | | Sep. 8 | Galveston | 125 | 10 | 20 | 27.64 | 936 | 30.00 | 6,000 |
| 1909 | | Jul. 21 | Velasco | 120 | 12 | 10 | 29.00 | 982 | 2.00 | 41 |
| 1910 | | Sep. 14 | S. Padre Island | 120 | ? | ? | 28.50 | 965 | (minimal) | ? |
| 1912 | | Oct. 15 | ? | 100 | ? | ? | ? | ? | 0.03 | ? |
| 1913 | | Jun. 27 | Padre Island | 100 | ? | 13 | ? | ? | ? | ? |
| 1915 | | Aug. 17 | Freeport | 120 | 11 | 16 | 28.06 | 950 | 50.00 | 275 |
| 1916 | | Aug. 18 | N. Padre Island | 130 | 11 | 9 | 28.00 | 948 | 1.80 | 20 |
| 1919 | | Sep. 14 | Corpus Christi | 120 | 10 | 16 | 27.99 | 948 | 20.30 | 284 |
| 1921 | | Jun. 22 | Port O'Connor | 110 | ? | 7 | 28.91 | 979 | (minimal) | ? |
| 1929 | | Jun. 28 | Port O'Connor | 90 | 17 | 3 | 28.62 | 969 | 0.70 | 3 |
| 1932 | | Aug. 13 | Freeport | 110 | 17 | 6 | 27.83 | 942 | 7.50 | 40 |
| 1933 | | Sep. 4 | Brownsville | 100 | 8 | 13 | 28.02 | 949 | 12.00 | 40 |
| 1934 | | Jul. 25 | Rockport | 70 | ? | 10 | 28.79 | 975 | 4.50 | 11 |
| 1936 | | Jun. 27 | Port Aransas | 80 | ? | ? | 29.15 | 987 | 0.60 | ? |
| 1940 | | Aug. 7 | Port Arthur | 80 | 8 | 21 | 28.87 | 978 | 1.80 | ? |
| 1941 | | Sep. 23 | Freeport | 90 | 13 | 10 | 28.31 | 959 | 6.00 | 4 |
| 1942 | | Aug. 22 | Gilchrist | 72 | ? | 7 | 29.35 | 994 | 0.60 | 0 |
| | | Aug. 30 | Matagorda Bay | 110 | 14 | 15 | 28.10 | 952 | 26.50 | 8 |
| 1943 | | Jul. 27 | Port Bolivar | 100 | 9 | 3 | 28.78 | 975 | 16.60 | 19 |
| 1945 | | Aug. 27 | Matagorda Bay | 130 | 4 | 15 | 28.57 | 967 | 20.10 | 3 |
| 1947 | | Aug. 24 | Galveston | 80 | ? | 4 | 29.30 | 992 | 0.20 | 1 |
| 1949 | | Oct. 3 | Freeport | 135 | 13 | 11 | 28.88 | 978 | 6.70 | 2 |
| 1957 | Audrey | Jun. 27 | Near Sabine Pass | 100 | ? | ? | 28.29 | 958 | 8.00 | 9 |
| 1959 | Debra | Jul. 25 | Dickinson | 80 | 6 | 3 | 29.07 | 984 | 7.00 | 0 |
| 1961 | Carla | Sep. 11 | Port O'Connor | 150 | 6 | 22 | 27.49 | 931 | 408.00 | 34 |
| 1963 | Cindy | Sep. 17 | High Island | 60 | 8 | 4 | 29.41 | 996 | 11.70 | 3 |
| 1967 | Beulah | Sep. 20 | Near Brownsville | 140 | 8 | 12 | 27.98 | 948 | 200.00 | 15 |
| 1970 | Celia | Aug. 3 | Corpus Christi | 130 | 12 | 9 | 27.80 | 941 | 453.00 | 11 |
| 1971 | Fern | Sep. 10 | Rockport | 78 | 9 | 6 | 29.04 | 983 | 30.00 | 2 |
| 1980 | Allen | Aug. 10 | Port Mansfield | 115 | ? | 12 | 27.91 | 945 | 600.00 | 2 |

*Sources*: Bureau of Economic Geology, University of Texas, *Natural Hazards of the Texas Coastal Zone, 1977*; and U.S. Department of Commerce, *National Summary, 1957–1982.*

[a] No names were assigned to hurricanes prior to 1954.
[b] At the time of landfall.

TABLE D-2.
*Number of Tropical Storms and Hurricanes, 1871–1982*

|  | Jun. | Jul. | Aug. | Sep. | Oct. | Total |
|---|---|---|---|---|---|---|
| **Storms making landfall in Texas** | | | | | | |
| Tropical storms | 7 | 6 | 2 | 9 | 2 | 26 |
| Hurricanes | 7 | 6 | 13 | 10 | 3 | 39 |
| *Total* | 14 | 12 | 15 | 19 | 5 | 65 |
| **Storms entering Texas via Mexico or Louisiana** | | | | | | |
| Tropical storms | 1 | 1 | 5 | 5 | 0 | 12 |
| Hurricanes | 1 | 0 | 2 | 1 | 0 | 4 |
| *Total* | 2 | 1 | 7 | 6 | 0 | 16 |
| **Total** | **16** | **13** | **22** | **25** | **5** | **81** |

TABLE D-3.
*Names for North Atlantic Tropical Storms and Hurricanes, 1983–1985*

| 1983 | 1984 | 1985 | 1986 | 1987 |
|---|---|---|---|---|
| Alicia | Arthur | Ana | Andrew | Arlene |
| Barry | Bertha | Bob | Bonnie | Bret |
| Chantal | Cesar | Claudette | Charley | Cindy |
| Dean | Diana | Danny | Danielle | Dennis |
| Erin | Edouard | Elena | Earl | Emily |
| Felix | Fran | Fabian | Frances | Floyd |
| Gabrielle | Gustav | Gloria | Georges | Gert |
| Hugo | Hortense | Henri | Hermine | Harvey |
| Iris | Isidore | Isabel | Ivan | Irene |
| Jerry | Josephine | Juan | Jeanne | Jose |
| Karen | Klaus | Kate | Karl | Katrina |
| Luis | Lili | Larry | Lisa | Lenny |
| Marilyn | Marco | Mindy | Mitch | Maria |
| Noel | Nana | Nicholas | Nicole | Nate |
| Opal | Omar | Odette | Otto | Ophelia |
| Pablo | Paloma | Peter | Paula | Philippe |
| Roxanne | Rene | Rose | Richard | Rita |
| Sebastien | Sally | Sam | Shary | Stan |
| Tanya | Teddy | Teresa | Tomas | Tammy |
| Van | Vicky | Victor | Virginie | Vince |
| Wendy | Wilfred | Wanda | Walter | Wilma |

TABLE D-4.
*Names for Eastern North Pacific Tropical Storms and Hurricanes,*
*1983–1985*

| 1983 | 1984 | 1985 | 1986 | 1987 |
|------|------|------|------|------|
| Adolph | Alma | Andres | Agatha | Adrian |
| Barbara | Boris | Blanca | Blas | Beatriz |
| Cosme | Christina | Carlos | Celia | Calvin |
| Dalilia | Douglas | Dolores | Darby | Dora |
| Erick | Elida | Enrique | Estelle | Eugene |
| Flossie | Fausto | Fefa | Frank | Fernanda |
| Gil | Genevieve | Guillermo | Georgette | Greg |
| Henriette | Herman | Hilda | Howard | Hilary |
| Ismael | Iselle | Ignacio | Isis | Irwin |
| Juliette | Julio | Jimena | Javier | Jova |
| Kiko | Kenna | Kevin | Kay | Knut |
| Lorena | Lowell | Linda | Lester | Lidia |
| Miriam | Manuel | Marty | Madeline | Max |
| Narda | Norbert | Nora | Newton | Norma |
| Octave | Odile | Olaf | Orlene | Otis |
| Priscilla | Polo | Pauline | Paine | Pilar |
| Raymond | Rachel | Rick | Roslyn | Ramon |
| Sonia | Simon | Sandra | Seymour | Selma |
| Tico | Trudy | Terry | Tina | Todd |
| Velma | Vance | Vivian | Virgil | Veronica |
| Winnie | Wallis | Waldo | Winifred | Wiley |

TABLE E-1.
*Deaths, Injuries, and Property Damage Exacted by Most Memorable Tornadoes*

| Date | Location | Deaths | Injuries | Damage ($ million) |
|------|----------|--------|----------|--------------------|
| Apr. 28, 1893 | Cisco | 23 | 93 | .40 |
| May 15, 1896 | In/near Sherman | 76 | ? | .23 |
| May 18, 1902 | Goliad | 114 | 230 | .05 |
| Apr. 26, 1906 | Bellevue/Stoneburg | 17 | 20 | .30 |
| Mar. 23, 1909 | Slidell | 11 | 10 | .03 |
| May 30, 1909 | Zephyr | 28 | "many" | .09 |
| Apr. 9, 1919 | Leonard/Ector/Rowena | 20 | 45 | .13 |
|  | Henderson/Van Zandt/Wood/Red River counties | 42 | 150 | .45 |
| Apr. 13, 1921 | Melissa | 12 | 80 | .50 |
| Apr. 15, 1921 | Wood/Cass/Bowie counties | 10 | 50 | .09 |
| May 4, 1922 | Austin | 12 | 50 | .50 |
| May 14, 1923 | Howard/Mitchell counties | 23 | 100 | .05 |
| Apr. 12, 1927 | Edwards/Real/Uvalde counties | 74 | 205 | 1.23 |
| May 9, 1927 | Garland | 11 | ? | .10 |
|  | Neveda/Wolfe City/Tigertown | 28 | 200 | .90 |
| May 6, 1930 | Bynum/Irene/Mertens/Ennis/Frost | 41 | ? | 2.10 |
|  | Kenedy/Runge/Nordheim | 36 | 34 | .13 |
| Mar. 30, 1933 | Angelina/Nacogdoches/San Augustine counties | 10 | 56 | .20 |
| Jun. 10, 1938 | Clyde | 14 | 9 | .09 |
| Apr. 28, 1946 | Crowell | 11 | 250 | 1.50 |
| Jan. 4, 1946 | Angelina/Nacogdoches counties | 13 | 250 | 2.05 |
|  | Palestine | 15 | 60 | .50 |
| Apr. 9, 1947 | White Deer/Glazier/Higgins | 68 | 201 | 1.55 |
| May 3, 1948 | McKinney | 3 | 43 | 2.00 |
| May 15, 1949 | Amarillo | 6 | 83 | 5.31 |
| Mar. 13, 1953 | Jud/O'Brien/Knox City | 17 | 25 | .60 |
| May 11, 1953 | Waco | 114 | 597 | 41.15 |
|  | Near San Angelo | 11 | 159 | 3.24 |
| Apr. 2, 1957 | Dallas (Oak Cliff) | 10 | 200 | 4.00 |
| May 15, 1957 | Silverton | 21 | 80 | .50 |
| Apr. 3, 1964 | Wichita Falls | 7 | 111 | 15.00 |
| Jun. 2, 1965 | Hale Center | 4 | 76 | 8.00 |
| Apr. 18, 1970 | Clarendon | 17 | 42 | 2.10 |
| May 11, 1970 | Lubbock | 26 | 500 | 135.00 |
| Mar. 10, 1973 | Hill County | 6 | 77 | ? |
| Apr. 10, 1979 | Wichita Falls/Vernon | 54 | 1,807 | 442.00 |
| Apr. 2, 1982 | Paris | 11 | 173 | 51.00 |

TABLE E-2.
*Number of Observed Tornadoes, 1953–1982*

| Year | Jan.–Feb. | Mar. | Apr. | May | Jun. | Jul. | Aug. | Sep. | Oct. | Nov.–Dec. | Total |
|---|---|---|---|---|---|---|---|---|---|---|---|
| 1953 | 2 | 2 | 3 | 9 | 4 | 5 | 11 | 0 | 3 | 10 | 49 |
| 1954 | 3 | 1 | 31 | 27 | 23 | 5 | 3 | 6 | 7 | 0 | 106 |
| 1955 | 2 | 8 | 19 | 76 | 57 | 14 | 17 | 8 | 3 | 0 | 204 |
| 1956 | 3 | 6 | 8 | 49 | 7 | 18 | 8 | 2 | 24 | 2 | 127 |
| 1957 | 1 | 21 | 69 | 33 | 18 | ? | 6 | 3 | 6 | 5 | 162 |
| 1958 | 2 | 7 | 12 | 15 | 16 | 10 | 8 | 1 | 1 | 8 | 80 |
| 1959 | 0 | 7 | 4 | 31 | 14 | 10 | 2 | 4 | 5 | 6 | 83 |
| 1960 | 5 | ? | 8 | 29 | 14 | 3 | 4 | 2 | 11 | 1 | 77 |
| 1961 | 1 | 21 | 15 | 24 | 28 | 9 | 2 | 12 | 0 | 10 | 122 |
| 1962 | 4 | 12 | 9 | 25 | 56 | 12 | 15 | 7 | 2 | 1 | 143 |
| 1963 | 0 | 3 | 9 | 19 | 24 | 8 | 4 | 6 | 4 | 3 | 80 |
| 1964 | 1 | 6 | 23 | 15 | 11 | 9 | 7 | 3 | 1 | 3 | 79 |
| 1965 | 7 | 3 | 7 | 43 | 24 | 2 | 9 | 4 | 6 | 3 | 108 |
| 1966 | 4 | 1 | 21 | 22 | 15 | 3 | 8 | 3 | 0 | 0 | 77 |
| 1967 | 2 | 11 | 17 | 34 | 22 | 10 | 5 | 124 | 2 | 5 | 232 |
| 1968 | 3 | 3 | 13 | 48 | 21 | 4 | 8 | 5 | 8 | 27 | 140 |
| 1969 | 1 | 1 | 16 | 65 | 16 | 6 | 7 | 6 | 8 | 1 | 127 |
| 1970 | 4 | 5 | 23 | 23 | 9 | 5 | 20 | 9 | 20 | 3 | 121 |
| 1971 | 20 | 10 | 24 | 27 | 33 | 7 | 20 | 7 | 16 | 27 | 191 |
| 1972 | 1 | 19 | 13 | 43 | 12 | 19 | 13 | 8 | 9 | 7 | 144 |
| 1973 | 15 | 29 | 25 | 21 | 24 | 4 | 8 | 5 | 3 | 13 | 147 |
| 1974 | 3 | 8 | 19 | 18 | 26 | 3 | 9 | 6 | 22 | 2 | 116 |
| 1975 | 7 | 9 | 12 | 50 | 18 | 10 | 3 | 3 | 3 | 2 | 117 |
| 1976 | 2 | 14 | 54 | 64 | 11 | 16 | 4 | 10 | 3 | 0 | 178 |
| 1977 | 0 | 3 | 34 | 50 | 4 | 5 | 5 | 12 | 0 | 10 | 123 |
| 1978 | 0 | 0 | 34 | 65 | 10 | 13 | 6 | 5 | 1 | 3 | 137 |
| 1979 | 3 | 24 | 33 | 39 | 14 | 12 | 10 | 4 | 15 | 3 | 157 |
| 1980 | 2 | 7 | 26 | 44 | 21 | 2 | 34 | 10 | 5 | 1 | 152 |
| 1981 | 7 | 7 | 9 | 71 | 26 | 5 | 20 | 5 | 23 | 3 | 176 |
| 1982 | 0 | 6 | 27 | 123 | 36 | 4 | 0 | 3 | 0 | 3 | 202 |
| **Average**[a] | **4** | **9** | **21** | **40** | **20** | **8** | **9** | **9** | **7** | **5** | **132** |

*Source:* Based upon data from the U.S. Department of Commerce, *Storm Data,* 1953–1982.

[a] Rounded off.

TABLE E-3.
*Most Notable Flash Floods*

| Date | Location | Deaths | Damage ($ million) |
|---|---|---|---|
| Jun. 27–Jul. 1, 1899 | Brazos River basin | 30–35 | 9.00 |
| Apr. 5–8, 1900 | Colorado River near Austin | 23 | 1.25 |
| May 22–25, 1908 | Trinity River near Dallas | 11 | 5.00 |
| Dec. 1–5, 1913 | Brazos River basin | 177 | 8.54 |
| Sep. 8–10, 1921 | Central Texas | 215 | 19.00 |
| Apr. 23–28, 1922 | Dallas–Fort Worth | 11 | 1.00 |
| May 24–31, 1929 | Near Houston | | 6.00 |
| Jun. 30–Jul. 2, 1932 | Upper Watersheds of Nueces and Guadalupe rivers | 7 | .50 |
| Sep. 15–18, 1936 | Concho River basin, including San Angelo | 2 | 1.80 |
| Sep. 27–30, 1936 | Brazos River basin, including Waco | ? | 3.00 |
| May 16–17, 1949 | Trinity River at Fort Worth | 10 | 6.00 |
| Sep. 8–10, 1952 | Pedernales River basin, including Kerrville, Blanco, and Boerne | 5 | "many millions" |
| Jun. 26–28, 1954 | Pecos River basin and Rio Grande | ? | ? |
| Apr.–May 1957 | Pecos River to Sabine River | 17 | ? |
| Oct. 28, 1960 | Near Austin | 11 | 2.50 |
| Sep. 21–23, 1964 | Trinity River near Dallas–Fort Worth | 2 | 3.00 |
| Jun. 11, 1965 | Sanderson | 26 | 2.72 |
| Apr. 22–29, 1966 | Northeast Texas | 19 | 12.00 |
| Apr. 28, 1966 | Dallas County | 14 | 15.00 |
| May 11–12, 1972 | South Central Texas near New Braunfels and Seguin | 17 | 17.50 |
| Jun. 12–13, 1973 | Southeastern Texas | 10 | 50.00 |
| Nov. 23–24, 1974 | Central Texas, including Austin | 13 | 1.00 |
| Jan. 31–Feb. 1, 1975 | Nacogdoches County | 3 | 5.50 |
| May 23, 1975 | Near Austin | 4 | 5.00 |
| Jun. 15, 1976 | Harris County | 8 | 25.00 |
| Mar. 27, 1977 | Near Dallas–Fort Worth | 5 | 1.00 |
| May 26, 1978 | Palo Duro Canyon | 4 | 5.00 |
| Aug. 1–4, 1978 | Edwards Plateau and Low Rolling Plains | 33 | 50.00 |
| May 24, 1981 | Austin | 13 | 40.00 |
| Aug. 30–31, 1981 | Lavaca and surrounding counties | 5 | 40.30 |
| Oct. 11–14, 1981 | North Central Texas | 5 | 105.00 |
| Jul. 29–30, 1982 | Amarillo | 0 | 25.00 |

TABLE F-1.
*Average Monthly Snowfall Accumulations (inches)*

| Region | Location | Jan. | Feb. | Mar. | Apr. | Nov. | Dec. | Total[a] |
|--------|----------|------|------|------|------|------|------|----------|
| High Plains | Amarillo | 3.2 | 4.3 | 2.6 | 0.3[d] | 2.2 | 2.2 | 15.2 |
| | Dalhart | 3.4 | 3.8 | 2.3 | 0.7[d] | 2.1[b] | 2.6 | 15.5 |
| | Lubbock | 1.9 | 3.6 | 2.0[b] | 0.1[d] | 1.5[b] | 1.7 | 10.9 |
| | Midland | 1.1 | 1.1[b] | 0.4[d] | 0 | 0.7[d] | 0.4[c] | 3.7 |
| | Muleshoe | 2.7 | 3.3 | 1.4[b] | 0.3[d] | 1.3[b] | 2.0 | 11.2 |
| | Plainview | 2.4 | 4.2 | 1.4[b] | 0 | 1.1[b] | 1.7 | 10.9 |
| | Seminole | 2.2 | 2.8 | 1.4[b] | 0 | 1.5[d] | 1.2[b] | 9.1 |
| | Spearman | 3.8 | 4.3 | 3.0[b] | 0.8[d] | 2.4[b] | 2.9 | 17.7 |
| Low Rolling Plains | Abilene | 1.8 | 1.6 | 0.7[d] | 0 | 0.7[d] | 0.7[d] | 5.4 |
| | Childress | 2.0 | 2.5 | 1.3[c] | 0 | 0.9[c] | 2.2[d] | 8.9 |
| | Clarendon | 1.9 | 2.6[b] | 0.9[c] | 0 | 1.2[c] | 1.0[c] | 7.6 |
| | Wichita Falls | 1.6 | 2.1 | 0.7[c] | 0 | 0.5[d] | 1.0[c] | 6.0 |
| North Central | Brownwood | 0.9[c] | 0.9[c] | 0 | 0 | 0.4 | 0.1 | 2.3 |
| | Dallas–Fort Worth | 1.7[b] | 1.2[c] | 0.2 | 0 | 0.2 | 0.2 | 3.5 |
| | Paris | 1.4[b] | 1.5[c] | 0.1 | 0 | 0.2 | 0.2 | 3.3 |
| | Sherman | 1.4[b] | 2.1[b] | 0.1 | 0 | | 0.3 | 4.0 |
| | Waco | 0.7[c] | 0.4[c] | 0 | 0 | 0.1[d] | 0.1[d] | 1.2 |
| East | Mount Pleasant | 0.8[c] | 0.6[c] | 0.1[d] | 0 | 0 | 0.2[d] | 1.8 |
| | Palestine | 0.6[c] | 0.4[c] | 0.3[d] | 0 | 0 | 0.1[d] | 1.3 |
| Trans Pecos | Alpine | 1.3[c] | 0.7[d] | 0 | 0 | 0.5[d] | 0.2[d] | 2.6 |
| | El Paso | 0.7[c] | 1.0[c] | 0.4[d] | 0 | 1.4[d] | 1.3[c] | 4.9 |
| | Fort Stockton | 0.3[d] | 0.7[c] | 0 | 0 | 0.3[d] | 0.2[d] | 1.6 |

*Source:* Based upon observations for the period 1951–1980.

[a] Monthly sums may not add up to the total due to rounding.
[b] Only one snow accumulation every other year.

[c] One snow accumulation in one out of every three years.
[d] One show accumulation in one out of every four years—or less frequent than that.

TABLE F-2.
*Average Wind Direction and Speed (mph)*

| Location | January | April | July | October |
|----------|---------|-------|------|---------|
| Abilene | s 12 | sse 14 | sse 11 | sse 11 |
| Amarillo | sw 13 | sw 16 | s 12 | sw 13 |
| Austin | s 10 | sse 11 | s 8 | s 8 |
| Brownsville | sse 12 | se 14 | se 12 | se 10 |
| Corpus Christi | sse 12 | se 14 | sse 12 | se 10 |
| Dallas–Fort Worth | s 11 | s 13 | s 9 | s 10 |
| El Paso | n 9 | wsw 12 | sse 9 | n 8 |
| Houston | nnw 8 | sse 9 | s 7 | ese 6 |
| Lubbock | sw 12 | sw 15 | s 11 | s 11 |
| Midland–Odessa | s 10 | sse 13 | sse 11 | s 10 |
| Port Arthur–Beaumont | n 11 | s 12 | s 8 | n 9 |
| San Angelo | sw 10 | s 12 | s 10 | s 9 |
| San Antonio | n 9 | se 11 | sse 9 | n 8 |
| Waco | s 12 | s 13 | s 11 | s 10 |
| Wichita Falls | n 11 | s 13 | s 11 | s 11 |

*Source:* Based upon data from U.S. Department of Commerce, *Local Climatological Data,* 1980.

*Note:* Averages are mean values for periods of about 25–35 years ending in 1980.

TABLE F-3.
*The Beaufort Wind Scale*

| Beaufort Force | Velocity mi./hr. | knots | NWS Terminology | Specifications (for use on land) |
|---|---|---|---|---|
| 0 | 1 | 1 | Calm | Calm; smoke rises vertically |
| 1 | 1–3 | 1–3 | Light air | Direction of wind shown by smoke drift but not by wind vanes |
| 2 | 4–7 | 4–6 | Light breeze | Wind felt on face; leaves rustle; ordinary vane moved by wind |
| 3 | 8–12 | 7–10 | Gentle breeze | Leaves and small twigs in constant motion; wind extends light flag |
| 4 | 13–18 | 11–16 | Moderate breeze | Raises dust and loose paper; small branches are moved |
| 5 | 19–24 | 17–21 | Fresh breeze | Small trees in leaf begin to sway; crested wavelets form on inland water |
| 6 | 25–31 | 22–27 | Strong breeze | Large branches in motion; whistling heard in telegraph wires; umbrellas used with difficulty |
| 7 | 32–38 | 28–33 | Near gale | Whole trees in motion; resistance felt in walking against wind |
| 8 | 39–46 | 34–40 | Gale | Breaks twigs off trees; generally impedes progress |
| 9 | 47–54 | 41–47 | Strong gale | Slight structural damage occurs (chimney pots and slate removed) |
| 10 | 55–63 | 48–55 | Storm | Seldom experienced inland; trees uprooted; considerable structural damage occurs |
| 11 | 64–72 | 56–63 | Violent storm | Very rarely experienced; accompanied by widespread damage |
| 12 | 73–82 | 64–71 | Hurricane | Maximum wind damage |
| 13 | 83–92 | 72–80 | Hurricane | Maximum wind damage |
| 14 | 93–103 | 81–89 | Hurricane | Maximum wind damage |
| 15 | 104–114 | 90–99 | Hurricane | Maximum wind damage |
| 16 | 115–125 | 100–108 | Hurricane | Maximum wind damage |
| 17 | 126–136 | 109–118 | Hurricane | Maximum wind damage |

TABLE F-4.
*Average Relative Humidity (%)*

| Location | January | | | | April | | | | July | | | | October | | | |
|---|---|---|---|---|---|---|---|---|---|---|---|---|---|---|---|---|
| | 6 A.M. | 12 M. | 6 P.M. | 12 P.M. | 6 A.M. | 12 M. | 6 P.M. | 12 P.M. | 6 A.M. | 12 M. | 6 P.M. | 12 P.M. | 6 A.M. | 12 M. | 6 P.M. | 12 P.M. |
| Abilene (17) | 72 | 55 | 51 | 66 | 73 | 47 | 41 | 64 | 70 | 44 | 38 | 57 | 75 | 51 | 48 | 67 |
| Amarillo (19) | 70 | 51 | 47 | 64 | 67 | 38 | 31 | 54 | 73 | 42 | 38 | 60 | 70 | 41 | 40 | 61 |
| Austin (19) | 79 | 61 | 58 | 73 | 83 | 58 | 54 | 76 | 87 | 50 | 46 | 74 | 83 | 54 | 53 | 75 |
| Beaumont–Port Arthur (20) | 89 | 70 | 77 | 87 | 91 | 63 | 71 | 88 | 94 | 65 | 71 | 92 | 90 | 57 | 72 | 88 |
| Brownsville (14) | 87 | 68 | 74 | 86 | 87 | 60 | 68 | 85 | 90 | 55 | 63 | 85 | 88 | 59 | 69 | 85 |
| Corpus Christi (16) | 89 | 71 | 73 | 86 | 90 | 64 | 69 | 87 | 92 | 58 | 63 | 87 | 89 | 60 | 68 | 85 |
| Dallas–Fort Worth (17) | 80 | 61 | 60 | 74 | 84 | 58 | 54 | 74 | 80 | 48 | 44 | 66 | 82 | 53 | 54 | 73 |
| Del Rio (16) | 76 | 55 | 46 | 66 | 77 | 52 | 40 | 62 | 79 | 52 | 39 | 59 | 82 | 57 | 50 | 71 |
| El Paso (20) | 63 | 44 | 34 | 54 | 36 | 22 | 15 | 26 | 60 | 38 | 29 | 45 | 61 | 36 | 29 | 50 |
| Galveston (63) | 85 | 77 | 80 | 83 | 86 | 74 | 79 | 85 | 81 | 70 | 73 | 80 | 80 | 65 | 71 | 75 |
| Houston (11) | 88 | 67 | 72 | 85 | 92 | 60 | 63 | 89 | 93 | 58 | 63 | 88 | 93 | 56 | 71 | 91 |
| Lubbock (33) | 72 | 50 | 47 | 65 | 68 | 40 | 32 | 54 | 74 | 47 | 39 | 61 | 77 | 47 | 44 | 65 |
| Midland–Odessa (17) | 70 | 47 | 41 | 62 | 67 | 34 | 27 | 53 | 70 | 41 | 33 | 54 | 79 | 45 | 41 | 69 |
| San Angelo (20) | 74 | 52 | 48 | 67 | 73 | 43 | 36 | 62 | 75 | 43 | 36 | 58 | 82 | 51 | 50 | 74 |
| San Antonio (38) | 80 | 60 | 58 | 76 | 83 | 56 | 52 | 76 | 87 | 51 | 45 | 75 | 84 | 53 | 52 | 77 |
| Victoria (19) | 87 | 66 | 69 | 84 | 88 | 60 | 64 | 85 | 91 | 57 | 60 | 87 | 89 | 56 | 64 | 85 |
| Waco (17) | 82 | 64 | 63 | 77 | 83 | 61 | 56 | 75 | 79 | 48 | 43 | 66 | 83 | 54 | 55 | 74 |
| Wichita Falls (20) | 81 | 57 | 57 | 74 | 81 | 49 | 47 | 72 | 77 | 42 | 38 | 65 | 83 | 50 | 53 | 73 |

*Source:* Based on data from U.S. Department of Commerce, *Local Climatological Data*, 1980.

*Note:* All times are local standard. The numbers in parentheses indicate the number of years of data, ending in 1980, used in calculating the averages.

TABLE F-5.
*Average Number of Cold Fronts*

| Location | January | April | July | October |
|---|---|---|---|---|
| **High Plains** | | | | |
| Amarillo | 6 | 8 | 6 | 7 |
| Lubbock | 6 | 8 | 5 | 7 |
| **Low Rolling Plains** | | | | |
| Abilene | 6 | 8 | 4 | 6 |
| Wichita Falls | 6 | 8 | 5 | 7 |
| **North Central** | | | | |
| Dallas–Fort Worth | 7 | 8 | 4 | 7 |
| Waco | 7 | 8 | 3 | 6 |
| **East** | | | | |
| Lufkin | 7 | 7 | 3 | 6 |
| Texarkana | 7 | 8 | 4 | 7 |
| **Trans Pecos** | | | | |
| Alpine | 6 | 9 | 3 | 5 |
| El Paso | 4 | 7 | 3 | 5 |
| **Edwards Plateau** | | | | |
| Del Rio | 6 | 8 | 2 | 5 |
| San Angelo | 6 | 9 | 3 | 6 |
| **South Central** | | | | |
| Austin | 6 | 8 | 2 | 5 |
| Corpus Christi | 5 | 6 | 1 | 4 |
| San Antonio | 6 | 7 | 2 | 5 |
| **Upper Coast** | | | | |
| Houston | 6 | 7 | 2 | 6 |
| **Southern** | | | | |
| Laredo | 5 | 6 | 1 | 4 |
| **Lower Valley** | | | | |
| Brownsville | 5 | 4 | 0 | 4 |

Source: Based upon data from Illinois State Water Survey, 1975, for the period 1961–1970.

TABLE F-6.
*Average Number of Days with Various Sky Conditions*

| Location | January Cr | PC | Cd | April Cr | PC | Cd | July Cr | PC | Cd | October Cr | PC | Cd |
|---|---|---|---|---|---|---|---|---|---|---|---|---|
| **High Plains** | | | | | | | | | | | | |
| Amarillo (39) | 12 | 7 | 12 | 12 | 8 | 10 | 13 | 12 | 6 | 17 | 7 | 7 |
| Lubbock (34) | 13 | 6 | 12 | 13 | 8 | 9 | 14 | 11 | 6 | 17 | 7 | 7 |
| Midland (32) | 12 | 7 | 12 | 13 | 8 | 9 | 13 | 11 | 7 | 17 | 6 | 8 |
| **Low Rolling Plains** | | | | | | | | | | | | |
| Abilene (41) | 11 | 6 | 14 | 11 | 8 | 11 | 14 | 10 | 7 | 15 | 7 | 9 |
| Wichita Falls (37) | 11 | 6 | 14 | 11 | 7 | 12 | 14 | 10 | 7 | 16 | 7 | 8 |
| **North Central** | | | | | | | | | | | | |
| Dallas–Fort Worth (27) | 10 | 5 | 16 | 9 | 7 | 14 | 15 | 9 | 7 | 14 | 8 | 9 |
| Waco (37) | 9 | 6 | 16 | 8 | 7 | 15 | 14 | 11 | 6 | 14 | 8 | 9 |
| **Trans Pecos** | | | | | | | | | | | | |
| El Paso (38) | 14 | 7 | 10 | 17 | 8 | 5 | 12 | 13 | 6 | 19 | 7 | 5 |
| **Edwards Plateau** | | | | | | | | | | | | |
| Del Rio (16) | 10 | 7 | 14 | 8 | 8 | 14 | 12 | 11 | 8 | 12 | 9 | 10 |
| San Angelo (32) | 12 | 6 | 13 | 11 | 8 | 11 | 14 | 10 | 7 | 16 | 7 | 8 |
| **South Central** | | | | | | | | | | | | |
| Austin (39) | 9 | 6 | 16 | 8 | 7 | 15 | 11 | 14 | 6 | 13 | 9 | 9 |
| Corpus Christi (38) | 7 | 7 | 17 | 5 | 10 | 15 | 11 | 14 | 6 | 13 | 10 | 8 |
| San Antonio (38) | 9 | 6 | 16 | 7 | 8 | 15 | 9 | 15 | 7 | 12 | 10 | 9 |
| **Upper Coast** | | | | | | | | | | | | |
| Houston (11) | 7 | 5 | 19 | 8 | 6 | 16 | 7 | 15 | 9 | 12 | 9 | 10 |
| Port Arthur (27) | 6 | 7 | 18 | 6 | 8 | 16 | 5 | 16 | 10 | 13 | 10 | 8 |
| Victoria (19) | 6 | 7 | 18 | 5 | 7 | 18 | 7 | 15 | 9 | 12 | 10 | 9 |
| **Lower Valley** | | | | | | | | | | | | |
| Brownsville (38) | 6 | 8 | 17 | 5 | 11 | 14 | 11 | 15 | 5 | 12 | 12 | 7 |

*Source:* Based upon data from U.S. Department of Commerce, *Local Climatological Data,* 1980.

*Note:* "Cr" denotes "Clear" skies; "PC," "Partly Cloudy"; and "Cd," "Cloudy." The numbers in parentheses indicate the number of years of data, ending in 1980, used in calculating the averages.

TABLE F-7.
*Average Amount of Sunshine (% possible amount)*

| Location | Dec. | Jan. | Feb. | Mar. | Apr. | May | Jun. | Jul. | Aug. | Sep. | Oct. | Nov. |
|---|---|---|---|---|---|---|---|---|---|---|---|---|
| **High Plains** | | | | | | | | | | | | |
| Amarillo (39) | 68 | 68 | 69 | 71 | 73 | 72 | 77 | 78 | 77 | 74 | 75 | 72 |
| Lubbock (8) | 71 | 63 | 70 | 78 | 77 | 76 | 81 | 78 | 80 | 70 | 79 | 70 |
| **Low Rolling Plains** | | | | | | | | | | | | |
| Abilene (32) | 66 | 62 | 65 | 70 | 70 | 70 | 78 | 79 | 77 | 68 | 72 | 69 |
| **North Central** | | | | | | | | | | | | |
| Dallas–Fort Worth | 55 | 51 | 54 | 59 | 57 | 61 | 73 | 77 | 75 | 69 | 65 | 62 |
| **Trans Pecos** | | | | | | | | | | | | |
| El Paso (38) | 79 | 77 | 82 | 85 | 88 | 89 | 89 | 80 | 81 | 82 | 84 | 83 |
| **South Central** | | | | | | | | | | | | |
| Austin (39) | 51 | 48 | 52 | 55 | 53 | 57 | 70 | 76 | 75 | 67 | 66 | 57 |
| Corpus Christi (38) | 47 | 46 | 52 | 56 | 57 | 62 | 74 | 82 | 78 | 68 | 69 | 58 |
| San Antonio (38) | 51 | 48 | 53 | 58 | 55 | 56 | 68 | 75 | 74 | 67 | 65 | 56 |
| **Upper Coast** | | | | | | | | | | | | |
| Galveston (90) | 49 | 48 | 51 | 55 | 61 | 68 | 76 | 72 | 71 | 68 | 72 | 60 |
| Houston (11) | 60 | 40 | 52 | 47 | 52 | 59 | 67 | 67 | 63 | 59 | 65 | 55 |
| Port Arthur (26) | 47 | 42 | 52 | 52 | 52 | 64 | 69 | 65 | 63 | 62 | 67 | 57 |
| **Lower Valley** | | | | | | | | | | | | |
| Brownsville (38) | 44 | 43 | 49 | 51 | 57 | 66 | 74 | 81 | 76 | 68 | 66 | 52 |

*Source:* Based upon data from U.S. Department of Commerce, *Local Climatological Data,* 1980.

*Note:* The numbers in parentheses indicate the number of years of data, ending in 1980, used in calculating the averages; the only exception is Dallas–Fort Worth (1942–1973).

TABLE F-8.
*Average Number of Days on Which Heavy Fog Occurs*

| Location | Dec. | Jan. | Feb. | Mar. | Apr. | May | Jun. | Jul. | Aug. | Sep. | Oct. | Nov. |
|---|---|---|---|---|---|---|---|---|---|---|---|---|
| **High Plains** | | | | | | | | | | | | |
| Amarillo (39) | 3 | 3 | 4 | 3 | 2 | 2 | 1 | 1 | 1 | 2 | 2 | 2 |
| Lubbock (34) | 2 | 3 | 3 | 1 | 1 | 1 | * | * | * | 1 | 1 | 2 |
| Midland (32) | 3 | 3 | 3 | 1 | 1 | * | * | * | * | 1 | 1 | 3 |
| **Low Rolling Plains** | | | | | | | | | | | | |
| Abilene (41) | 1 | 1 | 1 | 1 | * | * | * | * | * | * | 1 | 1 |
| Wichita Falls (37) | 2 | 2 | 2 | 1 | 1 | 1 | * | * | * | 1 | 1 | 1 |
| **North Central** | | | | | | | | | | | | |
| Dallas–Fort Worth (27) | 2 | 3 | 2 | 1 | 1 | * | * | 0 | * | * | 1 | 1 |
| Waco (37) | 2 | 3 | 2 | 1 | 1 | 1 | * | * | 0 | * | 1 | 2 |
| **Trans Pecos** | | | | | | | | | | | | |
| El Paso (41) | 1 | 1 | * | * | * | * | 0 | 0 | 0 | * | * | * |
| **Edwards Plateau** | | | | | | | | | | | | |
| Del Rio (16) | 4 | 3 | 2 | 1 | * | * | 0 | 0 | 0 | * | 1 | 3 |
| San Angelo (33) | 1 | 1 | 1 | 1 | * | * | * | * | * | * | 1 | 1 |
| **South Central** | | | | | | | | | | | | |
| Austin (39) | 4 | 5 | 3 | 3 | 1 | 1 | * | * | * | 1 | 2 | 3 |
| Corpus Christi (38) | 5 | 6 | 5 | 4 | 3 | 1 | * | * | * | * | 1 | 4 |
| San Antonio (38) | 5 | 5 | 3 | 3 | 2 | 1 | * | * | * | * | 2 | 3 |
| **Upper Coast** | | | | | | | | | | | | |
| Houston (11) | 5 | 7 | 4 | 4 | 3 | 2 | 1 | * | 1 | 2 | 4 | 5 |
| Port Arthur (27) | 7 | 8 | 6 | 6 | 3 | 1 | * | * | * | 1 | 3 | 5 |
| Victoria (19) | 6 | 7 | 5 | 5 | 4 | 2 | 1 | * | * | 1 | 3 | 6 |
| **Lower Valley** | | | | | | | | | | | | |
| Brownsville (38) | 5 | 6 | 5 | 4 | 2 | 1 | * | * | * | * | 1 | 3 |

*Source:* Based upon data from U.S. Department of Commerce, *Local Climatological Data,* 1980.

*Note:* The numbers in parentheses indicate the number of years of data, ending in 1980, used in calculating the averages.
* Less than one-half day.

TABLE F-9.
*Cooperative Weather Observers with Lengthy Service*

| Observer | City (County) | Years of Service | Date of First Observation |
|---|---|---|---|
| **Individuals Still Active as of September 1982** | | | |
| Edwin Ramey | Dimmitt (Castro) | 59 | Jan. 1923 |
| R. R. Traylor | Matagorda (Matagorda) | 56 | May 1926 |
| Addie L. Koenig | Runge (Karnes) | 50 | Feb. 1933 |
| Mrs. John G. Kenedy Jr. | Sarita (Kenedy) | 49 | Mar. 1934 |
| John Holdsworth | Crystal City (Zavala) | 42 | May 1940 |
| Roy Turner | Tascosa (Oldham) | 42 | Jan. 1941 |
| Samuel P. Herren | Haskell (Haskell) | 41 | Sep. 1941 |
| W. G. Rountree | Hamlin (Jones) | 40 | May 1942 |
| Lawrence Warburton | Freer (Duval) | 40 | Aug. 1942 |
| R. C. Davidson | Shamrock (Wheeler) | 39 | Jul. 1943 |
| Etoile H. Moore | Davilla (Milam) | 39 | Oct. 1943 |
| Lillie H. Fry | Houston: Barker (Harris) | 38 | Oct. 1944 |
| W. G. Stein | Brenham (Washington) | 38 | Oct. 1944 |
| Ross I. Davis | Waller (Waller) | 37 | May 1945 |
| Mavis M. Altmiller | Darrouzett (Lipscomb) | 37 | Jul. 1945 |
| John D. Isenhower | Putnam (Callahan) | 37 | Mar. 1946 |
| **Former Weather Observers** | | | Period of Service |
| Earle family | Hewitt (McLennan) | 85 | 1879–1963 |
| Kenedy family | Sarita (Kenedy) | 83 | 1899–present[a] |
| Newman family | Mexia (Limestone) | 71 | 1904–1974 |
| Stevens family | Coleman (Coleman) | 70 | 1910–1979 |
| Hubbard family | Kaufman (Kaufman) | 66 | 1904–1969 |
| Koenig family | Runge (Karnes) | 65 | 1918–present[a] |
| Hembree family | Bridgeport (Wise) | 63 | 1921–present[a] |
| R. M. Jones | Clifton (Bosque) | 60 | 1911–1970 |
| McCleary family | Honey Grove (Fannin) | 60 | 1916–1975 |
| Pearl Smith | Brownwood (Brown) | 55 | 1905–1959 |
| Michael Kangerga | Henderson (Rusk) | 55 | 1908–1962 |
| Judge H. E. Haass | Hondo (Medina) | 53 | 1899–1951 |
| Black family | Crowell (Foard) | 51 | 1932–present[a] |
| W. S. Ownsby | Cleburne (Johnson) | 49 | 1913–1961 |
| J. J. McMickin | Memphis (Hall) | 48 | 1919–1966 |
| R. J. Klump | Muleshoe (Bailey) | 48 | 1921–1968 |
| Daugherty family | Seymour (Baylor) | 46 | 1925–1970 |

*Source:* U.S. Department of Commerce, *Substation History: Texas,* 1982.

[a]This is a list of weather stations with longest family-run continuous weather observations; some are still being maintained by a family member kin to, but not identical with, the one who first started the records.

**TABLE F-10**
*Military Outposts Serving as Weather Observation Points*

| Outpost | Location (nearest present city) | Period of Observations | Percentage of Complete Data |
|---|---|---|---|
| Fort Houston | Palestine | 1842–1868 | 100 |
| Forth Worth | Fort Worth | 1849–1853 | 100 |
| Fort Croghan | Burnet | 1849–1853 | 100 |
| Fort Inge | Uvalde | 1849–1861 | 60 |
| Fort Bliss | El Paso | 1850–1876 | 70 |
| Fort Graham | Kopperl | 1850–1853 | 100 |
| Fort Chadbourne | Robert Lee | 1852–1861 | 100 |
| Fort McKavett | Menard | 1852–1883 | 70 |
| Fort Belknap | Graham | 1852–1858 | 95 |
| Fort Ewell | Cotulla | 1852–1854 | 100 |
| Fort Davis | Fort Davis | 1855–present | 60 |
| Camp Colorado | Coleman | 1857–1861 | 85 |
| Camp Hudson | Comstock | 1858–1861 | 100 |
| Fort Quitman | El Paso | 1859–1874 | 25 |
| Fort Stockton | Fort Stockton | 1870–present | 90 |
| Fort Clark | Brackettville | 1871–1920 | 90 |
| Fort Ringgold | Rio Grande City | 1871–1906 | 100 |
| Fort Duncan | Eagle Pass | 1871–1877 | 100 |
| Fort McIntosh | Laredo | 1871–1875 | 90 |
| Fort Elliott | Mobeetie | 1879–1890 | 90 |

*Source:* U.S. Department of Agriculture, *Summary of the Climatological Data for the United States,* 1925.

**TABLE F-11.**
*Earliest Weather Stations Operated by Cooperative Observers*

| Location | Period of Observations | Percentage of Complete Data |
|---|---|---|
| El Paso | 1850–1900 | 83 |
| Austin | 1856–1900 | 99 |
| Jacksboro | 1868–1883 | 83 |
| Albany | 1869–1882 | 90 |
| Brownsville | 1871–1900 | 99 |
| Galveston | 1871–1900 | 100 |
| San Antonio | 1871–1900 | 97 |
| Indianola | 1872–1886 | 100 |
| Clarksville | 1872–1885 | 88 |
| McKinney | 1872–1881 | 92 |
| Corsicana | 1874–1900 | 90 |
| Dallas | 1874–1900 | 86 |
| Denison | 1875–1883 | 86 |
| Fredericksburg | 1877–1900 | 74 |
| Graham | 1877–1900 | 52 |
| Eagle Pass | 1878–1900 | 75 |

*Source:* U.S. Department of Agriculture, *Summary of the Climatological Data for the United States,* 1925.

# Glossary

*absolute humidity*: the ratio of the mass of water vapor present in moist air to the volume occupied by the mixture; usually expressed in grams of water vapor per cubic meter (gm/m³).

*accretion*: the growth of a precipitation particle by the collision of a frozen particle (ice crystal or snowflake) with a supercooled liquid droplet, which freezes upon contact.

*acre-foot*: the volume of water required to cover one acre to a depth of one foot (i.e., 43,560 cubic feet).

*advection*: the transport of a property (such as temperature or moisture) solely by the motion of the atmosphere.

*advection fog*: a type of fog caused by the advection of moist air over a cold surface and the subsequent cooling of that air below its dew point.

*air mass*: a widespread body of air that is approximately homogeneous in its horizontal extent, especially with respect to temperature and moisture.

*albedo*: the ratio of the amount of radiation reflected by a body to the amount of radiation incident upon that body.

*altocumulus*: a type of middle-level cloud, white or gray in color, that occurs as a layer or patch with a waved aspect.

*altostratus*: a type of middle-level cloud that appears in the form of a gray or bluish sheet or layer of striated, fibrous, or uniform cloud elements.

*anemometer*: an instrument designed to measure the speed of the wind.

*aneroid barometer*: an instrument, containing no liquid, that measures atmospheric pressure.

*anticyclone*: a weather system having an anti-cyclonic (or clockwise, in the Northern Hemisphere) circulation pattern; also known as a high-pressure cell.

*atmosphere*: the envelope of air surrounding Earth and confined next to Earth due largely to Earth's gravitational influence.

*atmospheric pressure*: the pressure exerted by the atmosphere as a result of gravitational attraction imposed upon a "column" of air lying directly over the point in question.

*autumnal equinox*: the time (approximately September 22) when the sun's noon rays are directly overhead at the Equator and when the sun approaches the Southern Hemisphere; the official beginning of autumn.

*bar*: a unit of pressure equal to 1 million dynes per square centimeter (10 dynes/cm²), or 29.53 inches of mercury.

*barometer*: an instrument for measuring atmospheric pressure.

*barometric pressure*: same as atmospheric pressure.

*Bermuda high*: semipermanent subtropical high-pressure cell in the North Atlantic, whose circulation pattern is largely responsible for the warm and humid conditions that prevail in Texas in summer.

*black frost*: a killing dry freeze, with respect to its effect upon vegetation.

*blizzard*: a severe wintry condition typified by cold temperatures and strong winds (with speeds at least 32 mph) bearing a great amount of snow.

*blowing dust*: dust particles picked up from Earth's surface and blown about by the wind as clouds or sheets.

*blowing snow*: snow lifted from Earth's surface by the wind to a height of 6 feet or more and blown about to such an extent that horizontal visibility is restricted at or above that level.

*breeze*: a light wind (with speeds ranging from 4 to 27 knots, or 4 to 31 mph).

*ceiling*: the height ascribed to the lowest layer of clouds or other phenomenon (such as fog or smoke) that obscures visibility.

*Celsius*: a temperature scale (formerly known as centigrade) whose ice point is 0° and boiling point is 100°.

*cirrocumulus*: a type of high-level cloud that appears as a thin, white patch of cloud without shadows and is composed of very small elements in the form of grains, ripples, etc.

*cirrostratus*: a type of high-level cloud that appears as a whitish veil, usually fibrous but sometimes smooth, and sometimes produces the halo phenomenon around the sun or moon.

*cirrus*: a type of high-level cloud consisting of elements in the form of white, delicate filaments, of white patches, or of narrow bands.

*clear*: a sky condition when clouds are absent or when the cloud cover is less than 0.1.

*climatology*: the scientific study of climate or the long-term manifestations of the weather.

*cloud*: a visible aggregate of minute water and/or ice particles in the atmosphere above Earth's surface.

*cloud base*: the lowest level in the atmosphere at which the air contains a perceptible quantity of cloud.

*cloud seeding*: a technique performed to add to a cloud a certain amount of particles that will alter the natural development of that cloud.

*cloudy*: a sky condition in which clouds cover about 0.7 or more of the sky.

*coalescence*: the merging of two water drops into a single larger drop.

*cold front*: the leading edge of an advancing air mass that replaces a warmer air mass.

*cold wave*: a rapid fall in temperature within 24 hours that necessitates substantially increased protection to agriculture, industry, commerce, and social activities.

*condensation*: the process by which a vapor becomes a liquid or a solid.

*condensation nucleus*: a solid or liquid particle upon which condensation of water vapor begins in the atmosphere.

*continental air*: a type of air whose characteristics are developed over a large land area and that is often marked by a low moisture content.

*convection*: motion within the atmosphere that results in the transport and mixing of certain properties of the atmosphere (e.g., moisture).

*convective cloud*: a cloud that owes its vertical development to currents of convection.

*cooperative observer*: an unpaid volunteer weather observer who maintains a weather station for the NWS.

*coriolis force*: an apparent force exerted on moving particles that stems from the rotation of Earth on its axis.

*corona*: one or more prismatically colored rings that concentrically surround the sun or moon when veiled by a thin cloud layer.

*cumulonimbus*: a principal cloud type that appears as mountains or huge towers and often produces heavy rain of a showery nature; its popular name is *thundercloud* or *thunderhead*.

*cumulus*: a type of low-level cloud—made up of individual, detached elements that generally are dense—that develops vertically as rising mounds, domes, or towers.

*cut-off low*: a low-pressure cell that becomes displaced from the basic westerly current in the mid-latitudes.

*cyclogenesis*: the development or intensification of a cyclonic circulation pattern (e.g., a low-pressure cell).

*cyclone*: any weather system having a closed counterclockwise circulation pattern.

*deepening*: a decrease (or intensification) in the central pressure of a weather system (such as a low).

*degree day*: a measure of the departure of the mean daily temperature from a given standard (most often 65°F or 29°C).

*density*: the ratio of the mass of any substance to the volume occupied by it; usually expressed in grams per cubic centimeter ($gm/cm^3$).

*depression*: an area of low pressure; also known as a *low* or *trough*.

*dew*: water condensed onto grass and other objects near the ground whose temperatures have fallen below the dew point of the layer of air next to Earth's surface but are still above freezing (if temperature is below freezing, *hoarfrost* occurs; if temperature falls below freezing after dew has formed, the frozen dew is called *white dew*).

*dew point*: the temperature to which the air must be cooled in order for it to become saturated (assuming pressure and moisture content remain constant).

*diurnal*: daily, particularly with reference to processes that are completed within 24 hours and that recur every 24 hours.

*dog days*: period of greatest heat in summer, usually from mid-July to the end of August.

*downrush*: the strong downward-flowing current of air associated with a dissipating thunderstorm.

*downwind*: the direction toward which the wind is blowing.

*drifting snow*: snow raised by the wind from Earth's surface to a height of less than 6 feet above the surface and then deposited behind obstacles and irregularities of the surface in heaps referred to as *snow drifts*.

*drizzle*: very small, numerous, and uniformly dispersed water drops that may appear to float and that fall to the ground; classifications include *very light drizzle*, which does not completely wet an exposed surface, regardless of duration; *light drizzle*, the rate of fall ranging from a trace to 0.01 inch per hour; *moderate drizzle*, 0.01 to 0.02 inch per hour; and *heavy drizzle*, more than 0.02 inch per hour.

*drought*: a period of abnormally dry weather of sufficient length to cause a serious hydrologic imbalance (i.e., crop damage, water-supply shortage, etc.).

*dry freeze*: freezing of soil and objects on the ground due to a lowering in temperature when the air does not contain enough moisture for hoarfrost to form; with reference to vegetation, it is known as *black frost*.

*dry snow*: powdery snow from which a snowball cannot easily be made.

*dust*: solid materials suspended in the atmosphere that give a tannish or greyish hue to distant objects.

*Dust Bowl*: a region of the United States, including the Texas High and Low Rolling Plains, afflicted by extreme drought and dust storms in the decade of the 1930s.

*dust devil*: a small, vigorous, well-developed whirlwind rendered visible by the dust, sand, or other debris picked up from the ground.

*dust storm*: a severe weather condition marked by strong winds and dust-filled air that reduces visibilities to ⅝ mile or less (if lowered to 5/16 mile, a *severe dust storm*).

*easterly wave*: a migratory disturbance, imbedded within the broad easterly current that moves from east to west across the tropics, that occasionally evolves into a tropical cyclone.

*echo*: the appearance on a radar indicator of radio energy reflected or scattered back from a radar target.

*effective precipitation*: the portion of precipitation that reaches stream channels as runoff or that remains in the soil and is available for consumptive use.

*effective temperature*: the temperature at which motionless, saturated air would induce, in a sedentary worker wearing ordinary indoor clothing, the same sensation of comfort as that brought about by the actual conditions of temperature, humidity, and air movement.

*elevation*: a measure of the height of a point on Earth's surface above a reference plane (most often, mean sea level); usually expressed in feet (ft.) or meters (m).

*equinox*: the moment at which the sun passes directly above Earth's equator (*see also* autumnal equinox *and* vernal equinox).

*evaporation*: the process by which a liquid is transformed to the gaseous state; the opposite of condensation.

*evapotranspiration*: the combined processes by which water is transferred from Earth's surface to the atmosphere through evaporation of liquid or solid water plus transpiration from plants.

*extratropical*: typical of weather events that occur poleward of the belt of tropical easterlies.

*eye*: the roughly circular area of comparatively light winds and fair skies found at the center of an intense tropical cyclone (such as a tropical storm or a hurricane).

*Fahrenheit*: a temperature scale whose ice point is 32° and boiling point is 212°.

*fair*: a term generally descriptive of pleasant weather; it implies no precipitation, less than 0.4 sky cover of low clouds, and no other extreme conditions of cloudiness, visibility, or wind.

*first-order station*: any weather-observing facility staffed in whole or in part by NWS personnel; there are eighteen such installations in Texas.

*flash flood*: a flood that rises or falls quite rapidly with little or no advance warning, most often as a result of high-intensity rainfall over relatively small areas.

*flood*: the condition of water overflowing the natural or artificial confines of a stream or other water body; also the accumulation by drainage of water in low-lying areas.

*fog*: a visible aggregate of minute water droplets suspended in the atmosphere near Earth's surface (*see also* advection fog, frontal fog, ground fog, radiation fog, sea fog, steam fog, upslope fog).

*freeze*: the condition in which air temperature remains below freezing (32°F or 0°C) over a widespread area; if it cuts short a growing season, it is termed a *killing freeze*; if it is sufficiently cold and prolonged, it is known as a *hard freeze*, a phenomenon recognized by the destruction of seasonal vegetation, a ground surface frozen solid underfoot, and heavy ice on small water surfaces, such as puddles.

*freezing drizzle*: drizzle that falls in liquid form but freezes upon contact with an object to form a coating of glaze.

*freezing level*: the lowest altitude in the atmosphere, over a given location, at which the air temperature is 32°F or 0°C.

*freezing rain*: rain that falls in liquid form but freezes upon impact to form a coating of glaze on the ground and other exposed objects.

*front*: the interface or transition zone between two air masses having differing densities; also referred to as *frontal surface*, *frontal system*, and *frontal zone*.

*frontal fog*: fog associated with frontal surface and frontal passages; *prefrontal fog* results from rain falling through a cold stable air mass and raising its dew-point temperature, whereas *frontal-passage fog* stems from the mixture of warm and cold air within the frontal zone.

*frost*: a deposit of interlocking ice crystals on Earth's surface and earthbound objects when the temperature of the surface and those objects falls below freezing.

*gale*: an unusually strong wind, categorized as follows: *moderate gale*, 28–33 knots (32–38 mph); *fresh gale*, 34–40 knots (39–46 mph); *strong gale*, 41–47 knots (47–54 mph); and *whole gale*, 48–55 knots (55–63 mph).

*gale warning*: a storm warning for marine interests of impending winds with speeds from 28 to 47 knots (32–54 mph).

*gamma ray*: electromagnetic radiation having extremely short wavelength (between X rays and cosmic rays) that contributes to the ionization of the atmosphere, one manifestation of which is lightning.

*glaze*: a coating of ice formed on exposed objects by the freezing of a film of water deposited by rain, drizzle, fog, or even supercooled water vapor.

*greenhouse effect*: the effect of heating exerted by the atmosphere upon Earth as a result of the absorption and reemission of radiation by the atmosphere.

*ground clutter*: a type of radar echo that stems from the reflection of a radar signal by fixed ground targets (such as tall buildings).

*ground fog*: a fog that hides less than 0.6 of the sky and that does not extend to the base of any clouds that may be above it.

*growing season*: the period of the year when the temperature of cultivated vegetation remains sufficiently high to allow plant growth.

*gust*: a sudden brief increase in the speed of the wind; it is reported when peak wind speed reaches at least 16 knots (18 mph) and the variation in speed between peaks and lulls is at least 9 knots (10 mph); its duration is usually less than 20 seconds.

*hail*: precipitation in the form of balls or irregular lumps of ice always produced by convective clouds (such as thunderstorms).

*halo*: an atmospheric optical phenomenon that appears as colored or whitish rings and arcs around the sun or moon when seen through a layer or cloud of ice crystals.

*hard freeze*: a condition when seasonal vegetation is destroyed, the ground surface is frozen solid underfoot, and heavy ice is formed on small containers of water.

*haze*: dust or salt particles, so small they cannot be felt or seen individually with the human eye, that diminish horizontal visibility and give the atmosphere an opalescent appearance.

*heat*: a form of energy transferred between systems as a result of a difference in temperature.

*heating-degree day*: a popular indicator of fuel consumption; one heating-degree day is assigned for each degree that the daily mean temperature departs below the base of 65°F (19°C).

*heat wave*: a period of abnormally and uncomfortably hot and usually humid weather.

*high*: an expression for an area of "high" pressure, which refers to a maximum of atmospheric pressure.

*hoarfrost*: a deposit of interlocking ice crystals formed by direct sublimation on objects, most often those freely exposed to the air (e.g., tree branches, wires, poles, plant stems).

*humidity*: a measure of the water-vapor content of air (*see also* relative humidity).

*hurricane*: a severe tropical cyclone in the North Atlantic Ocean, having a sustained wind speed of 64 knots (74 mph) or greater; classifications include a *major hurricane*, when winds of 101–135 mph (88–117 knots) and a minimum central pressure of 28.01–29.00 inches Hg (711–737 mm Hg or 948.5–982.0 mb) are observed, and an *extreme hurricane*, when maximum winds of 136 mph (118 knots) or higher and a minimum central pressure of 28.00 inches Hg (711 mm Hg or 948.2 mb) or less are noted.

*hurricane warning*: an advisory of impending winds of hurricane force.

*hurricane watch*: an announcement for a specific area that hurricane conditions pose a threat.

*hydrologic cycle*: the composite picture of the interchange of water substances among Earth, its atmosphere, and the seas.

*hygrometer*: an instrument that measures the humidity, or water-vapor content, of the atmosphere.

*ice pellets*: a type of precipitation consisting of transparent or translucent fragments of ice, 0.2 inch (5 mm) or less in diameter, that bounce when hitting hard ground and make a sound upon impact; commonly known as *sleet*.

*ice storm*: a weather event characterized by the fall of freezing precipitation, which creates hazardous conditions by causing glaze to form on terrestrial objects.

*infrared radiation*: electromagnetic energy having a wavelength from about 0.8 microns to an indefinite upper boundary; or bounded on its lower limit by visible radiation and on its upper limit by microwave radiation.

*insolation*: solar radiation received at Earth's surface.

*instability*: an atmospheric condition in which certain disturbances, when introduced into the steady state, will increase in magnitude.

*intertropical convergence zone* (ITCZ): the broad trade-wind current of the tropics; the dividing line between the southeast and northeast trade winds.

*inversion*: a deviation from the usual decrease or increase with altitude of the value of an atmospheric property (e.g., temperature or moisture).

*ionosphere*: the layer of the upper atmosphere characterized by a high ion density, having a base at about 43–50 miles (70–80 km) and a ceiling of indefinite height.

*isallobar*: a line of equal change in atmospheric pressure for a specific time interval.

*isallotherm*: a line connecting points of equal change in temperature for a given time period.

*isobar*: a line of equal or constant pressure; on a weather map, a line drawn through all points having the same atmospheric pressure.

*isotherm*: a line of equal or constant temperature.

*jet stream*: a concentration of relatively strong winds within a narrow stream of Earth's atmosphere; commonly referred to as a current of maximum winds imbedded in the upper atmosphere within the mid-latitude westerlies.

*killing freeze*: a condition in which the surface temperature of the air remains below 32°F (0°C) for a sufficiently long time to destroy all but the hardiest herbaceous crops and to shorten the growing season.

*knot*: a unit of speed in the nautical system; one nautical mph; equal to 1.1508 statute mph (0.5144 m/sec).

*land breeze*: a coastal breeze blowing from land to sea, set up by a difference in temperature when the sea surface is warmer than the adjacent land surface; usually occurs at night.

*lapse rate*: the decrease of an atmospheric variable (most often temperature) with height.

*leader*: the streamer of electrical charge that initiates the first phase of each stroke of a lightning discharge.

*light freeze*: the condition when the surface air temperature drops to below freezing (32°F or 0°C) for a short time period, such that only the tenderest plants and vines are harmed.

*lightning*: any and all forms of visible electrical discharge produced by thunderstorms.

*lightning rod*: a grounded metallic conductor with its upper extremity extending above the structure that is to be protected from damage due to lightning.

*long wave*: a wave in the major belt of westerly winds high in the atmosphere that is characterized by large length and significant amplitude.

*low*: an expression for an area of "low" pressure, which refers to a minimum of atmospheric pressure.

*major trough*: a long-wave upper atmospheric low-pressure area.

*Marfa front*: a transition zone between moist air to the east and desertlike air to the west that oscillates eastward in daytime and westward at night across the Trans Pecos region of Texas; it is so named because a weather station at Marfa transmits data hourly that allow forecasters to monitor its movement.

*maritime air*: a type of air whose characteristics are developed over an extensive water body; customarily high in moisture content, at least in its lowest levels.

*mean sea level*: the average height of the sea surface, based upon hourly observations of the height of tides for all stages over a 19-year period.

*mesopause*: the boundary between the mesosphere and the thermosphere; at an altitude of about 50 miles (80 km), it is usually marked by a sudden change in the rate at which temperature drops with height.

*mesosphere*: that portion of the atmosphere extending from the top of the stratosphere to the mesopause; ranging in altitude between 12 miles (20 km) and 50 miles (80 km), it is characterized by a broad temperature maximum at about 30 miles (50 km).

*meteorology*: the science that deals with the phenomena of the atmosphere.

*millibar*: a measure of atmospheric pressure; equal to about 0.03 inch of mercury, or 33.86 millibars equal 1.0 inch.

*minor trough*: an atmospheric pressure area having a scale smaller than a major, or long-wave, trough; usually moves rapidly.

*mist*: an aggregate of microscopic water droplets suspended in the atmosphere that produces, generally, a thin, grayish veil over the landscape; intermediate between haze and fog.

*moist air*: air that is a mixture of dry air and any amount of water vapor.

*moisture*: a general term referring to the water vapor content, or total water substance, in a given volume of air.

*nautical mile*: the distance unit in the nautical system; its value is 1,852 m (6,076.103 feet or 1.1508 statute miles).

*nimbostratus*: a type of middle-level cloud that is gray colored and often dark, rendered diffuse by falling rain, snow, or sleet.

*normal*: the average value of a meteorological element over a fixed period of time (customarily 30 years) that is recognized as standard for the area and element concerned.

*norther*: a strong cold wind, from between the northwest and northeast, that accompanies a cold-air outbreak; characteristically a rushing blast that brings a sudden drop of temperature of as much as 25°F in one hour or 50°F in a 3-hour period; a phenomenon most often observed from November to April.

*nucleus*: a particle of any nature (salt, sand, soil, etc.) upon which molecules of water or ice accumulate as a result of a phase change to a more condensed state.

*occluded front*: a composite of two fronts, formed when a cold front overtakes a warm or quasi-stationary front; a common process in the late stages of the development of a strong surface low-pressure area.

*overcast*: a sky condition in which the sky cover is solid and at least a portion of the sky cover is attributable to clouds or some other obscuring phenomenon.

*overrunning*: a condition when an air mass is in motion aloft above another air mass of greater density at the surface; a common circumstance in Texas' coastal plain when moist Gulf air pours up over a more dense dome of cool polar or Arctic air.

*ozone*: a nearly colorless gaseous form of oxygen that occurs in trace quantities in Earth's atmosphere, primarily in the stratosphere, where it results from photochemical processes involving ultraviolet radiation.

*partly cloudy*: a sky condition typified by an average cloudiness from 0.4 to 0.7 for a 24-hour period; popularly regarded as the condition when clouds are conspicuously present but do not completely dull the day or the sky at any moment.

*peak gust*: the highest "instantaneous" wind speed recorded at a weather station for a specific time period (usually a 24-hour period).

*persistence*: the tendency for the occurrence of a specific weather event to be more probable, at a given time, if that same event has occurred in the immediately preceding time period.

*point rainfall*: the rainfall during a given time interval measured in a rain gauge.

*polar air*: the type of air whose traits are developed over high latitudes, especially within the subpolar regions (e.g., northern Canada).

*potential evapotranspiration*: the amount of moisture that, if available, would be withdrawn from a given land area by the process of evapotranspiration; often determined in dry regions by the amount of irrigation water used.

*precipitable water*: the total amount of water vapor contained in a vertical column of atmosphere (of unit cross-sectional area) between any two specified levels in the atmosphere.

*precipitation*: any and all forms of water particles, whether liquid or solid, that fall from the atmosphere and reach Earth's surface.

*prefrontal squall line*: a line of squalls, about 50 to 200 miles in advance of a cold front, that moves in about the same way as the cold front.

*pressure*: *See* atmospheric pressure.

*prevailing wind direction*: the wind direction most frequently observed during a given time interval.

*probability of precipitation*: the forecast likelihood that a precipitation event (e.g., rain, snow, sleet) will occur at a particular point during a given interval of time (for instance, a probability of 30% implies that, in 100 similar weather situations, any point within the local forecast area should observe measurable precipitation 30 times).

*quasi-stationary front*: a front that is stationary or nearly so; one whose speed of movement is less than about 5 knots (6 mph).

*radar*: an electronic instrument used for the detection and ranging of distant objects having a composition that scatters or reflects radio waves.

*radar echo*: *See* echo.

*radiation*: the process by which electromagnetic radiation is propagated through free space due to joint variations in the electric and magnetic fields.

*radiational cooling*: the cooling of Earth's surface and nearby air that results when the surface sustains a net loss of heat.

*radiation fog*: a major type of fog that results over a land area when radiational cooling drops the air temperature to or below its dew point.

*radiosonde*: an instrument, borne by a balloon, used for simultaneously measuring and transmitting meteorological data.

*radome*: a dome-shaped covering that houses the antenna assembly of a radar to protect it from wind and other foul weather.

*rain*: a kind of precipitation consisting of liquid water drops having a diameter larger than 0.02 inch (0.5mm) or, if widely scattered, even smaller than that; classifications of rainfall intensity include *very light rain*, when scattered drops do not completely wet an exposed surface, regardless of duration; *light rain*, the rate varying between a trace and 0.10 inch (2.5 mm) per hour, with the maximum rate of fall amounting to no more than 0.01 inch (0.25 mm) in 6 minutes; *moderate rain*, from 0.11 inch (2.8 mm) to 0.30 inch (7.5 mm) per hour, the maximum rate being no more than 0.03 inch (0.8 mm) in 6 minutes; and *heavy rain*, over 0.30 inch per hour or more than 0.03 inch in 6 minutes.

*rainbow*: one of a family of circular arcs consisting of concentric colored bands, with red on the inside to blue on the outside, that may be seen on a "sheet" of water drops (such as rain, fog, or spray).

*rain day*: a 24-hour period having measurable precipitation, with 0.01 (0.25 mm) the most often used minimum threshold amount.

*rain gauge*: an instrument designed to measure the amount of precipitation that has fallen.

*rainmaking*: a common term referring to all activities designed to increase, through any of an assortment of artificial means, the amount of precipitation released from a cloud.

*rawinsonde*: a method of observing upper-atmospheric weather conditions—notably wind speed and direction, temperature, pressure, and moisture content—by means of a balloon-borne radiosonde tracked by a radar or other electronic finding device.

*reflectivity*: a measure of the portion of the total amount of radiation reflected by a given surface.

*refraction*: the process in which the direction of energy propagation is changed due to a change in density within the propagating medium.

*relative humidity*: the ratio (usually expressed as a percentage) of the actual vapor pressure of the air to the vapor pressure of the air when saturated; a popular measure of the amount of moisture in the air.

*ridge*: an elongated area of relatively high atmospheric pressure; the opposite of a ridge is a trough.

*rime*: a white or milky and opaque deposit of ice formed by the rapid freezing of supercooled water drops as they impinge upon an exposed cold object; it is lighter, softer, and less transparent than glaze, but denser and harder than hoarfrost.

*sandstorm*: a severe weather condition marked by strong winds carrying sand through the air that reduces visibilities to 5/8 mile or less (if lowered to 5/16 mile, a *severe sandstorm*); usually confined to lowest 10 feet and rarely rises more than 50 feet above the ground.

*saturation*: a condition of the air in which any increase in the amount of water vapor will initiate within the air a change to a more condensed state.

*scattering*: the process by which small particles suspended in a medium diffuse a portion of the incident radiation in all directions.

*scud*: rugged low clouds, usually seen moving rapidly beneath a layer of nimbostratus clouds or a base of a thunderstorm.

*sea breeze*: a coastal breeze blowing from sea to land, caused by the temperature difference that exists when the sea surface is colder than the adjacent land.

*sea fog*: a type of advection fog formed when air lying over a warm water surface is carried over a colder water surface.

*sea level*: *See* mean sea level.

*sea-level pressure*: the atmospheric pressure at mean sea level.

*semiarid climate*: a type of climate where plant life is short, drought-resistant grasses; regions having this type of climate are highly susceptible to severe drought.

*severe storm*: generally, any destructive storm, but often used to describe intense thunderstorms that produce heavy rain, hail, tornadoes, and/or strong winds.

*shear*: most often used in meteorology to describe the variation of wind speed and direction with height above the surface of Earth.

*short wave*: a progressive wave in the horizontal pattern of air motion within Earth's atmosphere.

*shower*: a precipitation event characterized by a suddenness with which it starts and stops, by rapid changes in intensity, and usually by rapid changes in sky appearance.

*sleet*: *See* ice pellets.

*small-craft warning*: an advisory to marine interests warning them of impending winds up to 28 knots (32 mph).

*smog*: a natural fog contaminated by industrial pollutants (i.e., a mixture of smoke and fog).

*smoke*: foreign particulate matter in the atmosphere resulting from combustion processes.

*snow*: precipitation consisting of white or translucent ice crystals, often agglomerated into snowflakes; classifications of intensity include *very light snow*, when scattered flakes do not completely cover or wet an exposed surface, regardless of duration; *light snow*, when visibility is ⅝ mile or more; *moderate snow*, when the visibility is less than ⅝ mile but more than ⁵⁄₁₆ mile; and *heavy snow*, when the visibility is less than ⁵⁄₁₆ mile.

*snowfall accumulation*: the actual depth of snow on the ground at any instant during a snowstorm or after any storm or series of storms.

*solstice*: popularly regarded as the time at which the sun is farthest north or south

(*see also* summer solstice *and* winter solstice).

*spring equinox*: same as vernal equinox.

*squall*: a strong wind typified by sudden onset, a duration on the order of minutes, and a sudden decrease in speed; a sustained wind speed of 16 knots (18 mph) or higher for at least two minutes.

*squall line*: a line or narrow band of active thunderstorms, not associated with a front.

*stability*: an atmospheric condition in which a displaced parcel of air is subjected to a buoyant force opposite to its displacement.

*standard atmosphere*: a hypothetical vertical distribution of temperature, pressure, and density that is regarded as representative of the atmosphere for calibration of equipment, aircraft design, etc.

*stationary front*: same as quasi-stationary front.

*steam fog*: a type of fog that forms when very cold air drifts across relatively warm water by which water vapor is added to the air, which is much colder than the vapor's source.

*stepped leader*: the initial streamer of a lightning discharge.

*storm*: any disturbed state of the atmosphere that strongly implies destructive or otherwise unpleasant weather.

*storm surge*: an abnormal rise of the sea along a shore as a result, mostly, of the winds of a storm; also known as *storm tide*.

*storm warning*: a specially worded forecast of severe weather, intended to alert the public to impending dangers.

*stratocumulus*: a type of low-level cloud in the form of a gray and/or whitish layer or patch, which nearly always has dark parts and is nonfibrous; composed of small water droplets.

*stratosphere*: the layer of the atmosphere above the troposphere and below the mesosphere, from an altitude of about 5–7 miles (8–11 km) up to about 15–18 miles (24–29 km).

*stratus*: a type of low-level cloud in the form of a gray layer with a rather uniform base; usually does not produce precipitation of consequence.

*sublimation*: the transition of a substance from the solid phase to the vapor phase.

*subsidence*: the descending motion of air in the atmosphere.

*subtropical high*: one of the semipermanent high-pressure cells that lie over the Atlantic and Pacific oceans and have a profound impact on Texas weather, especially in summer; also called a *subtropical ridge*.

*subtropics*: the belt in each hemisphere between the tropics and the temperate zones, or roughly 35–40° N and S.

*sultriness*: an oppressively uncomfortable state of the weather that stems from the simultaneous occurrence of high temperatures and high humidities; some lower limits are 95°F (35°C) and 25%, 86°F (30°C) and 40%, and 77°F (25°C) and 65%.

*summer solstice*: the time (approximately June 21) when the sun's noon rays are directly overhead the point at 23½°N latitude; the official beginning of summer.

*sunrise*: the phenomenon of the sun's daily appearance when the upper limb of the sun first is seen on the sea-level horizon.

*sunset*: the phenomenon of the sun's daily disappearance; when the upper limb of the sun just vanishes below the sea-level horizon.

*sunspot*: a relatively dark area on the surface of the sun; usually occurs in pairs with a lifetime from a few days to several months.

*supercooled water*: liquid water drops whose temperature is reduced below their nominal freezing point without a change of status.

*synoptic*: affording an overall view; used with reference to weather data obtained simultaneously over a wide area for the purpose of presenting a comprehensive and nearly instantaneous picture of the state of the atmosphere.

*temperate climate*: the climate of the "middle latitudes," between the extremes of the tropics and the polar regions.

*temperature*: the degree of hotness or coldness as measured on some definite temperature scale.

*thermometer*: an instrument for measuring temperature.

*thermosphere*: the top portion of Earth's atmosphere extending from just above the mesosphere (at an altitude of about 50 miles [80 km]) to outer space and characterized by steadily increasing temperature with height.

*thunder*: the sound emitted by rapidly expanding gases along the channel of a lightning discharge.

*thunderhead*: a popular term for the cloud mass of a thunderstorm, or cumulonimbus cloud; also called a *thundercloud*.

*thunderstorm*: a local storm invariably produced by a cumulonimbus cloud and always accompanied by lightning and thunder and usually accompanied by strong wind gusts and heavy rain.

*thunderstorm day*: an observational day during which thunder is heard (precipitation need not have fallen).

*tide*: the periodic rising and falling of Earth's oceans and atmosphere.

*tornado*: a violently rotating column of air, pendant from a cumulonimbus cloud; the most destructive of all local atmospheric phenomena.

*trace*: a precipitation amount of less than 0.005 inch (0.125 mm).

*trade wind*: the current of air that is a major component of the general circulation of the atmosphere; it blows from the subtropical highs toward the equatorial trough and is northeasterly in the North Atlantic.

*transpiration*: the process by which water in plants is transferred as water vapor to the atmosphere.

*tropical air*: the type of air whose traits are developed over low latitudes; *maritime tropical air* is generated over tropical and subtropical seas and is therefore very warm and humid, while *continental tropical air* is produced over subtropical arid regions and is consequently hot and very dry.

*tropical cyclone*: a weather system having a closed counterclockwise circulation that originates over the tropical oceans; includes *tropical depressions*, *tropical storms*, and *hurricanes*.

*tropical depression*: a tropical cyclone having a sustained wind speed not greater than 34 knots (39 mph) and usually appearing, on a weather map, with one or more closed isobars.

*tropical disturbance*: a tropical cyclone with only a slight surface wind circulation and appearing on a weather map with only one closed isobar or none at all.

*tropical storm*: a tropical cyclone with a sustained wind speed of 34 knots (39 mph) to 63 knots (73 mph).

*tropopause*: the boundary between the troposphere and the stratosphere, usually marked by a sudden change in the rate at which temperature drops with height.

*troposphere*: the portion of the atmosphere between Earth's surface and the tropopause, or the lowest 6 to 13 miles (10–20 km) of the atmosphere.

*trough*: an elongated area of relatively low atmospheric pressure; often used to describe a surface front or an upper-atmospheric storm system; its axis is known as a *trough line*.

*twister*: a colloquial term for *tornado*.

*typhoon*: a severe tropical cyclone (including a hurricane) in the Pacific Ocean; its counterpart in the Atlantic is the *hurricane*.

*ultraviolet radiation*: electromagnetic energy having a wavelength shorter than visible radiation but longer than X rays.

*upper air*: generally the portion of Earth's atmosphere above 850 millibars (or about 5,000 feet above mean sea level).

*upslope fog*: a type of fog formed when air flows upward over rising terrain and is cooled to or below its dew point.

*vapor pressure*: the partial pressure of water vapor in the atmosphere.

*veering wind*: a change in wind direction in a clockwise sense (in the Northern Hemisphere).

*vernal equinox*: the time (approximately March 21) when the sun's noon rays are directly overhead at the Equator and when the sun approaches the Northern Hemisphere; the official beginning of spring.

*virga*: wisps or streaks of water or ice particles falling out of a cloud but evaporating before reaching Earth's surface as precipitation.

*visibility*: the greatest distance in a given direction at which it is just possible to see and identify with the unaided eye, (a) in the daytime, a prominent dark object against the sky at the horizon and, (b) at night, a known, preferably unfocused, moderately intense light source.

*warm front*: any nonoccluded front that moves in such a way that warmer air replaces colder air.

*waterspout*: a tornado or lesser whirlwind occurring over water.

*water vapor*: water substance in the form of a vapor.

*weather*: the state of the atmosphere, mainly with respect to its effects upon life and human activities.

*weather modification*: any effort to alter artificially the natural phenomena of the atmosphere (including rainmaking, fog dissipation, frost prevention, etc.).

*wet-bulb temperature*: the temperature an air parcel would have if cooled to saturation (at constant pressure) by evaporating water into it.

*white dew*: dew frozen as a result of the temperature falling below the freeze level after the dew originally formed.

*white frost*: a relatively heavy coating of hoarfrost.

*wind*: air in motion relative to Earth's surface.

*wind chill*: that part of the total cooling of a body caused by air motion.

*wind direction*: the direction from which the wind is blowing.

*wind vane*: an instrument used to indicate wind direction.

*winter solstice*: the time (approximately December 22) when the sun's noon rays are directly overhead the point at 23½°S latitude; the official beginning of winter.

*X ray*: electromagnetic energy of very short wavelength, lying between the wavelength interval of 0.1 and 1.5 angstroms, or between gamma rays and ultraviolet radiation.

*zonal flow*: the flow of air along a latitude circle (essentially a westerly or easterly wind).

# Suggested Readings

Anthes, Richard A., Hans A. Panofsky, John J. Cahir, and Albert Rango. *The Atmosphere.* Columbus, Ohio: Merrill Publishing Company, 1978.

Barry, R. G., and R. J. Chorley. *Atmosphere, Weather, and Climate.* New York: Holt, Rinehart & Winston, 1970.

Breuer, Georg. *Weather Modification: Prospects and Problems.* Cambridge: At the University Press, 1976.

Byers, Horace R. *Elements of Cloud Physics.* Chicago: University of Chicago Press, 1973.

———. *General Meteorology.* New York: McGraw-Hill Book Co., 1959.

Carr, John T. *Hurricanes Affecting the Texas Gulf Coast.* Report 49. Austin: Texas Water Development Board, 1969.

Critchfield, Howard J. *General Climatology.* Englewood Cliffs, N.J.: Prentice-Hall, 1974.

Gokhale, N. R. *Hailstones and Hailstone Growth.* Albany: State University of New York Press, 1975.

Griffiths, John F., and Greg Ainsworth. *One Hundred Years of Texas Weather, 1880–1979.* College Station: Office of the State Climatologist, Texas A&M University, 1981.

Henry, Walter K., Dennis M. Driscoll, and J. P. McCormack. *Hurricanes on the Texas Coast.* Rev. ed. College Station: Center for Applied Geosciences, Texas A&M University, 1980.

Mason, B. J. *Clouds, Rain, and Rainmaking.* Cambridge: At the University Press, 1975.

Morgan, Griffith M., David G. Brunkow, and R. C. Beebe. *Climatology of Surface Fronts.* Circular 122. Urbana: Illinois State Water Survey, 1975.

National Research Council. *Weather and Climate Modification: Problems and Progress.* Detroit: Grand River Books, 1980.

Petterssen, Sverre. *Introduction to Meteorology.* New York: McGraw-Hill Book Co., 1969.

Riehl, Herbert. *Climate and Weather in the Tropics.* London: Academic Press, 1979.

Simpson, R. H., and Herbert Riehl. *The Hurricane and Its Impact.* Baton Rouge: Louisiana State University Press, 1981.

Trewartha, Glenn T., and Lyle H. Horn. *An Introduction to Climate.* New York: McGraw-Hill Book Co., 1980.

U.S. Department of Commerce. *Operations of the National Weather Service.* Washington: Government Printing Office, 1980.

———. *Tropical Cyclones of the North Atlantic Ocean.* Technical Paper, no. 55. Washington: Government Printing Office, 1965.

Wallace, John M., and Peter V. Hobbs. *Atmospheric Science: An Introductory Survey.* New York: Academic Press, 1977.

# Index